Minerals, Rocks and Inorganic Materials

Monograph Series of Theoretical and Experimental Studies

4

Edited by

W. von Engelhardt, Tübingen · T. Hahn, Aachen

R. Roy, University Park, Pa.

J. W. Winchester, Tallahassee, Fla. · P. J. Wyllie, Chicago, Ill.

O. Braitsch

Salt Deposits
Their Origin and Composition

Translated by

P. J. Burek and A. E. M. Nairn

In Consultation with

A. G. Herrmann and R. Evans

With 47 Figures

Springer-Verlag New York · Heidelberg · Berlin 1971

ISBN 0-387-05206-2 Springer-Verlag New York Heidelberg Berlin
ISBN 3-540-05206-2 Springer-Verlag Berlin Heidelberg New York

Translation of BRAITSCH, Entstehung und Stoffbestand der Salzlagerstätten.
(Mineralogie und Petrographie i. E., Band III) 1962
ISBN 0-387-02877-3 Springer-Verlag New York Heidelberg Berlin
ISBN 3-540-02877-3 Springer-Verlag Berlin Heidelberg New York

Printed in Germany. Typesetting, printing and binding: Brühlsche Universitätsdruckerei, Gießen

Foreword

Professor OTTO BRAITSCH'S *Entstehung und Stoffbestand der Salzlagerstätten* has since its publication in 1962 enjoyed a reputation as the most complete and authoritative treatment, in any language, of the geochemistry of evaporites. The issue of this translation will now make it accessible to a much wider circle of readers. The years since first publication of the German edition have seen a great surge of discussion and research on evaporites among English-speaking geologists, and while BRAITSCH'S work was important to some of these it can now no longer be ignored by any.

It may be worth pointing out that, title notwithstanding, this volume takes as its subject not only the salts but all evaporite minerals and rocks, but that it makes no attempt to go beyond the geochemical evidence in discussing the origin of these evaporites. The reader should not expect to find here any extensive discussion of paleogeography and stratigraphy, or even petrographic detail. Within its scope of geochemistry and mineralogy the coverage is extensive.

OTTO BRAITSCH'S research on evaporites began with his dissertation, and the present volume is in some respects an elaboration of that first work, written while he was a Privatdocent at Göttingen. He did not regard it as final and complete in any sense, as will be evident in the many questions he raises here that are not much closer to solution now than they were then. Had it not been for BRAITSCH'S untimely accidental death in 1966, we might well have seen a *new* English edition in 1971. But it is a tribute to his mastery of the subject that, on all but a few points, the book is still a definitive survey of the field. And the admonition with which he concluded his book still needs to be listened to: "Today we can and must demand of geological investigations, insofar as they have pretensions to pose and answer questions concerning the genesis of salt deposits, adequate documentation of their qualitative and quantitative composition and, over and above this, an understanding of the underlying physicochemical principles." If we follow this admonition and so generate facts and concepts that will make this book outdated, we will probably have accomplished the purpose for which OTTO BRAITSCH wrote it.

WILLIAM T. HOLSER, Eugene, Oregon

From the Foreword to the German Edition

The present work stems from the greatly underestimated pioneer work of VAN'T HOFF, JÄNECKE, and BOEKE, the important researches of KURNAKOV and his school and the investigations of equilibrium conditions by D'ANS, AUTENRIETH and BRAUNE, and many others. If it succeeds in putting into some sort of order, and in elucidating the manifold processes and alterations taking place in the formation of salt deposits, it is thanks to these earlier workers. In addition valuable explanations and progress resulted from the diverging points of view of other researchers, which in part is recorded in the literature and in part was obtained from lectures and discussions and personal communications. To all those persons the author expresses his deepest thanks.

This work could not have been successfully carried out without the harmonious collaboration of my colleagues and without the favourable working conditions in the Mineralogische Anstalt. For his constant stimulation and encouragement as well as for the invitation to produce this monograph, I should like particularly to thank Professor CORRENS. For his important contributions which included the experimental investigation of the partition of bromine, and for reading the manuscript, I should also like to express my gratitude to Dr. A. G. HERRMANN. Mr. R. THIEL for technical and chemical analyses as well as Miss R. BLANK for her chemical analyses are both to be thanked. My thanks are also due to Dr. R. KÜHN, Hannover, for his valuable advice particucularly in the first year of my work on salt deposits, for references to the literature, and for samples. His unselfish support always readily given despite differences in opinion on many questions is gratefully acknowledged. I should also like to thank Dr. BAAR of Kassel for his contributions to certain problems.

The investigation of salt deposits is by no means exhausted, and many times the gaps in our knowledge must be apparent. There are certainly errors in some points and I would be grateful to all readers for indications of gaps and errors and for ideas for possible improvements.

By means of many inclusions during the proof stage, the attempt was made to bring this work as up to date as possible. For the possibility to do this and for their painstaking preparation for printing, I am deeply grateful to Springer-Verlag.

The help of the Deutsche Forschungsgemeinschaft in supporting my co-workers and materially providing for chemical analyses is acknowledged.

My greatest debt is to my wife for her patient and understanding consideration, protecting my peace for working and not least for laboriously copying the manuscript.

OTTO BRAITSCH

Translators' Preface

To the general geologist, the term "evaporites" is an all-embracing one covering a large variety of salts, often unspecified and certainly poorly-understood. At most, reference is made to thickness, presence or absence of potash salts, and the presence or absence of gypsum and anhydrite. Interest may reach as far as to attempt paleoclimatological or paleotemperature reconstruction using the sequence of salts given in the work of VAN'T HOFF. In the geological literature in English, however, there is an astonishing dearth of detailed investigations of the salts as geochemical systems, despite the fact that there exists a considerable volume of accessible data.

In the study of the stable and metastable equilibria of salt minerals, in the consideration of the physico-chemical conditions affecting the formation of evaporites, and in the mineralogical investigation of salt deposits, much of it carried out in, or stimulated by the Kaliforschungs-Institut Hannover, German scientists have reigned supreme. Among them, OTTO BRAITSCH was one of the leading figures.

The publication of BRAITSCH's monograph in English comes at a time when evaporites are the subject of an awakening of interest among English-speaking geologists. The publication of the three Northern Ohio Geological Society symposia on evaporites, and the appearance of the proceedings of the International Conference on Saline Deposits (Geol. Soc. Amer., Spec. Paper 88) are evidence of this interest.

Although this monograph was never regarded as the final word on the subject, and indeed BRAITSCH refers constantly to areas where further research is required, it is nevertheless the most useful single work on the geochemistry of evaporites. It is comprehensible and readable, and does not burden the reader with the mass of detail upon which many of the geochemical arguments are based. BRAITSCH wrote it for those geologists who wish to understand the formation of evaporite deposits.

We are glad to have had the opportunity to translate BRAITSCH's work and have tried to make the English version as clear and readable as the original. If we have failed and if there are misinterpretations in the text the fault is ours. We should like to record the friendly interest in the progress of the translation shown by Dr. K. F. SPRINGER, and the prodigious efforts of Mrs. B. M. CROOK of Springer-Verlag Heidelberg. As a result of her painstaking work not only were a number of mistranslations corrected but the form of the translation considerably improved. Dr. A. G. HERRMANN of the Geochemical Institute of the University of Göttingen also checked the translation, made a number of suggestions and factual corrections and updated BRAITSCH'S work by the inclusion of many references published since the latter's tragic death, and we would like to acknowledge his expert help. Dr. R. EVANS of Mobil Research and Development Corporation of Dallas, Texas, helped to ensure that the terminology conforms to English usage and we would like to thank him and the company for providing the time and facilities for this assistance. Our final wish is that this translation will stand as a fitting memorial to an outstanding earth scientist.

P. J. BUREK, Dallas, Texas
A. E. M. NAIRN, Cleveland, Ohio

Contents

A. Introduction

1. Outline of the Theme and Statement of the Problem

In the following monograph only those salts precipitated from seawater or from solutions of similar composition will be considered. As can be seen from Table 1 (from SVERDRUP et al., 1949) if water itself is excluded there are actually only six main components in seawater, namely

the cations Na^+, Mg^{++}, Ca^{++}, K^+,

and

the anions Cl^-, SO_4^{--}.

These are also the components of the salt minerals which form the major part of salt deposits derived from oceanic water.

Most salt deposits have a chequered history. This history is best understood by considering salt deposits in a genetic sequence, beginning with the conditions under which the first deposits were formed, the composition of the solution, temperature etc., and then their subsequent alteration during the course of diagenesis and metamorphism.

VAN'T HOFF (1896–1911), with the help of his colleagues and students, by the systematic investigation of the solution equilibrium of the marine salt system, succeeded in laying the foundations of an understanding of the formation of salt deposits. The object of the present work is to proceed from there to quantitative deductions on a physico-chemical basis. Comparisons must therefore be made between observed salt sequences and those calculated for various solution equilibrium conditions. Naturally, only the most important basic types can be considered. The comparison with natural salt sequences is, however, only significant when quantitative data on the latter are to hand, which unfortunately is seldom the case.

This objective was also that of VAN'T HOFF (1905–1909) and JÄNECKE (1923). The models of these pioneer investigators of salt deposits gave no satisfactory agreement between theory and observation. Subsequently various attempts were made to find an explanation for this divergence, e.g. precipitation under conditions of falling temperature (BORCHERT, 1940), metastable equilibria (the Kurnakov school – see VALYASHKO, 1958), bacterial reduction of sulphates (BORCHERT, 1940), solid state reactions and assimilation (LEONHARDT, BERDESINSKI, 1949/50, 1952). Although these attempts proceeded from accurate observation, they were concerned only with particular cases which, in the general problem of the formation of salt deposits, are of only minor importance. Nevertheless, the validity or applicability of physico-chemical principles should not be doubted for that reason, as has often been the case, but rather the formulation of the model should be rigorously checked and improved. The recognition of secondary alterations brought about by solutions, first quantitatively considered by D'ANS and KÜHN (1940, p. 79) but appreciated qualitatively much earlier (TSCHERMAK (1871, p. 306), NAUMANN (1913), RÓZSA (1915) and others), represented a significant step forward.

Likewise of significance is the alteration of the original solution by influx from other sources (STURMFELS, 1943; VALYASHKO, 1958, etc.). A particular case of this kind, is the alteration caused by the precipitation of specific minerals due to the influx of fresh seawater (BRAITSCH, 1961c). The influx of fresh seawater is the most important single factor in any hypothesis explaining of the origin of all major salt deposits with a salt thickness in excess of 10 m (Ochsenius' Bar theory and variants).

The geological hypotheses concerning the origin of salt deposits, including non-marine evaporites, are not considered here, since they are to be found in LOTZE'S book, *Steinsalz and Kalisalz* (1938, 1957). The geological conclusions which arise out of physico-chemical considerations will, however, be dealt with in this book.

The restriction of research to the main components in solution equilibria is both necessary and useful. This does not mean, however, that minor components should be ignored in the examination of the composition of salt deposits which forms such a large part of any mineralogical-petrographical investigation. Even at a comparatively early stage VAN'T HOFF had concerned himself with the rare element boron which forms accessory minerals in salt deposits. The subsidiary components provide in many cases important clues to the genesis of deposits. This is particularly true of bromine, and the principal features of the laws governing its distribution were known to BOEKE (1908). In general, however, the behaviour of the accessory components in the formation and alteration of salt deposits is problematic. The attempt will be made here to present the current state of knowledge and also the problems without abandoning an independent point of view.

2. On the History of Seawater

For the purposes of this work all discussion will be based on the present composition of seawater. An increase in the overall salt content of the oceans during the course of the Earth's history has no real effect on the discussions. The greater part of the developmental history of the oceans occurred in Precambrian times, and furthermore, it is far from certain that the salt content varies linearly with time. Through the repetition of major periods of salt formation since the Cambrian, salt has been repeatedly abstracted from seawater, so that for the Earth's history since the Cambrian an approximately constant average salt content of seawater may be assumed. Although the anions and cations of seawater had different origins (RUBEY, 1951; v. ENGELHARDT, 1959), they had already reached a state of balance within the Precambrian.

Table 1. *The composition of modern seawater, referred to a standard chlorinity*[a] *of 19‰ (after* SVERDRUP, JOHNSON, *and* FLEMING, *1949, p. 173)*

	‰	$\frac{\text{mol}}{1000 \text{ mol } H_2O}$	Converted to formal JÄNECKE units (see p. 30)			Hypothetical components in weight percent
Na^+	10.56	8.567	H_2O	70,690		78.03 % NaCl
Mg^{++}	1.27	0.976	Na_2Cl_2 } Na_2F_2 }	302.8		0.01 % NaF
Ca^{++}	0.40	0.186	K_2	6.41	} 100	2.11 % KCl
K^+	0.38	0.181	Mg	69.00	} 100	9.21 % $MgCl_2$
Sr^{++}	0.008[b]	0.002	SO_4	24.59	} 100	0.25 % $MgBr_2$
Cl^-	18.98	9.988	$CaCO_3$	1.51		6.53 % $MgSO_4$
SO_4^{--}	2.65	0.514	$CaSO_4$	11.65		3.84 % $CaSO_4$
HCO_3^-	0.14	0.043	$SrSO_4$	0.12		0.05 % $SrSO_4$
Br^-	0.065	0.015	B	0.55		0.33 % $CaCO_3$
F^-	0.0013	0.001				100.00 %
B	0.0045	0.008				

[a] Chlorinity = total Cl^-, Br^-, I^- per g/kg seawater, where Br^- and I^- are converted to equivalent quantities of Cl^-. Chlorinity may be determined directly through the determination of Cl by titration with silver nitrate. Salinity = total salt content per g/kg seawater. Calculated from chlorinity according to the empirical relationship: Salinity $= 0.03 + 1.805 \cdot$ Chlorinity.

[b] After SUGAWARA and KAWASAKI (1958).

The ionic ratio in seawater is of overwhelming significance in the formation of salt deposits. According to the results of the Challenger Expedition (1870), it is virtually constant, so that from the analytical determination of one component the overall salt content of the oceans can be calculated (see Table 1, Footnote a). Because of the different origins of cations and anions, and because of differences in the relative proportion of limestones and dolomites in different geological ages (DALY, 1909; RONOV, 1959), constant ionic ratios in the geological past cannot be presumed. As the possible changes were certainly slow and possibly had reached a balance, and as the time interval here considered represents only about one sixth of the age of the oceans, such changes may be neglected, particularly as, in the existing state of knowledge of the history of seawater, a numerical correction is not possible. The percentage of certain elements, e.g. bromine, in the total salt content can be estimated within certain limits (p. 203) and the value given for bromine in the Permian seas is the same as in the present ocean.

3. The Zechstein Salt Succession

The majority of the examples to be used later are taken from the Permian, the period of greatest salt deposition in the Earth's history, with particular emphasis upon the German Zechstein. Table 2 gives the succession (abbreviated after RICHTER-BERNBURG, 1953) accepted by the geological surveys, oil and salt industries.

The salt series are further subdivided and individual horizons from the base upwards denoted by Greek letters: e.g. Na 3α = "Basis"-salt, Na 3β = "Linien"-salt etc., although for present purposes the shortened succession is sufficient. The potash beds of the Werra and Leine Series are intercalated with rock salt (halite). In addition, locally potash beds are found in the Leine Series. The regional distributions of the four evaporite cycles do not coincide, and there are also marked thickness variations. A simplified sketch map of the central part of the Zechstein basin with the approximate original boundary of the Stassfurt, Ronnenberg and Riedel seams is shown in Fig. 1 (after FULDA, 1935; LOTZE, 1938) The Thuringia and Hesse seams occur only in the Hessen-Thuringian region. Further paleogeographic maps are to be found in LOTZE's book (1938).

The following districts can be distinguished (the numbers refer to Fig. 1, the more important mines being gives in parentheses):

North German plain: 1. Lübtheen (Friedrich Franz, Jessenitz), 2. Conow, 3. Wustrow, 4. Wolmirstedt.

Magdeburg-Halberstadt region = subhercynian basin: 5. Rothenfelde, 6. Oberes Allertal, 7. Schönebeck (Graf-Moltke-Schacht), 8. Dorm, 9. Asse, 10. Vienenburg, 11. Huy (Wilhelmshall I + II), 12. Stassfurt-Egelner-Sattel (Westeregeln; Neustassfurt; Achenbach; Berlepsch; Maybach; Ludwig II; Leopoldshall I–V), 13. Bernburger-Plateau (Gröna; Solvayhall; Solvay i. Pr.), 14. Aschersiebener-Sattel (Aschersleben I–VII; Kleinschierstedt).

North Hannover region: 15. Ahnebergen, 16. Grosshäuslingen, 17. Grethem, 18. Hope (Adolfsglück), 19. Steinförde, 20. Habighorst (Mariaglück), 21. Hänigsen (Niedersachsen, Riedel), 22. Wunstorf

Table 2. *Stratigraphic divisions of the German Zechstein Z1–Z4*

Cycle	Name	Symbol	Stratigraphic description
Overlying beds			Trias (Bundsandstein)
Z 4	Aller Series [a]	A 4 r	Grenzanhydrit (boundary anhydrite)
		Na 4	Aller rock salt (= "Youngest" rock salt)
		A 4	Pegmatitanhydrit
		T 4	Red Salzton
Z 3	Leine Series	Na 3 { K 3 Ri	with Riedel potash seam
		Na 3	Leine rock salt (= "Younger" rock salt)
		Na 3 { K 3 Ro	with Ronnenberg potash seam
		A 3	Hauptanhydrit
		Ca 3	Plattendolomit
		T 3	Gray Salzton [b]
Z 2	Stassfurt Series	K 2	Stassfurt seam
		Na 2 (K)	Kieseritic transition beds
		Na 2	Stassfurt rock salt (= "Older" rock salt)
		A 2	Basalanhydrit
		Ca 2	Hauptdolomit (Ca 2 d = Hauptdolomit Ca 2 st = Stinkschiefer)
		T 2	Brownish red "Salzton" [b]
Z 1	Werra Series	A 1 β	Upper Werra anhydrite
		Na 1 { K 1 H	with Hesse potash seam
		Na 1	Werra rock salt (= "Oldest" rock salt)
		Na 1 { K 1 Th	with Thuringia potash seam
		A 1 α	Lower Werra anhydrite (= Anhydrit-Knotenschiefer)
		Ca 1	Zechstein limestone
		T 1	Kupferschiefer
Underlying beds			Lower Permian (Rotliegendes)

[a] The Aller and Leine Series are sometimes together referred to as the Niedersachsen Series.

[b] Salzton = salt clay, but here the term has a stratigraphical significance and in this table the German is retained – Transl.).

(Sigmundshall), 23. Benthe (Hansa Silberberg; Ronnenberg I, Deutschland), 24. Sehnde-Sarstedt (Bergmannssegen; Hugo), 25. Ölheim, 26. Ölsburg, 27. Rolfsbüttel, 28. Ehmen (Einigkeit I), 29. Thiede, 30. Ösel, 31. Flachstöckheim, 32. Salzgitter.

South Hannover region: 33. Hildesheimer Wald (Hildesia; Mathildenhall; Salzdetfurth), 34. Oberes Leinetal (Desdemona), 35. Grossrhüden, 36. Salzderhelden, 37. Solling (Wittekind, Hildasglück = Volpriehausen), 38. Nörten (Königshall-Hindenburg = Reyershausen).

South Harz region: 39. Worbis (Bismarckshall b. Bischofferode), 40. Bleicherode (Craja: Kleinbodungen; Sollstedt), 41. Hüpstedt (and

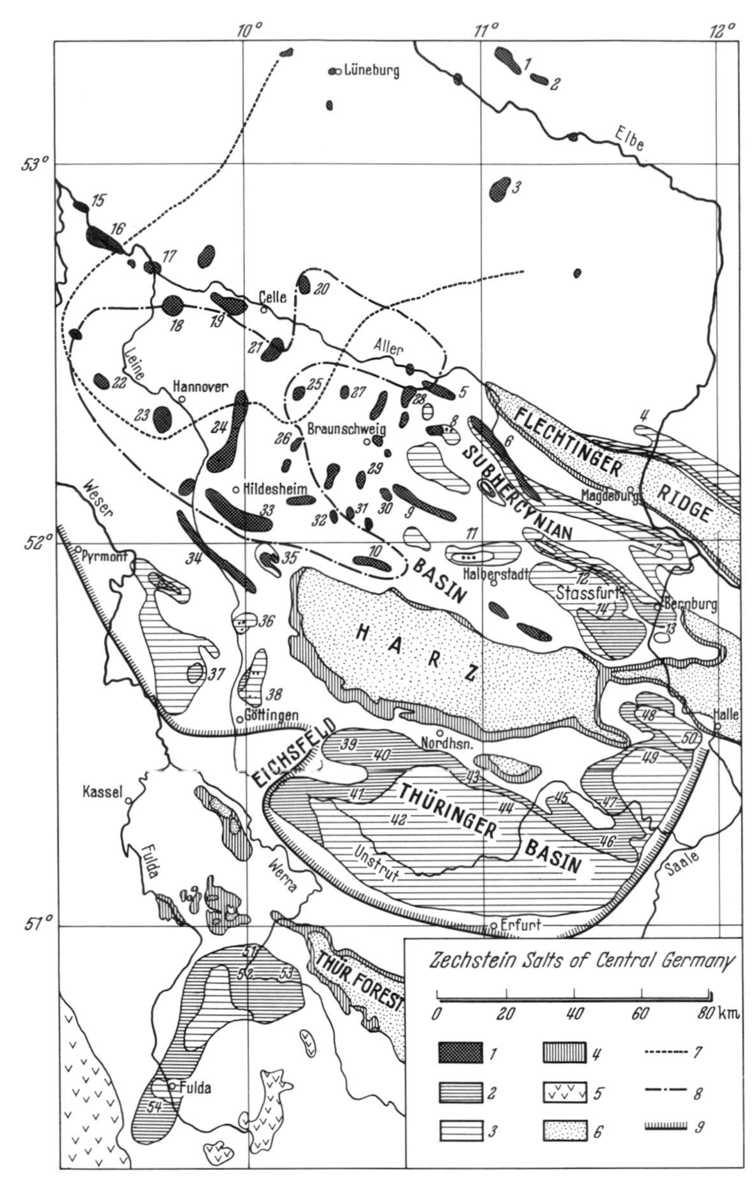

Fig. 1. Distribution of Zechstein salts in Central Germany (simplified after FULDA, 1935). 1. North German salt diapirs; 2. potash salts at depths of less than 1000 m, tectonically simple structures; 3. The same at depths greater than 1000 m; 4. Zechstein outcrops of gypsum, dolomite etc.; 5. Tertiary basalts of the Vogelsberg and Rhön; 6. Lower Permian and Paleozoic rocks; 7. Limits of the Riedel Seam (K 3 Ri); 8. Limits of the Ronnenburg Seam (K 3 Ro), after SYDOW, 1959; 9. South and southwestern limits of the Stassfurt seam. For numbers of potash districts, see text

Felsenfest bore), 42. Volkenroda-Pöthen. 43. Sondershausen-Wolkramshausen (Glückauf).

Unstrut-Saale region: 44. Hainleite, 45. Schmücke (Heldrungen), 46. Finne (Reichskrone), 47. Rossleben (Wendelstein; Georg; Unstrut), 48. Mansfelder Mulde, 49. Oberröblingen (Krügershall), 50. Halle/Saale.

Hesse-Thuringian region (only the Thuringia and Hesse seams occur): 51. Heringen/Werra (Alexandershall; Herfa-Neurode; Wintershall), 52. Philippstal (Hattorf; Sachsen-Weimar), 53. Salzungen (Grossherzog v. Sachsen; Kaiseroda), 54. Fulda (Neuhof-Ellers).

The English Zechstein can also be correlated with the German succession (LOTZE, 1958):

Upper Evaporite zone = Z 3,
Middle Evaporite zone = Z 2,
Lower Evaporite zone = Z 1.

4. The Minerals of Salt Deposits

a) The Main Components (Salt Minerals *sensu stricto*)

The mineralogy of salt deposits will not be dealt with exhaustively here. A detailed treatment by KÜHN is in preparation (a revision of an unpublished work of 1950). Similarly, despite their importance, the questions of sampling and research methods will not be considered.

The most important crystallographic data and physical properties of salt minerals are summarized in Table 3. The minerals are arranged according to their main components, even if the succession does not strictly conform to the pattern of the subsequent review. In the absence of newer results, the data given are taken principally from DANA II which has proven to be the most reliable source. The lattice constants are largely drawn from STRUNZ (1957), the Herrmann-Mauguin symbol of the space-group was checked, and where necessary revised, from the extinction law in the original description. A few synthetic compounds are also included. The mineral names are presented with a brief history and etymology from KÜHN (1959).

Further details on occurrence and mineralogical details of the various salt minerals will be found in the appropriate section on solid phases preceding the review of stability conditions. A discussion of the crystal chemistry cannot be entered into here, particularly as a clear understanding of solution equilibria from a crystal chemistry standpoint is not yet forthcoming (see p. 29). A key diagram for the microscopic analysis is given in Fig. 2. The optic axial angle and the optical characteristics, and where appropriate, the non-corrosive immersion media will be included here.

Explanation and data sources for Table 3.

Table 3. *Salt minerals*

1	2	3	4		5	6	7
No.	Mineral (Synonym) Formula	Crystal system Space group	Unit cell dim. a_0 b_0 c_0	 α β γ	common forms habit, twinning	Cleavage Translation and simple Shear. Fracture Hardness, Density	Lustre Colour
a) Ca-bearing Sulphates							
1	Anhydrite $4[CaSO_4]$	orthorhombic *Bbmm*	6.238 6.991 6.996		isometric tabular (100) or (001) (010) prismatic [010], rare [100] striated (h0 l), (0k l); twinning (101), in part lamellar; (012) rare	(001), perfect, pearly; (010) good, (100) distinct $T(001)t=[010]$ $T(012)t=[100]$ or $[0\bar{2}1]$ $K_1(101)$ $K_2(\bar{1}01)$ $H=3^1/_2$–4 $D=2.96$–2.98	Colourless bluish (violet, white, rose acc. to inclusions)
2	Gypsum $4[CaSO_4 \cdot 2\,H_2O]$	monocl. prismatic $A\frac{2}{a}$	5.68 15.18 6.29	 113.85°	(010) tabular with $(\bar{1}11)$ (120) reticulate [001] or [101] granular, fibrous porphyroblastic Twinning (100) swallowtail $(\bar{1}01)$ montmartre	(010) perfect, pearly (100) good, conchoidal (011) fibrous $T(010)t=[001]$ $H=2$ $D=2.317$	Pearly-vitreous colourless (may be white, gray, yellow from inclusions)
3	Bassanite (hemihydrate) $6[CaSO_4 \cdot {}^1/_2\,H_2O]$	ditrigonal scalenohedral $C\,3_2\,2$ (?)	6.84 — 12.72		fine acicular	$D=2.7$	white
3a	"soluble anhydrite" $\gamma CaSO_4$	$P\,6_2 22$	6.99 — 6.34			$D=2.58_7$	
4	Glauberite $4[Na_2Ca(SO_4)_2]$	monocl. prismatic $C\frac{2}{c}$	10.10 8.28 8.51	 112.2	tabular (001) (111) ± (110) (100) prismatic $[\bar{1}01]$ or [001]	(001) perfect, pearly (110) indistinct, vitreous Conchoidal fracture, brittle $H=2^1/_2$–3 $D=2.85$ (calculated 2.86_6)	gray, yellowish (reddish from inclusions)
5	Polyhalite $4[K_2MgCa_2(SO_4)_4 \cdot 2\,H_2O]$	triclinic $F\bar{1}$	11.69 16.33 7.60	91.6 90.0 91.9	tabular (010); (100) $(\bar{1}k\,1)$ prismatic [001], fibrous. Always twinned (010) and (100) lamellar	$(10\bar{1})$ perfect (010) indistinct $H=3^1/_2$ $D=2.78$	vitreous colourless white gray or red from inclusions
6	Syngenite $2[K_2Ca(SO_4)_2 \cdot H_2O]$	monocl. prismatic $P\frac{2_1}{m}$	9.72 7.16 6.21	 104.1	tabular (100); (hk0), (h01) common. prismatic [001]. Cryst. crust, lammella aggregates. Twinning (100)	(100) prefect (110) perfect, (106°) (010) distinct Conchoidal fracture	vitreous colourless pale yellow blue (from inclusions) milkywhite
7	Goergeyite (Mikheevite, pentasalz) $K_2Ca_5(SO_4)_6 \cdot H_2O$	monocl.	17.1 6.71 18.2	 113.2	tabular (001)	(100) distinct $H=3^1/_2$ $D=2.77$	vitreous colourless reddish from inclusions
b) Chlorides							
8	Rock salt (halite) $4[NaCl]$	Cubic-hex. oct. $F\frac{4}{m}\bar{3}\frac{2}{m}$	5.6404		{100} granular, occ. fibrous; "funnel-shaped" crystals (hopper crystals)	(100) perf., conch. fract. $T(011)t=[110]$ $T(001)t=[110]$ $H=2$ $D=2.168$	vitreous colorless (white, yellow, red, blue purple)
9	Hydrohalite $NaCl \cdot 2\,H_2O$	monocl. (pseudohexogonal)	n.k.	n.k.	tabular (001)	not determined $D=1.630$	colourless
10	Sylvite $4[KCl]$	cubic-hex. oct. $F\frac{4}{m}\bar{3}\frac{2}{m}$	6.293		{100}; rare {111} granular	(001) perfect $T(110)$ $t[1\bar{1}0]$ $H=2$ $D=1.99$	vitreous, white (opalescent) grey, bluish yellowish red with haematite inclusions)

(explanations p. 14–17 and 20)

	8	9	10								11
	Refraction, Opt. orientation Birefringence, Opt. angle char. Dispersion	Solubility (20° C) Varieties Taste	Powder diagram d-values (see expl.) Relative intensity hkl								Paragenesis (only in salt deposits)
1	$n\alpha = 1.5698$ $\parallel c$ $n\beta = 1.5754$ $\parallel b$ $n\gamma = 1.6136$ $\parallel a$ $+0.0438$ $2V\gamma = 43.7°$ $\varrho < v$	0.2% at 45° C	3.87 6 111	3.498 100 {002 020	3.118 3 200	2.849 33 210	2.328 22 {202 220	2.208 20 212	1.869 15 230	1.648 14 232	from gypsum with dolomite and halite, in Hartsalz; replaced by polyhalite, hydrated to gypsum
2	$n\alpha = 1.5207$ $n\beta = 1.5230$ $\parallel b$ $n\gamma = 1.5299$ $\lvert c + 52°$ $+0.0092$ $2V\gamma = 58°$ $\varrho > v$, inclined strongly temp. dep.	0.2% Selenite, Alabaster	7.59 100 020	4.29 100 120	3.80 15 {031 040	3.07 60 140	2.87 60 {002 $\bar{1}22$	2.68 50 {022 051	2.088 2.073 (together) 50 {122 $\bar{2}42$ {$1\bar{1}3$ $\bar{2}51$	1.900 25 {013 080	Primary ppt. from seawater, secondary from anhydrite, polyhalite. etc.
3	$n\omega = 1.558$ $n\varepsilon = 1.586$ $+0.028$ $(2V\gamma = 10–15)$	0.88%									Metastable, from dehydration of gypsum, mostly a synthetic product
3a	$n\omega = 1.501$ $n\varepsilon = 1.546$ $+0.045$		6.03 90 10.0	3.48 55 11.0	3.01_5 100 20.0	2.80 75 10.2	2.14_4 20	1.91_6 10 30.1	1.85_0 35	1.697 15 30.2	from dry dehydration of gypsum, metastable
4	$n\alpha = 1.515$ $n\beta = 1.535$ $\lvert c\ 12°$ $n\gamma = 1.536$ $\parallel b$ -0.021 $2V\alpha = 7°$ $\varrho > v$ very strong Strong λ. Temp. dep.	incongruent soln. with ppt. of gypsum weak salty taste	6.21 15 110	4.677 8 200	4.37_5 25 111	4.15 3 020	3.941 100 002	3.166 35 $\bar{2}21$	{3.114 3.101 35 {$\bar{3}11$ 220	2.668 45 {$\bar{1}13$ 221	in oceanic and terrestrial salts with halite, anhydrite, gypsum, polyhalite, thenardite, mirabilite, bloedite. Primary and as alteration product
5	$n\alpha = 1.547$ $\lvert c\ 12°$ $n\beta = 1.560$ $\approx \parallel [\bar{1}10]$ $n\gamma = 1.567$ -0.020 $2V\alpha = 64°$	incongruent soln. with ppt. of gypsum almost without taste Krugite = mixture anhydrite and polyhalite	6.00 8 $1\bar{1}1$	3.33 3 $13\bar{1}$	5.02 7 200	4.002 7 $2\bar{2}0$	3.106 8 {$3\bar{1}1$ $\bar{3}11$	3.176 45 {$\bar{2}02$ 202	{2.89 2.912 100 {$\bar{1}\bar{5}1$ 400	2.946 12 151	early diagenetic-diagenetic from anhydrite in rock salt, with kieserite, rare in carnallite, anhydrite, metam. from kieserite. Carnallite with kieseritic sylvite, langbeinite
6	$n\alpha = 1.501_0$ $\lvert c - 2.3°$ $n\beta = 1.516_6$ $n\gamma = 1.517_6$ $\parallel b$ -0.016_6 $2V\alpha = 28.3°$ $\varrho < v$ very strong Strongly Temp. dep.	incongruent soln. with ppt. of gypsum almost without taste	9.5 65 100	5.73 15 110	4.75 20 200	4.63 3 011	3.95_5 9 210	3.160 100 300	2.89_4 25 310	2.857 20 220	glaserite, mirabilite in soln. cavities, sec. with halite, anhydrite, recent precipitate from waste solutions
7	$n\alpha = 1.560$ $\lvert a\ 40^1/_2°$ $n\beta = 1.569$ $\parallel b$ $n\gamma = 1.584$ $+0.024$ $2V\gamma = 79°$	incongruent soln. with pptn. of gypsum	8.04 2	6.30_2 11	5.61 15	4.91 3	3.37_1 15	3.16_6 30	3.01_0 100	2.13_5 8	metamorphic with glauberite, polyhalite, halite
8	$n_D^{18°} = 1.5443$	26.4%, salty yellow flame	3.258 13 111	2.821 100 200	1.994 55 220	1.701 2 311	1.628 15 222	1.410 6 400	1.294 1 331	1.261 11 420	in all salt rocks, primary and from metamorphic solutions
9	not known	incongruent soln. at 0.1° C with ppt. of halite	not known								winter ppt. in recent salt seas, from evap. of seawater at low temp.
10	$n_D^{20°} = 1.4903$	25.6% exceedingly salty red-violet lustre	3.146 100 200	2.224 59 220	1.816 23 222	1.573 8 400	1.407 20 420	1.284 13 422	1.112_6 2 440	1.049_0 6 600	primary with halite and anhydrite, metamorphic with halite, kieserite ± anhydrite, langbeinite

Table 3

1 No.	2 Mineral (Synonym) Formula	3 Crystal system Space group	4 Unit cell dim. a_0 b_0 c_0	4 α β γ	5 common, forms habit, twinning	6 Cleavage Translation and simple Shear. Fracture Hardness, Density	7 Lustre Colour
11	Bischofite $2[MgCl_2 \cdot 6H_2O]$	monocl. prismatic (pseudotrigonal) $C\frac{2}{m}$	9.92 7.16 6.11	93.7°	(100)(110)(111), granular, flaky in part fibrous short prisms [001] K_1 irrat. $K_2(1\bar{1}1)$ $\eta[1\bar{1}2]$ $T(110)$	cleavage not known $T(110)$ perfect Uneven conchoidal fracture $H=1$–2 $D=1.604$	vitreous dull colourless, white
12	Carnallite $12[KMgCl_3 \cdot 6H_2O]$	orthorhombic (pseudotrigonal) *Pban*	9.56 16.05 22.56		thick tabular (001); (0kl) (111) pseudotrigonal. Pyramidal, granular, massive, fibrous Twinning (110) lamellar	cleavage not known conchoidal fracture K_1 (110) K_2 $(1\bar{3}0)$ $H=2^1/_2$ $D=1.602$	greasy dull lustre. Colourless to milky white often reddish with metallic sheen due to oriented haematite inclusions rarely yellowish, blue
13	Tachyhydrite $CaMg_2Cl_6 \cdot 12H_2O$	trigonal?	not known		$\{10\bar{1}1\}$; {0001} massive, granular pressure twinning lamellar	$(10\bar{1}1)$ distinct $H=2$ $D=1.667$	vitreous dull lustre pale reddish yellow also colourless
14	Chlorocalcite (Baeumlerite) $KCaCl_3$	orthorhombic (pseudocubic)	not known		pseudocubic and octahedral; prismatic [001], tabular (001). Twinning (lamellar) ∥ pseudo cubic faces	pseudo cubic 1 good, 2 clear $H=2^1/_2$–3	white (violet)
c) Alkali-sulphates							
15	Glaserite (variety Aphthitalite) $1[K_3Na(SO_4)_2]$	ditrigonal scalenohedral $P\bar{3}m1$	5.66 — 7.30		tabular (0001); $(10\bar{1}2)$; $(10\bar{1}0)$; rhombohedral fissile, botryoidal aggregates, massive. Twinning (0001) trilling $(11\bar{2}0)$	$(10\bar{1}0)$ distinct (0001) indistinct, uneven conchoidal fracture $H=3$ $D=2.65_6$	vitreous, lustre, white rarely colourless (gray, blue, green, reddish through haematite)
16	Mirabilite (Glaubers salt) $4[Na_2SO_4 \cdot 10H_2O]$	monocl. prismatic $P\frac{2_1}{c}$	11.51 10.38 12.83	107.75	prismatic or lathlike [001] or [010] thin tabular (100) or (001) massive, felted crusts. Twinning rare (001) penetration twins (100)	(100) perfect (001) (010) (011) distinct conchoidal fracture $H=1^1/_2$–2 $D=1.490$	vitreous, colourless transparent, white
17	Thenardite $8[Na_2SO_4]$	rhomb. bipyr. *Fddd*	9.821 12.304 5.863 (25° C)		bipyramidal (111) (011) (010) (101); tabular (010); rarely prismatic [100]; powdery encrustations and efflorescence cruciform twins (110)	(010) perfect (101) good (100) indistinct $H=2^1/_2$–3 $D=2.664$	glossy-resinous pure; colourless also gray white, yellow, yellow brown or reddish transparent, translucent
d) Mg sulphates							
18	Epsomite (Reichardtite, bitter salt) $4[MgSO_4 \cdot 7H_2O]$	rhombic disphenoidal $P2_12_12_1$	11.86 11.99 6.85_8 (25° C)		short prisms [001] (110) (010) (111) (101); mostly fibrous, ∥ [001], woolly efflorescence; bothryoidal-kidney masses, stalactites. Twin. (110) rare	(010) perfect, (101) distinct conchoidal fracture $T(100)$ $t[110]$? $T(100)$; (011); (101) $H=2$–$2^1/_2$ $D=1.677$	vitreous, fibrous silky lustre dull also pale red or green
19	Hexahydrite (Sakiite) $8[MgSO_4 \cdot 6H_2O]$	monocl. prismatic $C\frac{2}{c}$	10.06 7.16 24.39	98.6	(001); columnal, fine fibrous, rarely thick tabular. Twinning (001); (110)	(100) perfect conchoidal fracture H not known $D=1.75$	pearly lustre colourless white pale green not transparent generally

(continued)

	8	9	10	11
	Refraction, Opt. orientation Birefringence, Opt. angle char. Dispersion	Solubility (20° C) Varieties Taste	Powder diagram *d*-values (see expl.) Relative intensity hkl	Paragenesis (only in salt deposits)
11	$n\alpha = 1.495$ $\parallel b$ $n\beta = 1.507$ $\mid c$ $9^1/_2{}^\circ$ $n\gamma = 1.528$ $+0.003$ $2V\gamma = 79°$ $\varrho > v$ weak	35.1% $MgCl_2$ very hygroscopic exceedingly bitter alcohol soluble	5.8 4.10 3.57 2.98 2.88 2.72 2.65 2.23 5 10 5 6 9 8 10 8 110 111 020 310 220 $1\bar{1}2$ {$\bar{2}21$, 112} 401	final ppt. with a little carnallite, kieserite, halite, much tachhydrite Metam. with carnallite
12	$n\alpha = 1.466$ $\parallel c$ $n\beta = 1.475$ $\parallel b$ $n\gamma = 1.494$ $\parallel a$ $+0.028$ $2V\gamma = 66°$ $\varrho < v$ weak	incongruent soln. with sylvite pptn. (27.3% $MgCl_2$) hygroscopic, bitter. Altered in alcohol and glycerine with sylvite pptn.	7.70 5.53 4.776 4.667 3.596 3.316 2.923 1.740 5 14 20 14 100 35 65 35 111 113 200 130 {223, 134} 143 {152, 321} n. k.	primary with halite, anhydrite, ± kieserite (polyhalite), rare with sylvite, kainite, hexahydrite, etc. Sec. by cooling of metam. solns. and in salt clay
13	$n\varepsilon = 1.512$ $n\omega$ 1.520 -0.008	very hygroscopic, sharp, bitter, lustre weakly sol. in alcohol	7.75 5.75 5.04 4.67? 3.798 3.085 2.875 2.633 2.602 10 15 30 <5 45 43 40 30 100	primary? with bischofite. Metam. with carnallite (bischofite), mostly in clay. Not paragenetic with sylvite, kieserite.
14	$n\alpha$ = not known $n\beta = 1.52$ $n\gamma$ = not known − small $2V\alpha$ small	hygroscopic, bitter	not known	with tachhydrite and halite, concordant (?) in the Leine Rock salt (only in Desdemona, Alfeld)
15	$n\omega = 1.491$ $n\varepsilon = 1.499$ occ. weak $+0.008$ 2 axes, small $2V$	~11% K_2SO_4 weakly salty, bitter	4.902 4.07 3.645 3.129 2.834 2.431 2.325 2.035 10 30 30 75 100 25 15 40 10.0 10.1 00.2 ? 11.0 00.3 20.1 20.2	secondary with bloedite, schoenite, syngenite, mirabilite in cap rocks. Metam. with polyhalite, anhydrite
16	$n\alpha = 1.394$ $\parallel b$ $n\beta = 1.396$ $n\gamma = 1.398$ $\mid c + 31°$ -0.004 $2V\alpha = 76°$ $\varrho < v$ strong	16% Na_2SO_4, strongly temp. dep., weakly salty and bitter. Incong. sol. in glycerine with rapid transf. to thenardite	6.12 5.89_3 5.46_7 5.329 4.771 3.41_4 3.264 3.206 10 7 100 45 60 15 20 15 002 111 200 $\bar{1}12$ 021 221 $\bar{1}31$ $\bar{1}04$	in Na_2SO_4 rich recent salt lakes as a winter ppt. with gypsum, halite, epsomite, glauberite, glaserite, bloedite, often partly replaced by thenardite
17	$n\alpha = 1.471$ $\parallel c$ $n\beta = 1.477$ $\parallel b$ $n\gamma = 1.484$ $\parallel a$ $+0.013$ $2V\alpha = 82^1/_2{}^\circ$	(34.5% oversat. in mirabilite) very salty, bitter	4.66 3.84 3.178 3.075 2.783 2.646 2.329 1.864 73 18 51 47 100 48 21 31 111 220 131 040 311 022 222 351	in terrestrial salt deps. with bloedite, glauberite, epsomite, gypsum, soda, halite; often from mirabilite. Problematic in oceanic salt deposits.
18	$n\alpha = 1.432_5$ $\parallel a$ $n\beta = 1.455_4$ $\parallel c$ $n\gamma = 1.4609$ $\parallel b$ -0.0284 $2V_\alpha = 51^1/_2{}^\circ$ $\varrho < v$ weak	26.2% $MgSO_4$, sharp, bitter and salty. Incongruently sol. in absolute alcohol	5.99 5.95 5.35 4.48 4.21 2.880 2.677 2.659 22 6 26 14 100 20 24 22 020 011 120 201 121 {410, 212} 240 420	poss. primary ppt. at low temps. with halite, sylvite, kainite, carnallite, leonite, schoenite, bloedite. Surface alteration on kieserite and polyhalite
19	$n\alpha = 1.426$ $\mid c - 25°$ $n\beta = 1.453$ $\parallel b$ $n\gamma = 1.456$ -0.030 $2V\alpha = 38°$ (see note p. 16)	(30.3% $MgSO_4$, oversat. in epsomite) salty, bitter	5.46 5.12 4.89 4.40 4.04 2.92 2.77 2.68 28 24 24 100 32 60 28 24 {111, $\bar{1}12$} 112 $\bar{1}13$ {$\bar{1}14$, 202} 006 {206, $\bar{2}21$} 222 $\bar{2}08$ 026	primary with halite, carnallite in recent salt lakes. From the dehydration of epsomite, possibly as a surface alteration on kieserite

Table 3

1 No.	2 Mineral (Synonym) Formula	3 Crystal system Space group	4 Unit cell dim. a_0 α b_0 β c_0 γ	5 common forms habit, twinning	6 Cleavage Translation and simple Shear. Fracture Hardness, Density	7 Lustre Colour
20	Pentahydrite $MgSO_4 \cdot 5\,H_2O$	triclinic morph. $a:1:c$ 0.6021 : 1 : 0.5604	morph. $\alpha = 98.5$ $\beta = 109.0$ $\gamma = 75.1$	artificially prismatic [001] (100) (010) (110) ($\bar{1}$11), and others	$D = 1.718$	
21	Leonhardtite (Starkeyite) $4[MgSO_4 \cdot H_2O]$	monocl. prismatic $P\frac{2_1}{m}$	5.922 13.60_4 90.8° 7.90_5	artifical: short prisms (110), (120), (001); fibrous, efflorescence?	$D = 2.01$	
22	Kieserite $4[MgSO_4 \cdot H_2O]$	monocl. prismatic $C\frac{2}{c}$	6.88_6 7.61_0 116.3° 7.53_4	mostly massive, fine-coarse grained; crystals rare {110} {$\bar{1}$11} and others. Twinning ($\bar{5}$54), (55$\bar{4}$), lamellar (001) rare	(110) (111) perfect ($\bar{1}$11) ($\bar{1}$01) (011) indistinct $H = 3^1/_2$ $D = 2.571$	vitreous colourless, green white yellowish
e) NaMg-Double salts						
23	Bloedite (Astrakhanite) $2[Na_2Mg(SO_4)_2 \cdot 4\,H_2O]$	monocl. prismatic $P\frac{2_1}{a}$	11.09 8.20 100.65 5.50	short prism. [001] (110) (210) (011) (001) (111) granular, massive, nodular, fibrous	(110)? (uncertain) conchoidal fracture $H = 2^1/_2$–3 $D = 2.25$	vitreous colourless, bluish green reddish from inclusions
24	Loeweite $3[Na_{12}Mg_7(SO_4)_{13} \cdot 15\,H_2O]$	rhombohedral $R\bar{3}$ or $R3$	18.96 — 13.47	granular, short columnar, artifical rhombohedral	(0001)? indistinct conchoidal fracture $H = 2^1/_2$–3 $D = 2.37_6$	vitreous, colourless reddish yellow from inclusions
25	Vanthoffite $Na_6Mg(SO_4)_4$	monocl. prismatic $P\frac{2}{c}$	9.79_7 9.21_7 113.5 8.19_9	granular, massive pseudo-prismatic [$\bar{1}$01]	cleavage not known fracture uneven flat, conchoidal $H = 3^1/_2$ $D = 2.69_4$	vitreous (perlitic) colourless
26	D'Ansite $4[Na_{21}MgCl_3(SO_4)_{10}]$	Cubic. tetrahedral $I\bar{4}3d$(?) (pseudocubic rhombic?)	15.90	from partial cryst. forms aggregates, artificial: Tristetrahedron twinning {110} lamellar	$D = 2.59$	vitreous colourless
f) KMg Doubles salts						
27	Schoenite (Picromerite) $2[K_2Mg(SO_4)_2 \cdot 6\,H_2O]$	monocl. prismatic $P\frac{2_1}{a}$	9.06 12.26 104.8 6.11	Short prismatic [001] (110) (001) ($\bar{2}$01) (010) (100); massive, encrusting, powdery	($\bar{2}$01) perfect $H = 2^1/_2$ $D = 2.03$	vitreous, colourless white, reddish yellow gray from inclusions
28	Leonite $4[K_2Mg(SO_4)_2 \cdot 4\,H_2O]$	monocl. prismatic $C\frac{2}{m}$	11.78 9.53 95.4 9.88	tabular (100); elongate [001] (100) (hk0) (001); massive, granular. Twinning (100) lamellar	cleavage not known fracture chonchoidal $H = 2^1/_2$–3 $D = 2.20$	waxy-vitreous colourless also yellowish
29	Langbeinite $4[K_2Mg_2(SO_4)_3]$	Cubic tetrahedral pyritohedron $P2_13$	9.920	{111} {100} and others nodules, layers	cleavage not known fracture brittle conchoidal $H = 3^1/_2$–4 $D = 2.83$	birght vitreous colourl., in part yellowish rose, reddish-violet greenish-gray

(continued)

	8	9	10	11
	Refraction, Opt. orientation Birefringence, Opt. angle char. Dispersion	Solubility (20° C) Varieties Taste	Powder diagram d-values (see expl.) Relative intensity hkl	Paragenesis (only in salt deposits)
20	$n\alpha = 1.482$ almost ⊥ (010) $n\beta = 1.492$ $n\gamma = 1.493$ -0.011 $2V\alpha = 45.1°$ $\varrho < v$ (strong)	metastable		in recent salt lakes surface alteration on kieserite?
21	$n\alpha = 1.490$ $n\beta = 1.491$ $n\gamma = 1.497$ ∥ b $+0.007$ $2V\gamma - 22°$ (said to be) $\varrho < v$	metastable	6.83 br 5.429 5.156 4.706 4.47 br 3.953 3.216 2.95 br 40 60 5 28 100 60 40 100 {011, 020} 110 021 101 {120, 111} 002 $\bar{1}12$ {$\bar{1}22$, 140}	in recent salt lakes with penta- and hexahydrite Surface alteration on kieserite?
22	$n\alpha = 1.520$ | $c + 41°$ $n\beta = 1.533$ ∥ b $n\gamma = 1.584$ $+0.064$ $2V\gamma = 55°$ $\varrho > v$	forms metastable hydrates and epsomite, tasteless	4.79 3.397 3.358 3.315 3.050 2.553 2.526 2.048 100 100 25 80 35 30 40 30 110 111 $11\bar{2}$ 021 200 $\bar{2}21$ 022 221	in primary and diagenetic recryst. with halite, anhydrite, polyhalite, kainite, carnallite; metamorphic with sylvite and langbeinite, loeweite, [illegible]
23	$n\alpha = 1.483$ | c 37° $n\beta = 1.486$ ∥ b $n\gamma = 1.487$ -0.004 $2V\alpha . = .71°$ $\varrho < v$ strong	incongruent sol. with pptn. of mirabilite above 24.5° C congruent soln. weakly salty and bitter. Var.: Simonyite = Fe^{II} bearing, green	5.45 4.54_5 4.45? 4.275 3.280 3.250 2.962 2.728 3 90 3 30 100 90 35 65 200 210 ? $\bar{2}01$ {220, 021} {211, $\bar{1}21$} 221 {400, 002}	primary in Na_2SO_4 salt lakes, and saltpetre-deserts. In oceanic salts usually sec. from soln. metam. of kieserite. With halite, loeweite, kainite, Mg sulphates
24	$n\varepsilon = 1.471$ $n\omega = 1.490$ -0.019	weak bitter taste	10,4 9.48 7.01 6.23 4.312 4.062 3.471 3.191 2.704 75 20 40 20 90 05 100 100 100 10.1 11.0 20.1 10.2 31.1 (30.2) 11.3 30.3 50.1 {31.4, 15.2}	from soln. metam. of kieserite with vanthoffite, bloedite ± langbeinite ± glaserite (polyhalite, glauberite) anhydrite
25	$n\alpha = 1.485$ not known $n\beta = 1.488$ ∥ b $n\gamma = 1.489$ -0.004 $2V\alpha = 84°$ $\varrho < v$ weak	very weakly bitter, remains clear on heating	8.96 5.82 4.60 4.485 4.026 3.429 3.112 3.054 20 10 7 6 55 35 100 50 100 011 020 200 210 $\bar{2}12$ 211 102	from soln. metam. of kieserite, loeweite, with halite ± langbeinite D'Ansite
26	$n = 1.503$ weak double refraction poss. 2 axes		6.50 4.25 3.973 3.559 3.392 3.118 2.811 2.517 10 20 25 7 100 35 30 20 112 321 400 420 332 510 440 620	metam. in Vanthoffite; possibly also with thenardite, glauberite, glaserite, anhydrite
27	$n\alpha = 1.4607$ | $a - 1°$ $n\beta = 1.4629$ ∥ b $n\gamma = 1.4755$ $+0.0148$ $2V\gamma = 47.9°$ $\varrho > v$ weak	incongruent sol. bitter, dissolves slowly in glycerine with pm. of intermediate phase	7.16 5.36 5.03 4.40 4.156 4.073 3.705 3.066 2.968 85 8 5 30 65 100 100 50 60 110 011 120 200 111 $\bar{2}01$ 130 040 {002, 112}	from the soln. of kainite and leonite, with halite, bloedite, glaserite. Primary with sylvite possible at low temperatures
28	$n\alpha = 1.479$ $n\beta = 1.483$ ∥ b $n\gamma = 1.487$ | a small $+0.008$ $2V\gamma \approx 90°$	incongruent sol. weak, bitter	6.08 5.86 5.26 4.93 4.76 3.42 3.30 3.04 6 40 15 15 100 85 80 45 $\bar{1}11$ 200 $\bar{2}01$ 002 020 022 311 $\bar{2}22$	in kainite cap rock from kainite with kainitic halite(sylvite)(polyhalite) (bloedite)
29	$n = 1.5347$ (1.533)	congruent sol. above 61°. Tasteless Sol. very low	5.727 4.05 3.137 2.991 2.751 2.651 2.406 1.842 1.609 1 4 10 5 5 6 4 4 5 111 211 310 311 320 321 {322, 410} {432, 520} {532, 611}	metam. with kieserite ± sylvite (kainite), halite; with anhydrite, polyhalite

Table 3

1 No.	2 Mineral (Synonym) Formula	3 Crystal system Space group	4 Unit cell dim. a_0 α b_0 β c_0 γ	5 common forms habit, twinning	6 Cleavage Translation and simple Shear. Fracture Hardness, Density	7 Lustre Colour
30	Kainite $4[K_4Mg_4Cl_4(SO_4)_4 \cdot 11\,H_2O]$	monocl. prismatic $C\frac{2}{m}$	19.76 (?) 16.26 94.9° 9.57 (?)	thick tabular (100) Isometric, saccharoidal, massive, fibrous and in aggregates lamellar twinning	(001) perfect Fracture smooth, splintery, brittle $H = 2^1/_2$–3 $D = 2.15$	vitreous, colourless gray, blue violet yellowish reddish
g) Addition: Fe chloride						
31	Rinneite $2[K_3NaFeCl_6]$	ditrigonal scalar $R\bar{3}c$	11.98 — 13.84 ($a_{rh} = 8.31_4$ α 92.2°)	thick tabular (0001) short prism [0001] coarse grained, massive Twinning (0001)	(11$\bar{2}$0) good Fracture splintery, conchoidal $H = 3$ $D = 2.347$	diamantine-silky lustre-colourless (pure) usually rose, yellow violet. Surf. often brown
32	Douglasite $KFeCl_3 \cdot 2\,H_2O$ (?)	monocl. prismatic	β 104.8 (?)	coarse grained (?)	($\bar{2}$01) good	vitreous light green in the atmosphere is brown
33	Erythrosiderite $4[K_2FeCl_5 \cdot H_2O]$	Orthorhombic dipyramid *Pnma*	13.78 9.94 6.94	tabular (100) (110) (h01) twinning [21$\bar{1}$], [20$\bar{1}$]	(210) (011) perfect $D = 2.372$	vitreous ruby red, red brownish red

Data in columns 2, 3, 5–8 where no source is quoted, are taken from DANA, System of Mineralogy, 7th edition, Vol. II, New York, 1951, in column 4 (metric Å) from STRUNZ, Mineralogische Tabellen, 3rd edition Leipzig, 1957; these give further references on crystal structure. Data are in all cases referred to the unit cell given in column 4. The solubility data in column 9 are derived from D'ANS (1933), reactions with organic liquids are from KÜHN (1950). Column 10 lists specific X-ray characteristics (generally the first 4 reflections without regard to intensity; and the 4 or 5 strongest remaining lines in the range up to $d \gtrsim 1.5$ Å; large deviations are possible in the intensity data; measurements were made with a Philips Zählrohr Goniometer). The number preceding the formula in column 2 gives the number of formula units in the unit cell.

1. *Anhydrite.* Setting from HINTZE: Handb. d. Mineralogie **1**, 3, 2, p. 3735 (1930) (differs from DANA II). Lattice constants and powder diagram from SWANSON et al., Nat. Bureau of Standards, U.S. Dept. of Commerce, Circ. **539**, IV, 65 (1955). Transform. hkl (DANA II) → hkl (HINTZE) 001/100/010.

2. *Gypsum.* Setting from DE JONG and BOUMAN: Zeitschr. Krist. **100**, 275 (1938). Lattice constants calculated from ATOJI and RUNDLE: J. Chem. Phys., **29**, 1306 (1958). (Wooster structural data improved through

(continued)

	8 Refraction, Opt. orientation Birefringence, Opt. angle char. Dispersion	9 Solubility (20° C) Varieties Taste	10 Powder diagram *d*-values (see expl.) Relative intensity hkl	11 Paragenesis (only in salt deposits)
30	$n\alpha = 1.494$ $n\beta = 1.505$ ‖ b $n\gamma = 1.516$ | $c = 13°$ -0.022 $2V\alpha \approx 90°$ $\varrho > v$ weak	salty, weak bitter taste	8.15 7.78 7.39 6.28 4.756 3.133 3.08 3.035 30 80 40 11 25 13 100 40 020 $\bar{1}11$ 111 220 002 440 {042, $\bar{5}31$} {620, 113, $\bar{4}41$}	primary after halite, epsomite, hexahydrite (sylvite) (carnallite) mostly secondary from kieserite + sylvite or carnallite also with langbeinite
31	$n\omega = 1.5886$ $n\varepsilon = 1.5894$ $+0.0008$ abnormal colours strongly λ dep.	astrigent taste soluble in abs. alcohol with pptn. of KCl + NaCl	5.996 5.749 3.770 3.645 3.455 2.811 2.653 2.593 2.509 55 40 10 15 30 85 35 45 100 11.0 10.2 12.1 11.3 30.0 31.1 13.2 12.4 22.3	metamorphic, from the alteration of carnallite, with halite, sylvite, kieserite, anhydrite
32	$n\alpha = 1.488$ $n\beta = 1.488$ $n\gamma = 1.500$ $+0.012$ $2V\gamma \approx 0°$			uncertain; after halite, carnallite, sylvite
33	$n\alpha = 1.715$ ‖ a $n\beta = 1.75$ ‖ c $n\gamma = 1.80$ ‖ b $+0.085$ $2V\gamma = 62°$ $\varrho > v$ strong	very hygroscopic		weathering product on rinneite

neutron diffraction); ⚹ β morphology; Column 10d and I from HANAWALT et al., Ind. Eng. Chem. Anal. Ed., **10**, 9, 457—512 (1938); hkl pers. source.

3. *Bassanite* (Hemihydrate). Col. 4 after GALLITELLI, P., Periodico Mineralog., **4**, 1—42 (1933); pseudo-hexagonal. Col. 3 from FLÖRKE: Neues Jb. Mineral. Abh., A **84**, 189—240 (1952); possibly dimorphic, below about 45° C orthorhombic with double a_0. Powder diagram similar to γ $CaSO_4 \cdot \gamma$ $CaSO_4$ = Sol. anhydrite, cols. 3, 4, 8 from FLÖRKE (see Hemihydrate).

4. *Glauberite*. Col. 4 corrected from powder diagram (Glauberite of Villarubbia, Spain and from Neustassfurt).

5. *Polyhalite*. Setting from PEACOCK: Amer. Mineralogist **23**, 38 (1938). Col. 10 from BRAITSCH: Beitr. Mineral. u. Petrog. **8**, 84—91 (1961). Col. 4 from M. SCHLATTI, K. SAHL, A. ZEMANN, and J. ZEMANN: Tschermaks Min. Petr. Mitt. **14**, 75—86 (1970).

6. *Syngenite*. Col. 4 after LASZKIEWICZ from STRUNZ: Powder diagram indices from synthetic material (from the Kaliforschungsinstitut Hannover). The powder diagram in the ASTM index (1944) II 1244 and from CONLEY and BUNDY, 1958, differs greatly, but approximates to Kurylenko's lattice constants (KURYLENKO, C., Acta Cryst., **7**, 630 (1954). $a_0 = 9.55$; $b_0 = 7.13$; $c_0 = 5.66$; $\beta = 105°$). The lattice

constants give a cell content $Z = 1.77$ so that it is not conclusive for syngenite. Whether water-free $K_2Ca(SO_4)_2$, which would give $Z=2$, exists is not known. Crystal structure unknown.

7. *Goergeyite.* β after MEIXNER (in MAYRHOFER: Neues Jb. Mineral., 1953, 35—44). Lattice constants still uncertain, col. 10 from synthetic material (still uncertain because of incomplete indices).

8. *Halite* (rock salt). Col. 10 from SWANSON and FUYAT: Natl. Bur. Standards (U.S.) Dept. of Commerce Circ. **539**, II, 41 (1953).

9. *Hydrohalite.* Density after ADAMS and GIBSON (1930).

10. *Sylvite.* Cols. 4 and 10: SWANSON and TATGE: Natl. Bur. Standards (U.S.) Dept. of Commerce Circ. **539**, I, 65 (1953).

11. *Bischofite.* Col. 10: *d*-value and I (1—10) HANAWALT et al.: Ind. Eng. Chem., Anal. Ed., **10**, 9, 492 (1938). hkl pers. source.

12. *Carnallite.* c_0 (col. 4) from powder diagram, probably 0.1 Å too large.

14. *Chlorocalcite.* Paragenesis O. RENNER: Centr. Mineral. Geol. 1912, p. 106 (previous comm.).

15. *Glaserite.* Col. 10 from synthetic glaserite; line 3.129 I = 75 not indexed. Literature data (ASTM 1944 II, 1959, etc.) markedly different.

16. *Mirabilite.* Col. 4 RUBEN, H. W., D. H. TEMPLETON, R. D. ROSENSTEIN and T. OLOVSSON: J. Am. Chem. Soc., **83**, 820—824 (1961). (Crystal structure and entropy of sodium sulfate decahydrate). Col. 10 on synthetic material; rapid superficial decomposition to thenardite.

17. *Thenardite.* Cols. 4 and 10 after SWANSON and FUYAT: Natl. Bur. Standards (U.S. Dept. Commerce Circ. **539** II, 59 (1953), see DANA II, transformed to setting by ZACHARIASEN and ZIEGLER, Z. Krist. **81**, 92 (1932).

18. *Epsomite.* Cols. 4 and 10: Natl. Bur. Standards (U.S.) Dept. Commerce Circ., **539**, **7**, 30 (1957).

19. *Hexahydrite.* Col. 10 d_{hkl} after GARAVELLI, C., Atti soc. toscana sci. nat. Pisa Mem. Ser. A **64**, 6—15 (1957). hkl pers. source. The refraction data in col. 8 appear to be too low in comparison with those of epsomite.

20. *Pentahydrite.* Occurrence in the Hansa mine near Empelde, Hannover, described without detailed information (LEONHARDT and BERDESINSKI: Z. anorg. u. allgem. Chem., **265**, 284—287 (1951). Nomenclature see FLEISCHER, M., Amer. Mineralogist **36**, 641 (1951), and W. BERDESINSKI ("Allenite") Neues Jb. Mineral. Monatsh., 1952, 28.

21. *Leonhardtite.* Col. 4 after W. H. BAUR pers. comm. Col. 10 d_{hkl} after GARAVELLI (see Hexahydrite); I: ASTM 1—0341; improved d-values by calculation from lattice constants, hkl checked against single crystal data (W. H. BAUR). Col. 8 from ROBSON, H. L., J. Amer. Chem. Soc., **49**, 2777 (1927).

22. *Kieserite*. Col. 4 from LEONHARDT, J., and R. WEISS: Naturwiss. **44**, 338 (1937), converted to the reduced cell corresponding to the setting from DANA II. Transformations see WEINERT, G., Neues Jb. Geol. Mineral. Paleontol. A. Beil. B., **75**, 297—314 (1939). Twinning and optical orientation FRIEDRICH, K., et al., Kali u. Steinsalz **3**, 221—227 (1961).

23. *Bloedite*. Structure: RUMANOVA and MALITSKAYA: Kristallografiya **4**, 510—525 (1959); Giglio, M., Acta Cryst., **11**, 789—794 (1958). Col. 4 corrected with respect to RUMANOVA and MALITSKAYA with powder photos, hkl checked with the help of structural data. Col. 10 Bloedite from Leopoldshall near Stassfurt. (Reflections in MIKHEEV: Rentgenometricheskiy opredelitel mineralov, p. 535, Moscow, 1957 not indexed).

24. *Loeweite*. Formula: KÜHN, R., and H. RITTER: Kali u. Steinsalz **2**, 238—240 (1958). Col. 4 and density: SCHNEIDER, W., and J. ZEMANN: Beitr. Mineral. u. Petrog., **6**, 201—202 (1959). Col. 10 synthetic material.

25. *Vanthoffite*. Col. 3 after FISCHER, W., and HELLNER, E.: Fortschr. Mineralogie 340 (1961). Col. 4 FISCHER and HELLNER: Acta cryst. **17**, 1613 (1964). The assumed triclinic setting from KÜHN, R., Kali u. Steinsalz **2**, 217 (1958) is converted on rounding angles α and γ to 90° to the new setting using the transformation matrix KÜHN → FISCHER (101/010/101) (FISCHER pers. comm.). Col. 10 synthetic material.

26. *D'Ansite*. Col. 4 and physical properties: AUTENRIETH and BRAUNE: Naturwiss. **45**, 362 (1958); GÖRGEY 1909. Col. 10. STRUNZ, H.: Neues Jb. Mineral. Monatsh. 1958, 152—155. Supplementary information on optics, twinning and pseudo space group: BURZLAFF and HELLNER, Z. Elektrochem. **65**, 565 (1961).

27. *Schoenite*. Col. 10 synthetic material.

28. *Leonite*. Col. 4 SCHNEIDER, W., Acta Cryst., **14**, 784—791 (1961). Col. 10 Leonite from Wittmar/Asse; analysis p. 65. n_β increased with respect to SCHALLER and HENDERSON 1932 as $2V_\gamma \sim 90°$.

29. *Langbeinite*. Col. 4. ZEMANN, A. and J., Acta Cryst. **10**, 409 to 413 (1957) (structure). Col. 10: RAMSDELL, L. S.: Amer. Mineralogist **20**, 569 (1935), *d*-values corrected using col. 4 (111 observed in langbeinite from Hessen seam, Hattorf).

30. *Kainite*. Formula: KÜHN, R., and H. RITTER: Kali u. Steinsalz **2**, 238—240 (1958). Col. 4 LINSTEDT, H.: Naturwiss. **38**, 476 (1951), setting not certain; see EVANS, R. C., in DANA II, p. 594. LINSTEDT'S lattice constants give a better agreement between theoretical and observed *d*-values. Col. 10 Kainite from Stassfurt. Intensity varies markedly between different samples.

31. *Rinneite*. Col. 4 BELLANCA, A.: Periodico mineral. (Rome) **16**, 199 (1948), corrected using powder photos and BOEKE'S α angle

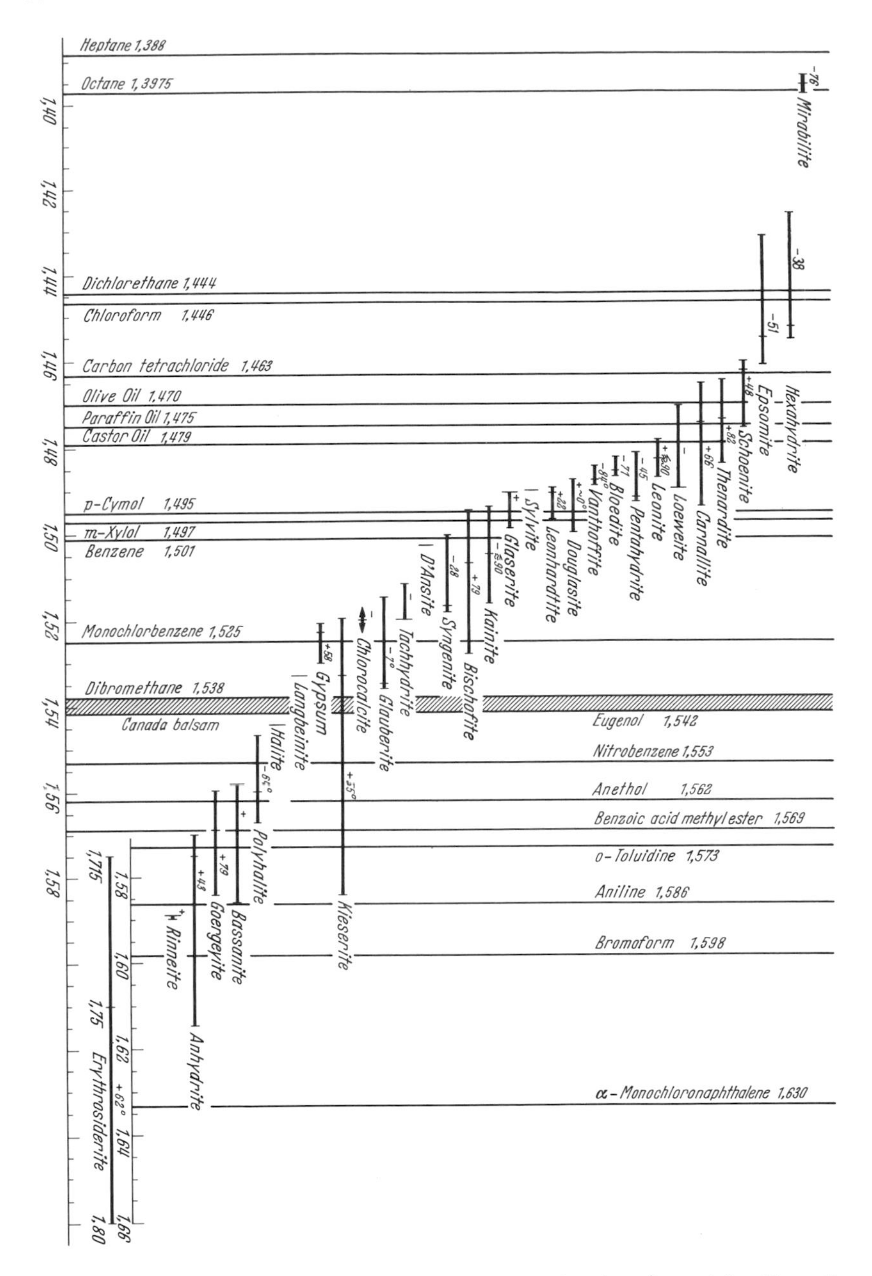

Fig. 2. Refractive indices and axial angles of the salt minerals and liquid media. The values for hexahydrite in Table 3 are probably too low (cf. p. 11)

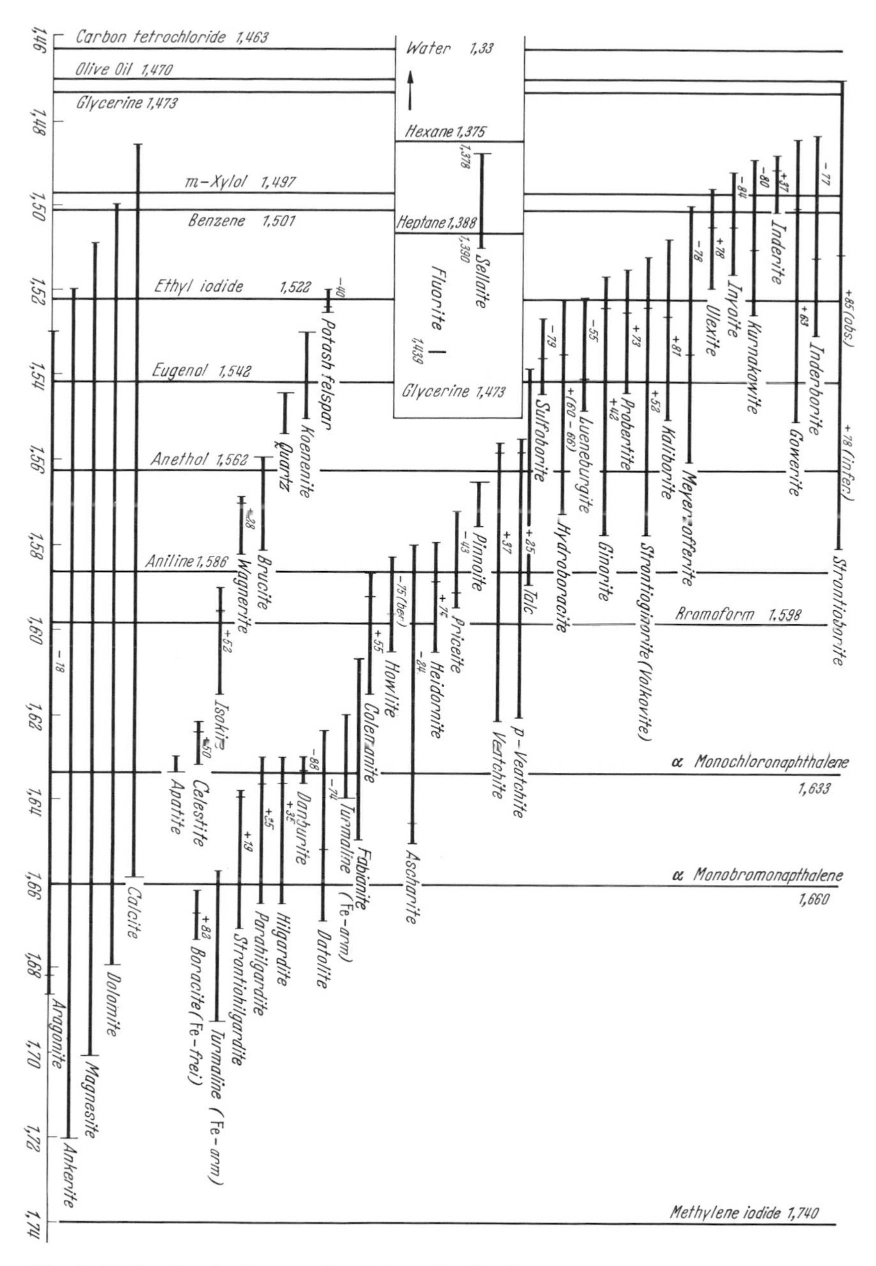

Fig. 3. Refractive indices and axial angles for borates and other water-insoluble minor minerals in salt deposits. Refractive indices for koenenite after KÜHN (1961) are higher ($n_w = 1.55$, $n_\varepsilon = 1.58_5$). Strontiaborite is problematic. Addition: fabianite ($n_\alpha = 1.608_5$; $n_\beta = 1.637_5$; $n_\gamma = 1.650_0$)

(Centr. Mineral. Geol., 1909, 72). Original data from BELLANCA: $a_0 = 11.89$, $c_0 = 13.89$ Å.

32. *Douglasite*. Formula from BOEKE (Neues Jb. Mineral. Monatsh. 1909 II, 44); Natural occurrence questionable.

b) Accessory and Authigenic Minerals

Certain of the minor constituents of seawater such as boron, fluorine, phosphorus, strontium, etc., form distinctive minerals which may be found in salt deposits. Of course, the composition of the minerals does not always depend, or only depends in part, upon the composition of seawater. They form a transitional sequence with impurities whose composition is predominantly derived from material of continental origin or from secondary solutions. Only a small part of the foreign matter is in its original form; these are such non-saline impurities as pyroxene, hornblende, rutile, zircon and many others, and none of them is of concern here. The greater part of the foreign material is altered either in the concentrated brines, or after deposition by pore solutions or percolating solutions. LEONHARDT and BERDESINSKI (1952) introduced the term "semisaline" for such transformations; syngenetic semi-saline for authigenic forms originating during primary salt precipitation, and epigenetic semi-saline for mineral phases formed subsequently. As there is a complete transition between minerals formed under saline and non-saline conditions, and as in many cases it is difficult to assign minerals to one or other of the genetic groups — indeed, some minerals may belong to both saline and non-saline groups — this distinction cannot be used as a means of classification.

Knowledge of the accessory minerals and their formation in salt deposits is still far from complete, as this area, excluding a few old studies (e.g. MÜGGE, 1913), has only recently received close attention. Detailed crystallographic, mineralogical and chemical data will not be listed here, as fuller descriptions of the minerals in the succeeding list appear elsewhere.

Up to the present time the following accessory and authigenic minerals are known in salt deposits.

Elements and Sulphides

Sulphur S, inclusions in halite, carnallite, etc.

Pyrite FeS_2 found in both haematite-free and haematite-bearing salts generally present as well-formed crystals.

Marcasite FeS_2 (rhomb.) }
Hauerite MnS_2 } occasionally found in Louisiana salt residues (TAYLOR, 1937)

Pyrrhotite FeS rare, known in the kainite cap at Aller-Nordstern (HARBORT, 1915).

Copper pyrites (chalcopyrite) $CuFeS_2$ occurrence as an authigenic mineral in sylvine from Bleicherode, Southern Harz, doubtful (see GLÖCKNER, 1914), grown on kieserite from Hallstadt (TSCHERMAK, 1871, p. 321).

Galena PbS as inclusions in anhydrite from the Alpine Salzgebirge (MAYER, 1912).

Halides and Oxychlorides

Sellaite MgF_2 in the Bleicherode Hauptdolomit (HEIDORN, 1932), and in the Hauptdolomit of West and Northwest Thuringia[1] in the carnallite "Begleitflöz" of the Ronnenberg seam at Salzdetfurth near Hildesheim (KÜHN, 1951) and also in the Ronnenberg sylvite (SIEMEISTER, 1961), Inder salt diapir (LOBANOVA and YARZHEMSKY, 1956).

Fluorite CaF_2 in biogenic dolomite (FÜCHTBAUER, 1958); in secondary anhydrite in Stassfurt halite (BRAITSCH, 1960).

Koenenite $Mg_5(Na_2Mg)_3Al_2Cl_6(OH)_{11} \cdot x\,H_2O$ (?) as a secondary joint filling in the gray Salzton and Hauptanhydrit, fine intergrowths with other salts, particularly in Zechstein 3 and 4 (KÜHN, 1961).

Zirklerite probably a Fe^{II} rich koenenite from the Adolfsglück potash mine some 30 km north of Hannover (HARBORT, 1928).

Oxides and Hydroxides

Haematite Fe_2O_3 the most important iron mineral in all salt deposits, particularly in sylvite and carnallite rocks.

Magnetite Fe_3O_4 subordinate to, or in place of haematite.

Rutile TiO_2 rare, not certain if truly authigenic.

Brookite TiO_2 in black carnallite from Hildesia (MÜGGE, 1913). Identification not verified.

Anatase TiO_2 also uncertain identification, in clay residues.

Quartz SiO_2 in the insolubles of pratically all salts. Authigenic and detrital; rarely as chalcedony, e.g. in the Tertiary gypsum of Sicily (L. OGNIBEN, 1955, p. 277).

Opal $SiO_2 \cdot x\,H_2O$ rare, e.g. in the gypsum cap of the Inder salt diapir (YARSHEMSKY, 1953).

---

[1] PENSOLD, G., THEILIG, F., THOMASER, P.: Auftreten von Sellait im Hauptdolomit des Zechsteins von Thüringen. Z. angew. Geol. **10**, 75—77 (1964).

(Periclase MgO) only synthetic! ("Stufen" of Neustassfurt). Possibly may occur as inclusions in wagnerite.

Goethite α FeO OH rare, in many brown salts; as an alteration product of rinneite (ARMSTRONG et al., 1951).

Brucite $Mg(OH)_2$ single report, diagnosis not yet confirmed (BRAITSCH, 1960).

Carbonates

Calcite $CaCO_3$ in marl bands. Rare in typical marine salt deposits.

Dolomite $CaMg(CO_3)_2$ chiefly in anhydrite and gypsum. In salt rocks often Fe-bearing (ankerite) (GÖRGEY, 1912; TAYLOR, 1937).

Magnesite $MgCO_3$ principal authigenic carbonate of chloride salt-rocks (OCHSENIUS, 1890).

Aragonite $CaCO_3$ rhomb., occurs rarely in gypsum residues of the Permian of Oklahoma.

Borates and Borosilicates

Ulexite $NaCaB_5O_9 \cdot 8\,H_2O$ rare, in Oklahoma gypsum (HAM et al., 1961), Nova Scotia, Inder salt diapir (GODLEVSKY, 1937)[2].

Probertite $NaCaB_5O_9 \cdot 5\,H_2O$ concretion in Oklahoma gypsum (HAM et al., 1961).

Kaliborite $KMg_2B_{11}O_{19} \cdot 9\,H_2O$ (?) rare in kainite cap rock; Inder salt diapir (YARZHEMSKY, 1945).

Ascharite $MgHBO_3$ widely distributed in rocksalt and many kieserite-sylvite-rocks, also occasionally in carnallite, and gypsum cap rock (synonyms szaibelyite, camsellite).

Pinnoite $MgB_2O_4 \cdot 3\,H_2O$ with kaliborite in kainite cap rock, rare (BOEKE, 1910).

Inderite $Mg_2B_6O_{11} \cdot 15\,H_2O$ monoclinic } in the Inter salt diapir.
Kurnakovite $Mg_2B_6O_{11} \cdot 15\,H_2O$ triclinic }

Lesserite identical with Inderite (SCHALLER and MROSE, 1960).

Preobrazhenskiite $Mg_8B_{10}O_{13} \cdot 5\,H_2O$ (?) Inder salt diapir (YARZHEMSKY, 1956).

Paternoite $MgB_8O_{13} \cdot 4\,H_2O$.

Inderborite $CaMgB_6O_{11} \cdot 11\,H_2O$ Inder salt diapir.

Hydroboracite $CaMgB_6O_{11} \cdot 6\,H_2O$ Inder salt diapir (GODLEVSKY, 1937).

---

[2] See also: KORITNIG, S.: Ulexit-Konkretionen ("cotton balls") im Zechsteingips der Werra-Serie. N. Jb. Miner. Abh. **103**, 31—34 (1965).

Fabianite CaB_3O_5 (OH) Werra rock salt, Rehden near Diepholz N.W. Germany (GAERTNER, et al., 1962); KÜHN et al., 1962[3]; KÜHN and MOENKE, 1963[4]).

Kurgantaite $(Sr, Ca)_2B_4O_8 \cdot H_2O$ Inder salt diapir (YARZHEMSKY, 1952).

Priceite $Ca_4B_{10}O_{19} \cdot 7\,H_2O$ synonym Pandermite, rare in gypsum (HAM et al., 1961); Inder salt diapir (GODLEVSKY, 1937).

Inyoite $Ca_2B_6O_{11} \cdot 13\,H_2O$ Inder salt diapir (GODLEVSKY, 1937); Nova Scotia, Canada.

Colemanite $Ca_2B_6O_{11} \cdot 5\,H_2O$ Inder salt diapir (GODLEVSKY, 1937).

P-Veatchite $SrB_6O_{10} \cdot 2\,H_2O$[5], Stassfurt rock salt Na 2 γ, Königshall-Hindenburg (BRAITSCH, 1959); Yorkshire (BEEVERS and STEWART, 1960).

Ginorite $Ca_2B_{14}O_{23} \cdot 8\,H_2O$ concretion in gypsum, Nova Scotia.

Strontioginorite $(Sr, Ca)_2B_{14}O_{23} \cdot 8\,H_2O$ Stassfurt rock salt Na 2 β, Königshall-Hindenburg (BRAITSCH, 1959).

Strontioborite $\approx [4(Sr, Ca)O \cdot 2\,MgO \cdot 12\,B_2O_3 \cdot 9\,H_2O]$ Permian salts of the Caspian region (LOBANOVA, 1960); the mineral is problematic (heterogeneous?).

Boracite $(Mg, Fe, Mn)_3ClB_7O_{13}$ the most common borate mineral in the German Zechstein 2 and other series.

Hilgardite $Ca_2B_5O_8(OH)_2Cl$ monocl. Choctaw salt dome Louisiana (HURLBUT and TAYLOR, 1937); Inder salt diapir (YARZHEMSKY, 1945).

Parahildgardite $Ca_2B_5O_8(OH)_2Cl$ triclinic, Choctaw salt dome Louisiana (HURLBUT, 1938).

Strontiohildgardite $(Sr, Ca)_2B_5O_8(OH)_2Cl$ sylvinite of the Stassfurt Seam, Königshall-Hindenburg (BRAITSCH, 1959), rare.

Heidornite $Ca_3Na_2(SO_4)_2B_5O_8(OH)_2Cl$, joint filling in the Werra anhydrite (v. ENGELHARDT et al., 1956). The mineral was also found in the Werra anhydrite by FABIAN and KLENERT (1962)[6].

Sulphoborite $Mg_3(SO_4)(BO_2OH)_2 \cdot 4\,H_2O$ in carnallite, Stassfurt seam; Inder salt diapir (BÜCKING, 1893; LOBANOVA et al., 1959).

Lueneburgite $Mg_3(PO_4)_2B_2O(OH)_4 \cdot 6\,H_2O$ in gypsum from Lüneburg (NOELLNER, 1870); New Mexico (SCHALLER and HENDERSON, 1932); Zechstein 2 at Königshall-Hindenburg (BRAITSCH, 1961 a).

[3] KÜHN, R., ROESE, K.-L., GAERTNER, H.: Fabianit $CaB_3O_5(OH)$, ein neues Mineral. Kali und Steinsalz 285—290 (1962).

[4] KÜHN, R., MOENKE, H.: Ultrarotspektroskopische Charakterisierung des neuentdeckten Borminerals Fabianit $CaB_3O_5(OH)$ und seines Begleiters Howlith. Kali und Steinsalz, 399—401 (1963).

[5] $Sr_2[B_5O_8(OH)]_2 \cdot B(OH)_3 \cdot H_2O$ reported by GANDYMOV, O., RUMANOVA, I. M., BELOV, N. V., in Doklady of the Academy of Sciences U.S.S.R., Earth Science Sections **180**, 152—155 (1968).

[6] FABIAN, H. J., KLENERT, G.: Der Bor-Gehalt der Zechstein-Anhydrite. Erdöl und Kohle, Erdgas, Petrochemie **15**, 603—606 (1962).

Braitschite $7(Ca, Na_2)O \cdot RE_2O_3 \cdot 11\, B_2O_3 \cdot 7\, H_2O$ has been found in the marine evaporites in the Paradox Member of the Hermosa Formation of Pennsylvanian age in southeastern Utah. The mineral occurs in white to reddish-pink nodules in a 6-inch zone in anhydrite which immediately overlies the potash deposit in the Cane Creek mine of the Texas Gulf Sulphur Co. near Moab, Utah[7,8].

Datolite $CaBSiO_4OH$ basalt contact in Buggingen Tertiary salt (BRAITSCH et al., 1964)[8a].

Danburite $CaB_2Si_2O_8$ Basalanhydrit in the Bleicherode potash mine (KÜHN and BAAR, 1955); South Harz (BUDZINSKI et al., 1959); N. W. Germany (FABIAN et al., 1961; FABIAN and KLENERT, 1962[6], see footnote p. 23); Nova Scotia; Choctaw salt dome Louisiana.

Howlite $Ca_2SiB_5O_9(OH)_5$ in gypsum from Windsor, Nova Scotia; in the transition layer halite-anhydrite, Werra Series, Rheden near Diepholz, N.W. Germany (KÜHN et al., 1962, p. 23).

Tourmaline $Na(Mg, Fe)_3Al_6(OH)_4(BO_3)_3Si_6O_{18}$ in the Hildesia black carnallite (MÜGGE, 1913); countless other occurrences; strong local enrichment in the Stassfurt seam at the Riedel and Niedersachsen mines (LOHSE, 1958).

Sulphates and Phosphates

Celestite $SrSO_4$ the most important Sr mineral in chloride salts.

Kalistrontite $K_2Sr(SO_4)_2$ Stassfurt seam of the bore-hole Pleismar 2/64[9].

Jarosite $KFe_3(SO_4)_2(OH)_6$ from the Stassfurt-carnallite[10].

Wagnerite Mg_2FPO_4 Stassfurt rock salt and Stassfurt seam of Königshall-Hindenburg (BRAITSCH, 1960); Thuringia seam of Herfa-Neurode (BRAITSCH in WEBER, 1961).

Isokite $CaMgFPO_4$ Stassfurt rock salt and Stassfurt seam of Königshall-Hindenburg (BRAITSCH, 1960).

---

[7] RAUP, O. B., GUDE, 3d, A. J., DWORNIK, E. J., CUTTITTA, F., ROSE, H., JR.: Braitschite, a new hydrous calcium rare-earth borate mineral from the Paradox Basin, Grand County, Utah. Am. Mineralogist **53**, 1081—1095 (1968).

[8] RAUP, O. B., GUDE, 3d, A. J., GROVES, H. L., JR.: Rare-earth mineral occurrence in marine evaporites, Paradox Basin, Utah. U.S. Geol. Survey Prof. Paper **575-C**, c38—c41 (1967).

[8a] BRAITSCH, O., GUNZERT, G., WIMMENAUER, W., THIEL, R.: Über ein Datolithvorkommen am Basaltkontakt im Kaliwerk Buggingen (Südbaden). Beiträge zur Mineralogie und Petrographie **10**, 111—124 (1964).

[9] BADER, E., BÖHM, G.: Kalistrontit im Flöz Staßfurt des Roßleben-Unstrut-Reviers. Chemie der Erde **25**, 253—257 (1966).

[10] BRAITSCH, O., KEIL, K.: Jarosit $KFe_3(SO_4)_2(OH)_6$ aus Staßfurter Carnallit. Beiträge zur Mineralogie und Petrographie **11**, 247—249 (1965).

Apatite $Ca_5(F, OH, Cl)(PO_4)_3$ Stassfurt rock salt and Stassfurt seam of Königshall-Hindenburg (BRAITSCH, 1960); in the insoluble residues of the Cambrian rock salt of Hormuz Island in the Persian Gulf.

Goyazite (= Hamlinite) $SrAl_3H(OH)_6(PO_4)_2$ Ukrainian salt dome contact with diabase intrusions (PITROVSKAYA, 1939).

Silicates

Potash feldspar $KAlSi_3O_8$ authigenic in all German Zechstein salt rocks and in the Tertiary salts of the upper Rhine valley (BRAITSCH, 1960; FÜCHTBAUER and GOLDSCHMIDT, 1959).

Albite $NaAlSi_3O_8$ in many Zechstein anhydrites and dolomites (FÜCHTBAUER and GOLDSCHMIDT, 1959, p. 330).

Muscovite $KAl_2(OH)_2AlSi_3O_{10}$ and illite generally detrital, occasionally authigenic (see HOFFMANN, 1961).

Chlorite a) amesite form, e.g. $Mg_{4.7}Fe^{II}_{0.2}(Al_{0.85}Fe^{III}_{0.2}) \cdot Al_{1.05}Si_{2.95}O_{10}(OH)_8$ (analysis by R. BLANK, "insoluble" residues $< 2\mu$ of carnallite, Königshall-Hindenburg): predominantly in carnallite and potash salt residues, confused by older authors with kaolinite (unpubl.),
b) penninite form, e.g. $Mg_{5.2}Fe^{II}_{0.3}Al_{0.1}Fe^{III}_{0.4}Al_{0.5}Si_{3.5}O_{10}(OH)_8$ (analysis by R. BLANK, kieseritic sylvite-halite from the Thuringia seam, insoluble residues $< 6\mu$): predominant in barren halite horizons and occasionally in the Stassfurt rock salt, South Harz, Werra.

Corrensite-2 $Na_{0.12}Mg_{8.25}Fe_{0.24}Al_{2.7}Si_{6.2}O_{20}(OH)_{10} \cdot 5\,H_2O$ (analysis by Fresenius and Schneider, Wiesbaden, insoluble residues $< 0.6\mu$ from the Stassfurt rock salt, Königshall-Hindenburg. Variety: chlorite-vermiculite, mixed layer structure (see BRADLEY and WEAVER, 1956). In Stassfurt rock salt and in barren halite horizons, South Harz; Werra.

Talc $Mg_3(OH)_2Si_4O_{10}$ in anhydrite, gypsum (SENFT, 1861) and rock salt of all formations (BAILEY, 1949; STEWART, 1949; MAYRHOFER and SCHAUBERGER, 1953; BRAITSCH, 1958; FÜCHTBAUER and GOLDSCHMIDT, 1959; DREIZLER, 1962).

Serpentine $Mg_6(OH)_8Si_4O_{10}$ in the Younger rock salt of Königshall-Hindenburg, identification not yet verified (unpubl.).

Talc serpentine: a mixture of talc and poorly crystallized serpentine? Werra anhydrite, Emsland (FÜCHTBAUER and GOLDSCHMIDT, 1956).

There are in addition important non-mineralic components associated with salt deposits such as:

Salt Solutions

NaCl, $MgCl_2$, $CaCl_2$ solutions, etc. (BAUMERT, 1928; HERRMANN, 1961 b).

Organic Matter

Hydrocarbons, bitumen, and also fossil remains (spores, etc.) (LÜCK, 1913; KLAUS, 1953; MÜLLER and SCHWARTZ, 1953).

Gases

CO_2 (TAMMAN and SEIDEL, 1932); CO; H_2S, H (PRECHT cited by JOHNSEN, 1909; SAVCHENKO, 1958); He (HAHN, 1932); radon, neon (KOCZY in KARLIK, 1939); argon (see ZÄHRINGER, 1960).

For an account of the gases see KÜHN (1955 a) and BAAR (1960, p. 132).

Fig. 3 with optic angles and refractive index is a key diagram for the recognition of the mineralogically identifiable authigenic minerals.

B. The Stability Conditions of Salt Minerals

Introduction

Salt deposits are particularly amenable to a physico-chemical consideration for in comparison to other rock series they form a simple system. Thanks to the pioneer work of J. H. VAN'T HOFF (1896–1911), a sound basis for research has been established. It was based upon conditions existing in nature, but considering only the main components, and these in controlled concentrations which could be examined systematically, changing values of temperature, pressure and time. VAN'T HOFF'S fundamental observations are still valid, and since his time, apart from minor numerical corrections, there have been few basic advances.

From the outset VAN'T HOFF concentrated his attention upon the stable mineral phases. He was aware that as a result of the evaporation of a solution metastable mineral phases could develop in many instances. As these have a greater solubility than the stable phases, the study of solubility conditions provides a means of investigating stability conditions. As a direct indication of the existence of metastable forms, VAN'T HOFF often made use of measurement of differential vapour pressure. This arises out of the fact that the stable phases, which occasionally possess lower vapour pressures, are generally dehydration products of originally hydrated compounds. The less hydrated compounds must, however, be in equilibrium with a solution which has a lower vapour pressure than the more highly hydrated compounds. As an example, a solution at 25° C saturated with respect to hexahydrite, carnallite and bischofite has a lower vapour pressure than hexahydrite. Solution brings about a loss of water from the hexahydrite. It follows that hexahydrite is not a stable phase in such a solution.

The methods of examining solubility relations and the efforts and necessary precautions to obtain reliable data (D'ANS, 1933; AUTENRIETH, 1952–1960) will not be entered into here. The results themselves for the different partial systems will be presented insofar as is necessary in the discussion of their origin. In solubility data the chief concern is the saturation concentration of the dissolved salts in a solution which is in equilibrium with crystalline solid phases whose crystal size is sufficiently great that the saturation concentration is demonstrably no longer dependent upon them (VALETON, 1915).

The old solubility data have been tabulated by D'ANS (1933) and the majority of applications given here are derived from that source. SDANOVSKY and his co-workers (1953, 1954) have prepared a modern, comprehensive summary of experimental data, which was unfortunately not available during the preparation of this monograph.

In the last few years an increasing volume of new data on solubility conditions has been obtained, particularly in the Kaliforschungsinstitut Hannover. Unfortunately only a small part of it has been published (AUTENRIETH, AUTENRIETH, and BRAUNE, 1952–1960). In the following discussions the most recent available data will be used as far as is possible.

A qualitative appreciation of solution equilibria is possible from the application of Gibb's Phase Rule:

$$P + F = C + 2$$

that is, the sum of the phases and degrees of freedom is equal to the number of components +2. The explanation for this is to be found in all the newer petrology textbooks (e.g. CORRENS, 1969) and for that reason will be taken as proven here. A comprehensive account is given by FINDLAY (1958) and also by ZERNIKE (1957). A simple example in the system KCl–NaCl–H_2O will be introduced later (p. 46, 47).

The phase rule, however, only provides generalized information on the number of phases present under given conditions and with given components. In addition to the solid phases, in solution equilibria both the solution and the vapour are further phases which must be considered. The degrees of freedom to be considered are the mixture conditions (i.e. concentration)[11] of the components and temperature. When vapour is present, pressure is no longer a degree of freedom,

[11] In concentrated solutions, data given as g/l, mol/l, etc., are unique (that is with respect to the H_2O content) only when the density of the solution is also given. In contrast, data in the form g/100 g H_2O, g/100 g solution (%), mol/1000 mol H_2O are always clear and will be used here. In the case of solutions with many ions, use is made of mol/1000 mol H_2O, as this unit permits the representation of reciprocal salt pairs. To convert the latter data to g/100 g H_2O, the following conversion factors (from D'ANS, 1933) can be used:

$\frac{1\ \text{mol}}{1000\ \text{mol}\ H_2O}$		$\frac{\text{g}}{100\ \text{g}\ H_2O}$	$\frac{1\ \text{mol}}{1000\ \text{mol}\ H_2O}$		$\frac{\text{g}}{100\ \text{g}\ H_2O}$
Na_2Cl_2	=	0.6489	$MgCl_2$	=	0.5286
Na_2SO_4	=	0.7885	$MgSO_4$	=	0.6682
K_2Cl_2	=	0.8277	$CaCl_2$	=	0.6161
K_2SO_4	=	0.9673	$CaSO_4$	=	0.7556
			$SrSO_4$	=	1.0196

rather it is fixed in terms of vapour pressure in vacuum. The converse also holds; if the pressure exceeds a threshold value, the vapour phase disappears. In the application of the phase rule to solution equilibria, it is customary (JÄNECKE, 1908) to consider concentration and temperature in the presence of vapour as condition variables whereby the vapour pressure is fixed. In practice all solubility data are determined at atmospheric pressure. Strictly speaking, in this event, the vapour phase is not present.

Vapour pressure measurements are, however, often used to check the results. Volume, too, is a degree of freedom which depends upon the other factors. It also has a part to play in the investigation of salt systems to determine the transition points, since volume effects during phase changes or the formation of new phases are often easily measurable (dilatometer).

The phase rule permits no specific statement as to which phases will appear or their stability fields. This information has to be obtained experimentally. Nevertheless in a few simple cases, using the free energy of the stable phase it is possible to calculate the transition points. This will be shown for the gypsum-anhydrite transition (p. 37). Noteworthy research in this context has been carried out by SAHAMA (1945) on the stability of sellaite.

In other cases it is possible at least to estimate the anticipated stable phases with the help of lattice energy, for at the low temperatures at which salts form this provides the greater part of the free energy (ZEMANN, 1958).

These energy considerations and the awaited development of the theory of concentrated solutions (degree of association, solvation, etc.) open up new ways for petrology. The following considerations, based upon purely empirical data, are still far from a true understanding of the events, for nothing will be said about crystallization kinetics or crystal growth. Above all, they will indicate the directions for further study upon which further progress is dependent. A start in this direction has been made by TODES (1958). Nevertheless the present empirical data are adequate to deduce the naturally occuring paragenetic processes and the associated compositional conditions.

The following discussion will commence with a consideration of the system $CaSO_4$–H_2O $\pm$ additional ions. Next come the quaternary partial systems NaCl–KCl–$MgCl_2$–H_2O, NaCl–KCl–Na_2SO_4–H_2O, NaCl–Na_2SO_4–$MgCl_2$–H_2O of the five-component system, but only in the particular case of NaCl saturation. Then follows the five-component system, NaCl–KCl–$MgCl_2$–Na_2SO_4–H_2O, later to be expanded through the inclusion of Ca compounds. Systems with $CaCl_2$ and with $FeCl_2$ will also be briefly considered.

The Graphic Presentation of Solubility Data

In simple systems the saturation concentrations are generally given in absolute units, referred to a constant water proportion (see, for example, Fig. 8). As a consequence several isotherms can be projected on to the concentration plane or a concentration-temperature diagram (polythermal diagram) can be constructed (Fig. 9). In the case of reciprocal salt pairs and in the five-component system, it is advantageous for many purposes to depict the absolute concentration at different isotherms in the form used by von KARSTEN (1950), and AUTENRIETH, AUTENRIETH and BRAUNE (1953–1960). Here the concentrations of two components are plotted along the ordinate, e.g. $MgCl_2$ along the x axis and $MgSO_4$ along the y axis, of an orthogonal co-ordinate system, whilst the saturation concentrations of KCl and NaCl are shown as tie lines crossing the contours, from which the exact composition at any point can be ascertained by interpolation (see Fig. 9). As knowledge of the absolute concentration of the solution is necessary for the calculation of the precipitation processes, this type of diagram is often preferred.

On the other hand, the diagram showing relative concentrations, i.e. referred to the sum of total dissolved salts, may often be easier to follow, as in this case one dimension fewer is utilized (two salts being shown along one axis, with absolute concentrations along the horizontal).

The five-component system can even be reduced to three variable concentrations, Mg, K_2 and SO_4, as JÄNECKE (1923 and earlier papers) has shown, for there is always saturation with respect to NaCl; moreover, only neutral salts are formed so that these conditions determine the valence-equivalent quantity of Cl or Na for every concentration of the given components. The reduction to three components makes it possible to use a Gibbs ternary diagram. The corners of the figure are represented by 100 $MgCl_2$, 100 K_2Cl_2 and 100 Na_2SO_4. As molar composition is indicated, $MgSO_4$ lies mid-way between Mg and SO_4, and K_2SO_4 mid-way between K_2 and SO_4. The tie line K_2SO_4–$MgSO_4$ divides the diagram into two systems, Na_2SO_4–K_2SO_4–$MgSO_4$ and $MgCl_2 + K_2SO_4 \rightleftharpoons MgSO_4 + K_2Cl_2$. As there is simultaneous NaCl saturation, the two systems continuously change from one to the other.

To calculate the ionic percentages required in the JÄNECKE diagram from the concentration data given in the tables in mol/1000 mol H_2O, the sum of total Mg ($MgCl_2$ and $MgSO_4$), total SO_4 ($MgSO_4$ and Na_2SO_4) and K_2Cl_2 is obtained and then converted to 100. The conversion factor multiplied by 1000 yields the water content for $Mg + K_2 + SO_4 = 100$. The NaCl content can be computed in the same way. In a similar manner the salt minerals can be recalculated as ionic percentages (Table 4) and their positions in the ternary diagram fixed.

Table 4. *Salt minerals (common minerals italicized)*

Symbol	Name	Formula	Ternary diagram position [a]				Mol
			$[K_2]$	[Mg]	$[SO_4]$	H_2O	weight
a	*Anhydrite*	$CaSO_4$	—	—	100	—	136.15
bi	Bischofite	$MgCl_2 \cdot 6\,H_2O$	—	100	—	600	203.33
bl	Bloedite	$Na_2Mg(SO_4)_2 \cdot 4\,H_2O$	—	33.3	66.7	133	334.51
c	*Carnallite*	$KMgCl_3 \cdot 6\,H_2O$	33.3	66.7	—	400	277.88
cc	Chlorocalcite	$KCaCl_3$	100	—	—	—	185.54
da	D'Ansite	$Na_{21}MgCl_3(SO_4)_{10}$	—	9.1	90.9	—	1574.29
e	Epsomite	$MgSO_4 \cdot 7\,H_2O$	—	50	50	350	246.50
g	*Gypsum*	$CaSO_4 \cdot 2\,H_2O$	—	—	100	—	172.18
gs	Glaserite	$K_3Na(SO_4)_2$	42.9	—	57.1	—	332.42
gb	Glauberite	$Na_2Ca(SO_4)_2$	—	—	100	—	278.21
goe	Goergeyite	$K_2Ca_5(SO_4)_6 \cdot H_2O$	14.3	—	85.7	14	873.03
hx	Hexahydrite	$MgSO_4 \cdot 6\,H_2O$	—	50	50	300	228.49
k	*Kainite*	$KMgClSO_4 \cdot {}^{11}/_4\,H_2O$ [b]	20	40	40	110	244.48
ks	*Kieserite*	$MgSO_4 \cdot H_2O$	—	50	50	50	138.41
lg	Langbeinite	$K_2Mg_2(SO_4)_3$	16.7	33.3	50	—	415.04
lh	Leonhardtite	$MgSO_4 \cdot 4\,H_2O$	—	50	50	200	192.45
le	Leonite	$K_2Mg(SO_4)_2 \cdot 4\,H_2O$	25	25	50	100	366.71
loe	Loeweite	${}^1/_7[Na_{12}Mg_7(SO_4)_{13} \cdot 15\,H_2O]$ [b]	—	35	65	75	280.76
m	Mirabilite	$Na_2SO_4 \cdot 10\,H_2O$	—	—	100	1000	322.22
5h	Pentahydrite	$MgSO_4 \cdot 5\,H_2O$	—	50	50	250	210.47
p	Polyhalite	$Ca_2K_2Mg(SO_4)_4 \cdot 2\,H_2O$	16.7	16.6	66.7	33	602.98
sh	Schoenite	$K_2Mg(SO_4)_2 \cdot 6\,H_2O$	25	25	50	150	420.75
n	Rock salt (Halite)	NaCl	—	—	—	—	58.454
sy	*Sylvite*	KCL	100	—	—	—	74.553
sg	Syngenite	$K_2Ca(SO_4)_2 \cdot H_2O$	33.3	—	66.7	33	328.43
ta	Tachhydrite	$CaMg_2Cl_6 \cdot 12\,H_2O$	—	100	—	600	517.65
t	Thenardite	Na_2SO_4	—	—	100	—	142.06
vh	Vanthoffite	$Na_6Mg(SO_4)_4$	—	20	80	—	546.57

[a] Calculation based on $K_2 + Mg + SO_4 = 100$.

[b] The formula is written with the water content in non-integral figures, in order to have molecular weights comparable with the other salt minerals, as this is required for subsequent calculations of salt precipitation.

In the Gibbs ternary diagram, salt precipitation can be ascertained by a simple graphical technique (p. 70). The four bounding systems of the five-component system can be reduced to two variable concentrations, so that the temperature dependence can also be represented in the plane of the diagram (Fig. 15).

It must be observed, however, that the water content, which is not directly reproducible on the JÄNECKE figure, plays an essential part, particularly as solubility is only determined with respect to the water content. This water content can be obtained from the contours. The contours at high Mg concentrations, and in the kieserite field in particular, are irregular, and do not decrease evenly in the direction of the crystallization end point (see Fig. 31). In conclusion then, the JÄNECKE diagrams lead to errors, as the example of "Stufenmetamorphose" (geothermal metamorphism[12]) (p. 114) will show, unless the effects of the water content are taken into account. In addition, points representing the salt minerals do not lie in the saturation plane but always below it, as they contain much less water than the saturated solutions. The salt minerals occur only in the projection plane in the stability fields. This projection is, however, independent of water content and consequently may only be applied to constructions which are equally independent of water content.

I. System $CaSO_4$–H_2O ($\pm$ NaCl, etc.)

1. Solid Phases

Anhydrite $CaSO_4$. In pure anhydrite beds or in thick anhydrite lens the crystals are xenomorphic and markedly interlocking, although also often elongate with an orientation more or less clearly parallel to the bedding. Usually several anhydrite generations are observed. The younger crystals are larger, either idiomorphic (commonly prismatic after [010]) or in enriched nests or lens. At the contact with, or as inclusions within, chloride salts the anhydrite crystals are always somewhat larger than in pure anhydrite beds. Also of importance are the pseudomorphs after gypsum (p. 157) which have been recorded on many occasions. Notwithstanding this, anhydrite is itself frequently replaced by polyhalite (p. 161).

Gypsum $CaSO_4 \cdot 2\,H_2O$ (orientation according to de Jong and BOUMAN, see explanation to Table 3, reduced monoclinic cell). In salt deposits well-formed gypsum crystals are found most commonly in druses where they have developed from the incongruent dissociation of other minerals, such as polyhalite and glauberite. Primary precipitation of gypsum should result in elongated crystals of [101] type, according to G. OGNIBEN (1955). In such a case the form must be inferred from the structural arrangement for as a result of diagenetic recrystallization the gypsum has acquired secondarily a granular, polygonal appearance. Fibrous gypsum in joints of shaly rocks is generally formed with fibres parallel to [001] (R. SCHMIDT, 1914).

[12] "Normal progressive geothermal metamorphism" (in German „Stufenmetamorphose“) defined by H. BORCHERT and R. MUIR, Salt Deposits. Van Nostrand Company 1964.

Hemihydrate $CaSO_4 \cdot \frac{1}{2}H_2O$ as a mineral (bassanite) in salt deposits is dubious. It is clear from laboratory experiments that hemihydrate plays an important role and in a fine crystalline form it may be mistaken for anhydrite.

2. *Temperature Dependence*

As a result of the numerous studies of the $CaSO_4$–H_2O system, reliable data on temperature are available. The results of the investigation of solubility by various workers are shown in Fig. 4, taken from KELLEY et al. (1941, Fig. 10) and D'ANS et al. (1955, Fig. 1). The principal results are:

1. The solubility of gypsum changes slowly with temperature and passes through a very weak maximum at about 40° C.

2. The solubility of anhydrite decreases with rising temperature.

3. The solubility of hemihydrate, which as represented in the figure occurs in two forms, is higher than gypsum below 100° C and higher than that of anhydrite at all temperatures. It is accordingly always unstable. Its solubility decreases very rapidly with increasing temperature.

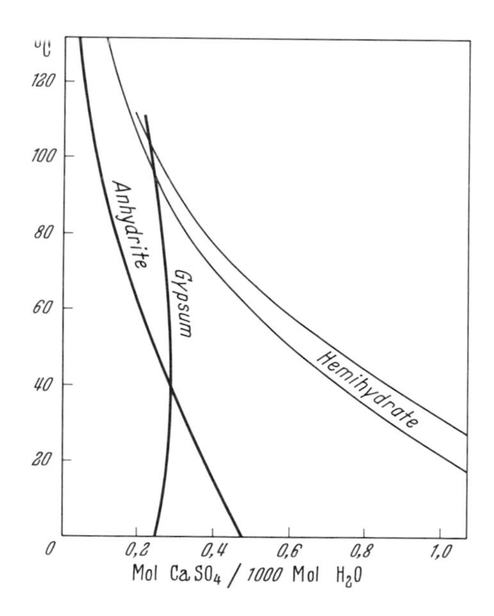

Fig. 4. Solubility of anhydrite, gypsum, and hemihydrate in water at different temperatures

4. The solubility curves of gypsum and anhydrite intersect at 42° C[13] and below that temperature gypsum is the stable form; above it, anhydrite is stable. The unsaturated solutions lie to the left of the curves. Solutions between the gypsum and anhydrite curves are saturated with respect to gypsum below 42° C and with respect to anhydrite above 42° C. In aqueous solutions gypsum then can only alter to anhydrite when the temperature exceeds 42° C. Above 100° C gypsum can change first to hemihydrate and then to anhydrite. At no temperature can the reverse occur and anhydrite change to give hemihydrate.

3. The Common Ion Effect (Excluding the Formation of Compounds)

The solubility of gypsum and anhydrite is in general influenced by the presence of additional ions, increasing through the addition of foreign ions, and conversely being reduced through the addition of the same ions. In salt deposits the most important additional ions are Na^+, Cl^-, and the common ion SO_4^{--}. The data from different workers appear to be in reasonably good agreement. The effects, using the data available up to 1961, are illustrated in Fig. 5.

In NaCl solution the following occurs:

1. The solubility of gypsum and anhydrite is first greatly increased, passes through a weak maximum and then diminishes gradually.

2. The solubility of gypsum is very nearly independent of temperature, although this assertion has not yet been adequately verified by systematic research on the effects of additional ions. The solubility of anhydrite, as in NaCl-free solutions, decreases markedly with increasing temperature.

[13] New results of investigations of the gypsum anhydrite equilibrium are published by the following authors:

D'ANS, J.: Der Übergangspunkt Gips↔Anhydrit. Kali und Steinsalz, Heft 3, 109—111 (1968).

HARDIE, L. A.: The gypsum-anhydrite equilibrium at 1 atm. pressure. Am. Min. **52**, 121—200 (1967).

KINSMAN, D. J. J.: Gypsum and anhydrite of Recent age, Trucial Coast, Persian Gulf. 2nd Symposium on Salt, **1**, 302—326, Northern Ohio Geological Society, Cleveland, Ohio (1966).

ZEN, E-AN.: Solubility measurements in the system $CaSO_4$–NaCl–H_2O at 35°, 50° and 70° C and 1 atm. pressure. J. Petrology **6**, 124—164 (1965)

The new data do not alter the *principal* conclusions published in this monograph.

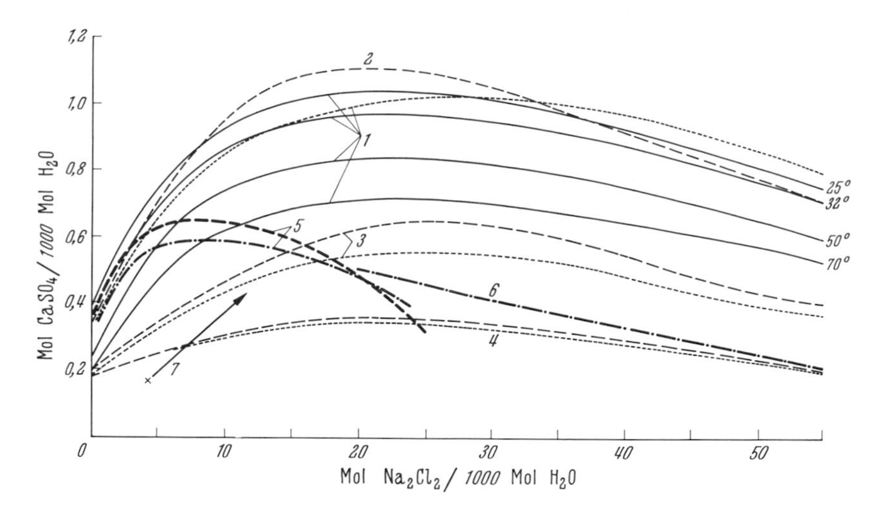

Fig. 5. Solubility of gypsum and anhydrite in saline solutions. Solid phases; anhydrite = solid and broken lines; gypsum = dotted lines (heavy and light). *1.* after D'ANS et al. (1955); *2–4.* After MADGIN and SWALES (1956) at 25° C; *3.* the system $CaSO_4$–NaCl–Na_2SO_4–H_2O with 1.4 mol Na_2SO_4/1000 mol H_2O; *4.* the same with 3.45–3.36 mol Na_2SO_4/1000 mol H_2O; *5.* in seawater at 30° C after POSNJAK (1940); *6.* in seawater extrapolated from *3* and *4*; *7.* seawater and its alteration in concentration

3. On account of the different temperature coefficients, the point of intersection of the gypsum and anhydrite solubility curves is displaced towards lower temperatures as the NaCl content of the solution increases.

With an additional SO_4^{--} content the solubility of gypsum and anhydrite is reduced and the maximum becomes flatter. The solubility curves of gypsum and anhydrite are very similar to one another.

In artificial seawater POSNJAK (1940) has shown that the foreign ion content (Na^+, Mg^{++}, Cl^-) strongly increases solubility at first, but that subsequently through the SO_4^{--} content solubility is similarly reduced. The gypsum and anhydrite curves intersect. This point requires closer investigation. POSNJAK's series of experiments broke off as soon as the point of intersection was reached. According to the data of MADGIN and SWALES (1956) for the NaCl + Na_2SO_4 system, the isotherms do not intersect. In any event the difference in solubility between gypsum and anhydrite at 30° C under conditions of high NaCl concentration, etc., appears to be small. The temperature effect in NaCl solutions containing SO_4 has not been experimentally investigated, but they must almost certainly correspond to conditions in NaCl solutions. From MADGIN and SWALES' data (1956) the saturation concentrations of $CaSO_4$, at the point where the precipitation of NaCl from seawater begins, can be found by extrapolation. The figure obtained, 0.2 mol $CaSO_4$/1000 mol H_2O, is probably a maximum value although it is far less than VAN'T HOFF's figure of 0.72 mol $CaSO_4$/ /1000 mol H_2O (VAN'T HOFF, 1909, p. 40). This fact has the important consequence that in regions where rock salt appears as a result of static evaporation considerably less $CaSO_4$ must occur than was formerly thought.

The concentration of $CaSO_4$ and NaCl in normal seawater is indicated by the point of origin of the arrow (Fig. 5). As long as only water is evaporating, the salt ratio remains constant. The composition moves away linearly from the origin (pure water). At the point where the arrow intersects the solubility curve of $CaSO_4$ in seawater (POSNJAK, 1940) gypsum begins to precipitate. This is at about $3^1/_2$ times the concentration of normal seawater. The composition of the solution then changes along the solubility curve. At about 5 times the concentration the intersection of the gypsum—anhydrite curve is reached. At higher NaCl concentrations at 30° C, anhydrite is the stable solid phase. These data assume normal seawater. If there is a reduction in the amount of $MgSO_4$ in seawater, either prior to or at the onset of gypsum precipitation, the amount of dissolved $CaSO_4$ lies between those found for normal seawater and those determined in NaCl solutions. Experimental data specifically for reduced $MgSO_4$ in seawater are still lacking.

It is not permissible to deduce from the solubility data evidence for the formation of primary anhydrite. Numerous crystallization experiments under simulated natural conditions have always produced gypsum, or at higher temperatures hemihydrate, as the first precipitate. This phenomenon, observed in the formation of many new phases (Ostwald's rule), arises out of the differing nucleation energies (VOLMER, 1939, p. 200), which are evidently substantially higher for anhydrite than for gypsum. Anhydrite nucleation arises secondarily in or upon gypsum, in a solution where, because of the rapid crystallization of metastable gypsum, the necessary oversaturation is not reached (CONLEY and BUNDY, 1958). In addition it is known that the nucleation energy of heterogeneous nuclei (that is to say, on a foreign base) as opposed to homogeneous nuclei is lower and, particularly in solution, favours the separation of oriented substances (LACMANN, 1961). This qualitative interpretation fits all the experimental data. Only at temperatures well above the transition temperature and with strong oversaturation can primary anhydrite be formed. A quantitative description of this is still impossible as the nucleation energy is unknown. In the first evaporation phase (Ca sulphate) at temperatures below 50° C the formation of primary anhydrite is unlikely.

4. Pressure Dependence

Solubility data at high pressure are experimentally difficult to obtain, and experiments on salt systems have only been carried out in isolated cases (see ADAMS, 1931, with references to the older literature; RAWITSCH, 1958; SOURIRAJAN and KENNEDY, 1962). Nevertheless, the pressure effects can be calculated from thermodynamic considerations. According to Le Chatelier's principle, pressure favours the development of the phase with the smaller volume. In the reaction

$$\text{Gypsum} \rightleftharpoons \text{Anhydrite} + 2\,H_2O$$

the sum of the specific volumes on the right-hand side of the equation is the greater by about 10%. With the same pressure on all phases, the formation of gypsum is therefore favoured.

From purely thermodynamic data the transition point Gypsum $\rightleftharpoons$ Anhydrite + 2 H_2O can be calculated (MACDONALD, 1953) for at this point the difference in free energy ΔG between gypsum and anhydrite vanishes.

According to KELLEY et al. (1941)[14] the free energy for the reaction Gypsum $\rightleftharpoons$ Anhydrite + 2 H_2O is dependent upon absolute temperature T.

$$\Delta G = 163.89\,T + 0.0215\,T^2 - 65.17\,T \cdot \log_{10} T - 2495\,\frac{\text{cal}}{\text{mol}}\,.$$

When $\Delta G = 0$, the temperature calculated is 40° C which agrees remarkably well with the transition temperature obtained from solubility data (42° C). Because of the relationship

$$\Delta G = \Delta H - T\Delta S$$

(where ΔH = heat of reaction at constant pressure; in the preceding case heat of transformation) the temperature dependence of the free energy corresponds to the negative entropy change ΔS:

$$\frac{d\,\Delta G}{d\,T} = -\Delta S\,.$$

From the preceding reaction

$$\Delta S = -163.89 - 0.043\,T + 65.17\,(0.43429 + \log_{10} T)\,\frac{\text{cal}}{\text{mol degree}}$$

and at the transition point ($T = 313.1$° K)

$$\Delta S = 13.59\,\frac{\text{cal}}{\text{mol degree}} = 580\,\frac{\text{cm}^3\text{ at}}{\text{mol degree}}\,.$$

[14] New data see ZEN, E-An.: Solubility measurements in the system $CaSO_4$–$NaCl$–H_2O at 35°, 50°, and 70° C and one atmosphere pressure. Journal of Petrology **6**, 124—164 (1965).

Substituting this value in the Clausius-Clapeyron equation (with $\Delta V = 6.8\ \mathrm{cm^3/mol}$ anhydrite) when the pressure on all phases is the same, results in the following pressure dependence of the transition temperature (as $\Delta S = \Delta H/T$ at the transition point):

$$\frac{dT}{dP} = \frac{T \cdot \Delta V}{\Delta H} = \frac{\Delta V}{\Delta S} = +0.0118 \frac{\text{degrees}}{\text{atmosphere}}$$

or, expressed in another way: the transition temperature from gypsum to anhydrite (in the absence of additional ions) is raised 1° C by a pressure of about 85 atmospheres. The pressure effect is thus small.

Under conditions of differential pressure (liquid phase under hydrostatic pressure with unrestricted connections to the surface, solid phases under the pressure of the overlying rocks which has an estimated average density of $2.4\ \mathrm{g/cm^3}$) the formation of anhydrite is favoured. A calculation analogous to the preceding gives:

$$\frac{dT}{dP_{(\text{diff.})}} = -0.0247 \frac{\text{degree}}{\text{atmosphere}}$$

or a reduction of about 1° in the transition temperature per 40 atmospheres pressure increase.

The effect of additional ions has also been investigated by MacDonald (1953) who demonstrated the dependence of the free energy of the dehydration upon the NaCl concentration. This can be done in an approximate way using the reduction in the vapour pressure which is a consequence of the NaCl concentration increase. The resulting reduction of the free energy of the reaction can be taken into consideration through an additional term $RT \cdot 2.303 \cdot \log_{10} \frac{p}{p_0}$ ($p =$ vapour pressure of the salt solution, $p_0 =$ vapour pressure of water, $R =$ gas constant). The calculation covers the known terms, but provides only approximate values because of certain simplifications in the calculation.

The dependence of the gypsum-anhydrite transition temperature upon the salt content and pressure (taking into account the thickness of the overlying rocks and the depth of the solution) is illustrated in Fig. 6. The oblique ruling indicates the region in which gypsum is stable assuming a normal geothermal gradient (3°/100 m) and a mean annual temperature of 11° C.

The transition temperatures at atmospheric pressure obtained from the solubility data of d'Ans et al. (1955) are also shown in Fig. 6. This curve has the opposite curvature to the others, which may be due to small inaccuracies in the solubility curves and to neglecting minor factors (further lowering of the vapour pressure due to the presence of $CaSO_4$ in solution; deviations from the ideal gas law, etc.).

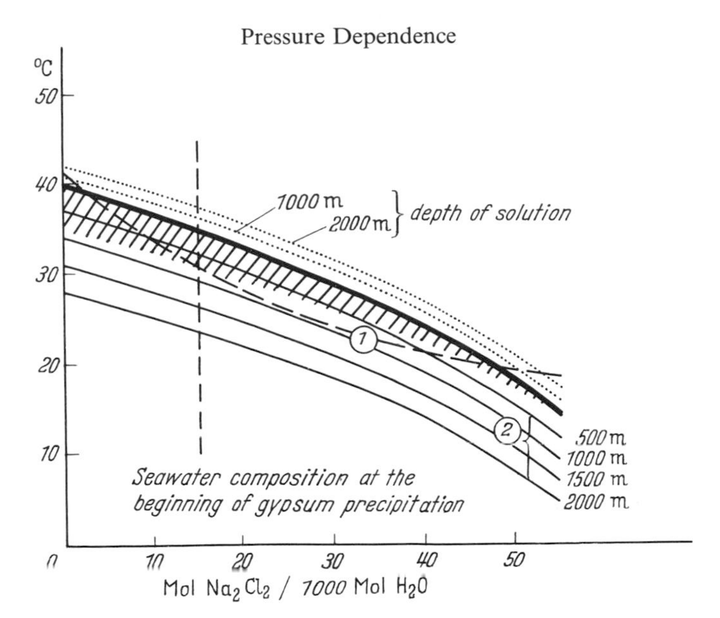

Fig. 6. Dependence of the gypsum-anhydrite transition temperature upon salt content, depth of solution (dotted curves) and thickness of sedimentary cover (*2* continuous curves, differential pressure). Calculation after MACDONALD (1953). *1* stability boundary determined from solubility data (after D'ANS et al., 1955)

The essential result is not affected by this uncertainty.

1. The transition temperature falls sharply with increasing concentration of NaCl.

2. Hydrostatic pressure causes an insignificant increase, and differential pressure a noticeable reduction, in the transition temperature. The transition temperature at the onset of precipitation of gypsum from seawater is 35° C (36° C at a depth of 870 m), at the onset of NaCl precipitation 14° C (15° C at a depth of 770 m). Thus, at the higher temperatures expected for evaporation of solutions saturated with halite, gypsum is no longer stable. Nevertheless it does precipitate in the metastable region, as has previously been established.

3. With a normal geothermal gradient, and in the absence of saline solutions, gypsum is stable to depths of $\gtrsim 800$ m. At higher thermal gradients (5°/100 m) close to active magma chambers the limit of the gypsum stability field is reached at a depth of about 500 m in a NaCl-free system, in the presence of aqueous solutions. (The heating of completely dry gypsum is likely to produce an alteration to hemihydrate, although at considerably greater depths than the above. The geological significance of such a change is still not known, but is probably small; hemihydrate is metastable and should alter rapidly to anhydrite.)

Observations do not contradict this. OGNIBEN (1957, p. 200) has reported gypsum in pure gypsiferous rocks at depths of 735 m. The secondary alteration of anhydrite to gypsum in weak saline solutions has been observed at depths down to about 600 m (L. OGNIBEN, 1957; VON GAERTNER, 1932).

II. System $NaCl–KCl–MgCl_2–H_2O$

1. Solid Phases

Rock salt or *halite* is the most important halide. In crystallization from natural salt solutions, scarcely any foreign ions in quantities greater than 0.05% are included in its lattice. The amount is influenced by the presence of other compounds in solution, but, in addition to the dominant cube faces, there occur very rarely {110}, {111} (rarely a determining habit, specifically from Na_2SO_4-rich solutions), and {012} (possibly only a solution form).

As crystal growth curiosities the following are noted:

Microlithic (hopper) crystals with step-like pseudofaces or depressions as the result of preferred growth at the edges at the surface of a statically evaporating solution (GAHM and NACKEN, 1954). The dip angles of the plane tangential to the growth steps depend upon the rate of evaporation and are not necessarily constant. Recent examples have been found in desert salt lakes. The probability of finding such forms persisting in fossil salt deposits is small because of recrystallization.

Fibrous rock salt in joints in porous rocks such as the saliferous clays (Salzton)[15]. The preferred long axis orientation is parallel to [110], less commonly parallel to [001] and still less frequently in other directions (R. SCHMIDT, 1914, but see also STURMFELS, 1943). The growth results from the simultaneous growth of the elongate crystal and expansion of the joint (MÜGGE, 1928; for the Br distribution in fibrous halite of the potash salt mine "Marie-Louise", Alsace/France see Herrmann, 1964 [16]).

Clayey rock salt (Tonwürfelsalz) is rock salt with distorted crystal growth, usually showing pseudo-rhombohedral form but with normal cubic cleavage (GÖRGEY, 1912; STURMFELS, 1943). In argillaceous saliferous rocks the short diagonal of the pseudorhomb is commonly found normal to the bedding because of lower growth potential in this direction.

[15] Salzton = salt or saliferous clay, but the term often has a stratigraphic context.

[16] HERRMANN, A. G.: Geochemische Untersuchungen an einem Vorkommen von Fasersteinsalz. Beiträge zur Mineralogie und Petrographie **10**, 374—378 (1964).

Halite takes part in paragenesis with all other salts which may occur as inclusions or occasionally as oriented growths within them, eg. sylvite. Moreover, liquid and gas inclusions may also occur, usually zonally arranged parallel to {100}. This zonal structure is not of itself sufficient evidence for primary cubic crystals, as it is just as common in the secondary precipitation of halite.

From the tectonic point of view the most important physical property of salt masses is their plastic deformation, which is explained in all the newer mineralogy textbooks. In NaCl rocks plastic deformation begins at stresses as low as 100–200 kp/cm^2 (STÖCKE, 1936), although according to H. BORCHERT (1959, p. 158) it may begin even as low as about 10 kp/cm^2 although plastic flow in single crystals does not begin until much higher stresses are applied (DOMMERICH, 1934).

Many of the properties of NaCl are not explicable on the basis of an ideal crystal structure. One example of this is the blue colour. The colour is related to lattice deformations of various kinds (colour centres) and is brought out by radiation. It can also be demonstrated artificially by radioactive radiation or by X-rays with initially a yellow colour resulting, and only after long exposure (W. BORCHERT, 1958), or under pressure (HOWARD and KERR, 1960), or by means of stabilizing foreign ions such as Pb (BORN, 1934) does the blue colour develop (see also VINOKUROV, 1958). The source of natural radiation is still problematic. If it were due to γ radiation from ^{40}K, as is often assumed, the blue colour should be pronounced in the sylvinites. Yet it is absent, for example, in the pure white sylvinite of the Ronnenberg seam (K 3 Ro). Hence it seems better to look for the source of radiation in the solutions from which the NaCl itself crystallizes (BORN, 9134, 1959).

Hydrohalite $NaCl \cdot 2\,H_2O$ is occasionally found in the Siberian salt lakes, but only as a periodic, winter precipitate (DZENS-LITOWSKI, 1955).

Sylvite KCl is industrially the most valuable and important potash mineral (equivalent to 63.17% K_2O). In natural outcrop it contains more impurities than rock salt, in particular Br (up to 0.5%) whilst other impurities such as Pb (HERRMANN, 1958) etc., are several orders of magnitude less. Sylvite often contains haematite inclusions, usually in the form of fine flakes strongly concentrated at the crystal margin so that macroscopically a strong red rim is observed and a weakly opalescent blue to turbid milky core. Oriented plates of haematite with $(0001)_{Haem.}$ parallel to {001}, {111} or {110} of sylvite have also been reported (LEONHARDT and TIEMEYER, 1938). Commoner is the oriented intergrowth of sylvite with NaCl (GÖRGEY, 1912, p. 418; STURMFELS, 1943, p. 173, D'ANS and KÜHN, 1938).

Although sylvite very rapidly develops a strong violet blue colour when exposed to X-ray radiation, it does not show any radiation colour in nature. In any case the artificially induced luminescence disappears relatively rapidly. In the so-called blue sylvite of the Werra-district, the blue colour is exclusively limited to the rock salt component (ROTH, 1953).

Bischofite $MgCl_2 \cdot 6\,H_2O$ is relatively rare in salt deposits and generally of secondary origin. Large primary bischofite outcrops are mentioned as occurring in the Lower Permian of the Caspian basin (FIVEG, 1961) and in the Gaboon province of the former French Congo where it occurs with primary tachhydrite (!) (MAYRHOFER, pers. comm.). Recent formation of bischofite is reported by SEDELNIKOV (1958) in isolated brine pools of Kara Bugas. It crystallizes in monoclinic form notwithstanding a distinct pseudo-trigonal form. It is strongly hygroscopic. Bischofite, too, shows plastic deformation (on (110), parallel $[1\bar{1}2]$), as well as simple shear (twinning planes) (MÜGGE, 1906).

Carnallite $KMgCl_3 \cdot 6\,H_2O$ with an equivalent of 16.95% K_2O is an important primary potash mineral, but much less common as a secondary product. In the natural material there is commonly up to 0.5% Br and other elements in trace amounts, in particular Rb, Cs, Tl, NH_4 (for K) and Fe^{II} (for Mg). Iron is most generally found in the form of oriented haematite scales (JOHNSEN, 1909), more uniformly distributed and much flatter than in sylvite. As a result of this orientation the mineral shows a characteristic brownish-red irridescence.

Secondary carnallite plays a minor role, e.g. as fibrous carnallite in joints of the Hartsalz (potash salts), in the grey Salzton (salt clay) and also in cavities in the Hauptanhydrit, and in ascharite and boracite nodules, etc. It is characterized by colour differences as compared with the adjacent, primary carnallite. In the Salzton it is not always colourless, but often has a yellowish or reddish tint in the form of cloudy bands within the uncoloured carnallite. Nevertheless some major carnallite deposits are presumably secondary (see p. 191, 194).

Carnallite also displays pressure twinning lamellae. These may develop at comparatively small stresses (the familiar crunching heard when the point of the hammer is inserted is due to shear).

2. *Stability Conditions*

a) System $NaCl–H_2O$

In this two-component system the only solid phases are NaCl, $NaCl \cdot 2\,H_2O$ (hydrohalite) and ice (Fig. 7). There are three invariant points (of which one is metastable), according to the phase rule for four phases at constant temperature and concentration:

+ 0.1° C: NaCl + NaCl · 2 H_2O + saturated solution + vapour

– 21.2° C: NaCl + NaCl · 2 H_2O + ice + saturated solution + vapour

– 25.2° C: NaCl + ice + saturated solution + vapour (metastable).

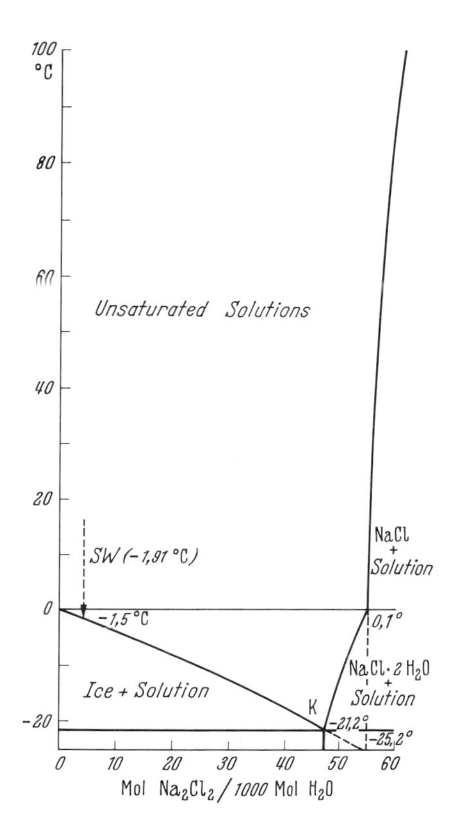

Fig. 7. The system NaCl–H_2O (essentially after D'ANS, 1933): The metastable areas are indicated by broken lines. Below *K* (= cryohydric point) all phases are solid with stable paragenesis (ice + hydrohalite), left of *K* with ice inclusions, right of *K* with hydrohalite inclusions

They form the intersection points of the univariant curves corresponding to the saturated solutions.

Halite has a very low positive temperature coefficient of solubility. In cooling a saturated solution from 100° C to 20° C, 5.5 mol out of

60.5 mol Na_2Cl_2 [17] will be precipitated. Precipitation of rock salt must therefore largely be the result of evaporation.

Hydrohalite has a marked positive temperature coefficient of solubility, and so can be precipitated from a saturated solution by cooling. It has nevertheless a relatively high nucleation energy so that it only appears after strong supercooling. It can develop from over-saturated NaCl solutions (particularly in the presence of other ions, as in seawater) above the transition temperature. It is then rapidly altered (see below).

The ice curve, which at low salt concentrations follows Raoult's Law for complete salt dissociation (reduction of freezing point proportional to salt concentration), has no real significance as far as the formation of salt deposits is concerned. Use can be made of this property technologically (DZENS-LITOWSKI, 1955). A halite solution with the same NaCl content as seawater of normal chlorinity, freezes at $-1.5°$ C, while seawater (19‰ Cl) because of the presence of other ions freezes at $-1.91°$ C. Thus the composition of the solution alters along the ice curve in the direction of a higher NaCl concentration. The ice curve for seawater has a much more complicated path for, with further cooling after ice has formed, other solid phases may also be precipitated (see SVERDRUP, et al., 1949, p. 216) before $-50°$ C is reached, when the residual solution freezes.

The solubility of NaCl increases slightly with increasing pressure, reaching a maximum at about 4000 atmospheres (at 25° C 59 mol Na_2Cl_2/1000 mol H_2O) and decreasing slowly until at around 12,500 atmospheres it has the same value as at atmospheric pressure (ADAMS, 1931). At 25° C between 7900 and 11,600 atmospheres hydrohalite is the stable form. In the study of the formation of salt deposits the pressure dependence of solubility can be ignored (see p. 129).

The invariant point at 0.1° C is an incongruent melting point (which was not the case at the transition temperature, 42° C, of gypsum to anhydrite, Fig. 6). In salt systems, incongruent melting points are common so that the simple example of hydrohalite will be enlarged upon. This example is also suitable for a demonstration experiment (BRAITSCH, 1961 b). At the incongruent melting point the solid phase melts with the formation of a new solid phase and a solution. At no point is there complete fluidity, the solution therefore never has the same composition as the original solid phase (hence the term "incongruent"). The newly

[17] Monovalent ions are calculated as double molecules so that they have the same bonding as the divalent ions in the solution. This formulation should not be confused with the composition of the soluble components. It refers to hydrated ions and says nothing about the degree of dissociation, possible association, complexes, ionic pairs and the "structure" of the solution (of GARRELS and THOMPSON, 1962).

forming crystals are at the same time surrounded by steadily enlarging droplets of solution.

From the solubility diagram the proportions of the different phases for the equilibrium composition can be deduced, as the solution in equilibrium with the two solid phases must have the same composition as at the incongruent melting point. Under such conditions the change can be represented by the equation[18]

	Hydrohalite	$\overset{0.1^\circ\,C}{\rightleftharpoons}$ x halite	+ y equilibrium solution
$NaCl$	1	1	110
H_2O	2	0	1000

from which

$$1 \text{ mol hydrohalite} \rightleftharpoons 0.78 \text{ mol NaCl} + 0.22\,(55 \text{ mol } Na_2Cl_2 + 1000 \text{ mol } H_2O)\,.$$

A graphic solution is also possible, but this is less precise and no quicker. With three unknowns the graphic solution becomes cumbersome and with further unknowns is generally impossible.

The temperature of incongruent melting is raised by pressure (ADAMS and GIBSON, 1930). Over the first 1000 atmospheres the relationship is essentially linear and is about 0.007°/at. At higher pressure the increase in the melting curve diminishes rapidly and at about 9400 atmospheres passes through a maximum of 25.8° C.

The simple system $NaCl$–H_2O is never reached by evaporating sea-water (cf. pp. 84 onwards). Such a simple system may correspond to the dissolving of exposed salt stocks in Iran by rainwater. From the salt water streams pure secondary rock salt can crystallize, often as microlithic crystals (p. 40). This is then collected for domestic use whilst the primary rock salt is much too impure to be used directly.

b) System $NaCl$–KCl–H_2O

In the system KCl–H_2O there are only sylvite and ice as solid phases. KCl has a very strong positive temperature coefficient of solubility of

[18] Calculations of this kind will be used later. They correspond to the solution of linear equations at an acceptable level using determinants (Kramer's rule). As many independent quantities are required as there are unknowns. The concentrations of components in each phase serve as coefficients, and these for simplicity are usually given in mols. The calculation of determinants can be found in any textbook of algebra. The calculation method is not essential for the understanding of the preceding result, the method of formulating the problem is. The method is a synoptic evaluation of the solution equilibria of VAN'T HOFF, D'ANS, AUTENRIETH and others, and it goes without saying that it leads to the same result, usually in a shorter time.

$+0.34$ mol K_2Cl_2/1000 mol H_2O per degree. This technically important property remains even in the presence of NaCl (Fig. 8). New solid phases (compounds of the two components) do not occur. Through the addition of similar ions, the solubility is considerably reduced although the temperature coefficient is hardly altered ($+0.31$ mol K_2Cl_2/1000 mol

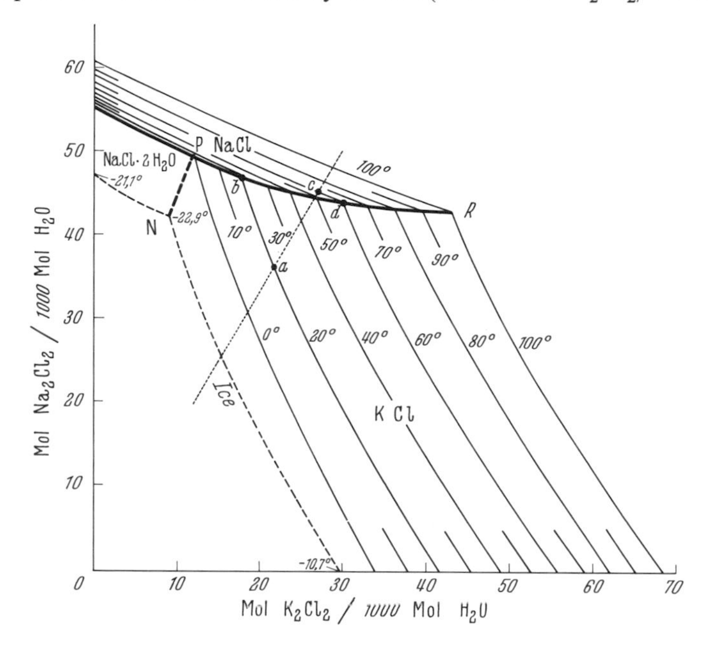

Fig. 8. The system NaCl–KCl–H_2O at different temperatures (simplified after D'ANS, 1933)

H_2O per degree in NaCl-saturated solutions). As a result the cooling of an NaCl + KCl saturated solution yields nearly as much KCl as can be obtained by precipitation from a cooled KCl-saturated NaCl-free solution. The solubility of NaCl in the presence of KCl is likewise reduced. Furthermore the temperature coefficient of solubility of NaCl below 100° C becomes negative (-0.1 mol Na_2Cl_1/1000 mol H_2O per degree between 0 and 50° C and somewhat less above 50° C, cf. curves for 0 mol $MgCl_2$/1000 mol H_2O in Figs. 10 and 11), and as a result halite precipitates from a KCl–NaCl saturated solution upon heating.

The precipitation data are simple. Several isothermal lines are shown in Fig. 8 (from D'ANS, 1933). The absolute solution concentration can be read directly. At any given temperature the unsaturated solutions lie between the 0 point and the appropriate isotherm. A solution with (25 mol Na_2Cl_2 + 15 mol K_2Cl_2)/1000 mol H_2O is unsaturated at all temperatures. Upon evaporation the composition changes linearly from the 0 point which corresponds to pure water. The progress is dependent upon temperature. Sylvite precipitates first when at 20° C the 20° isotherm is reached (at point a, Fig. 8). With further evaporation, and further precipitation of sylvite, the composition of the solution changes along the isotherm until point b on curve *RP* is reached. Thereafter NaCl and KCl precipitate simultaneously and the composition of the solution remains unchanged (constant solution). At 60° C, however, NaCl precipitates (point c) and the solution composition changes along the 60° isotherm in the direction of a lower NaCl concentration. After reaching the curve *RP* (at point d) the concentration remains constant until completely evaporated with the simultaneous precipitation of KCl and NaCl.

This simple system can serve as an example of Gibb's phase rule (p. 28). There are three independent components. The isotherms lie in the saturation planes where there are three phases (solid, liquid, vapour) and two degrees of freedom (temperature and concentration). Along the lines *RP*, *PN*, *PM* etc. two saturation planes intersect and here there are two solid phases. At a given temperature the concentration is fixed and the converse is also true, there thus remains one degree of freedom and the system is univariant. At points *P* and *N* there are three solid phases:

P: halite, hydrohalite, sylvite,
N: hydrohalite, sylvite, ice

and, in addition, solution and vapour. Temperature and concentration are both fixed and the system is invariant at these points.

If the system could be considered in the absence of the gas phase (see p. 29), then pressure can be counted as an additional degree of freedom and in this case points *P* and *N* are also univariant. Because of the relatively low pressure dependence of solution equilibria, this degree of freedom can be neglected in questions concerning the genesis of salt deposits.

c) System $NaCl–KCl–MgCl_2–H_2O$

This four-component system is particularly important for certain types of salt deposits. It is still quite simple. It is characterized by the appearance of a ternary compound, carnallite $KCl \cdot MgCl_2 \cdot 6\,H_2O$. In this system only the condition of an NaCl-saturated solution is of interest in the present context. This means that one solid phase is present and the NaCl concentrations are determined by the NaCl saturation requirement for each pair of values for KCl and $MgCl_2$. Thus the system can be regarded as a special case of a three component system.

Fig. 9 contains the solubility isotherms for KCl at various temperatures interpolated from data obtained by D'ANS (1933) and the NaCl isotherm at 20° C. The values have been smoothed by means of a difference method but differ from the original data by not more than 0.5 mol, particularly for the 60° isotherm. The carnallite isotherms are extrapolated using the saturation concentrations at the boundaries between the sylvite and bischofite fields. The curvature of the carnallite isotherms is certainly real, for the straight line connection of these end-points would fall under the solubility curves for sylvite. This is not possible since the stable compound must have a lower solubility. VAN'T HOFF (1905, p. 15) very early on assumed a curvature of the carnallite isotherms. A conclusive statement, however, must await the publication of the new experimental data of AUTENRIETH and BRAUNE (pers. comm.).

α) Isothermal Evaporation and Solution

This follows much the same course as in the system KCl–Na Cl–H_2O The solid phases which precipitate are those whose solubility curves are intersected by the line from the origin passing through the actual solution composition. If the first precipitates are assumed to be sylvite and NaCl, the consequence is that the solution is enriched in $MgCl_2$ up to point *E* (Fig. 9). At this point, however, there is no joint precipitation of carnallite and sylvite, there is rather a so-called "incongruent saturated solution". Although it remains constant so long as both solid phases are present, a reaction point, not a crystallization endpoint, is reached. That is to say, one of the solid phases will dissolve and the other precipitate. Upon evaporation sylvite will be reabsorbed and carnallite will form. If sylvite is reabsorbed or is removed from contact with the solution, as for example when a carnallite crust forms, as constantly happens in experiments involving evaporation or stirring, further carnallite will precipitate with further enrichment of the solution in $MgCl_2$ along the carnallite curve until point *D* is reached. Here, upon final evaporation the residual solution can yield bischofite, carnallite and halite. Whether, upon evaporation, point *E* is passed or not depends upon the composition of the initial solution. Only when it contains more KCl than corresponds to the molar ratio in carnallite can the solution be exhausted before all the sylvite has been dissolved at point *E* with the formation of carnallite. If the sylvite is removed from contact with the solution, point *E* can again be passed.

Thus it is not possible to produce incongruently saturated solutions by simultaneous dissolution of both the solid phases in equilibrium with them. It is much more likely that sylvite will separate with dissolution of carnallite after the solution has reached saturation in KCl until its composition reaches point *E*. The incongruent decomposition of carnallite is very important for the understanding of many sylvite rocks (p. 118).

As a special example, the evaporation of a solution with the molar ratio of carnallite will be calculated at 20° C. The composition of the initial solution at the beginning of sylvite precipitation and the constant solutions E_{20} and D_{20} can be found from Fig. 9.

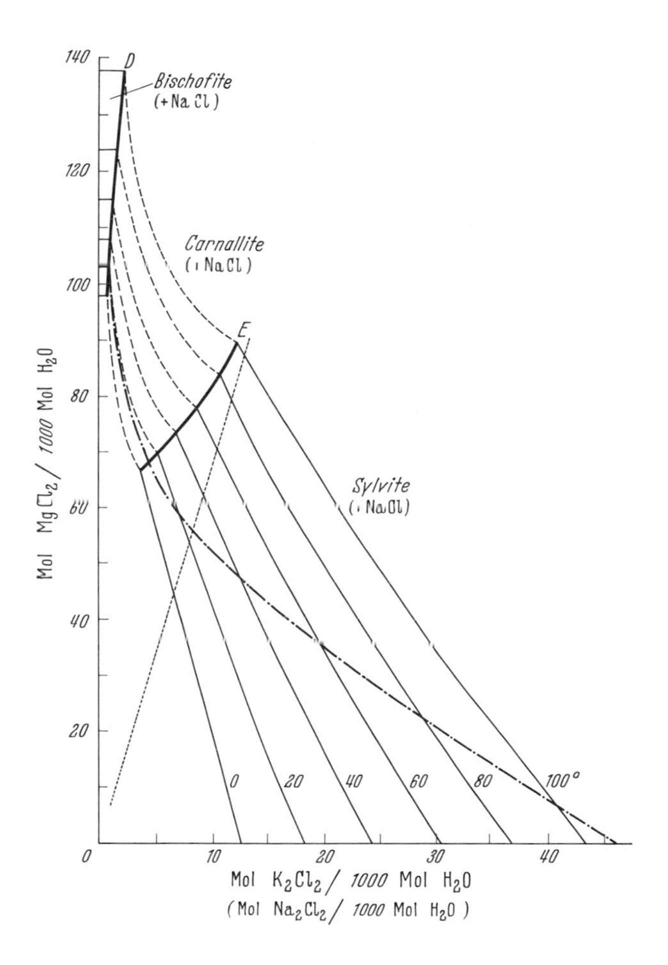

Fig. 9. The system NaCl–KCl–$MgCl_2$–H_2O at different temperatures, saturated with respect to NaCl. KCl saturation (solid lines) interpolated and corrected from D'ANS (1933). Curvature of the carnallite isotherms (broken lines) extrapolated (p. 48). The dot-dash line represents NaCl saturation concentration at 20° C. The dotted line represent the $MgCl_2/K_2Cl_2$ ratio in evaporating seawater ($MgSO_4$-free) during precipitation of pure NaCl

One obtains:

original solution → x_1 sylvite + y_1 halite + z_1 solution $E_{20}{}^{\circ}$

$MgCl_2$	26	0	0	69.5
KCl	26	1	0	10.4
NaCl	52	0	1	8

This gives 22.11 mol sylvite, 49.01 mol halite, 0.3741 solution E_{20}. For further development there are two possibilities, depending upon whether the sylvite formed up to this point is removed from solution or not.

In complete reaction, the solution changes, using up the previously formed sylvite and changing it to carnallite with the simultaneous precipitation of the remaining NaCl:

0.3741 solution E_{20} → x_1' sylvite + y_1' carnallite + z_1' halite

$MgCl_2$	26.00	0	1	0
KCl	3.89	1	1	0
NaCl	2.992	0	0	1

This gives −22.11 mol sylvite + 26 mol carnallite + 2.99 mol halite.

The total precipitate corresponds to the initial solution content. Sylvite is shown with a negative sign which indicates it is used up in the reaction. The occurrence of a negative sign is a purely formal indication of a simple incongruent saturated solution.

If the solid phases upon reaching E_{20} are completely removed from the solution, then the further precipitation is as follows:

0.3741 solution E_{20} → x_2 carnallite + y_2 halite + z_2 solution D_{20}

$MgCl_2$	26.00	1	0	103
KCl	3.89	1	0	1.6
NaCl	2.992	0	1	1.2

that is 3.541 mol carnallite, 2.730 mol halite, 0.2180 solution D_{20}, and similarly from the residual solution D_{20}, complete precipitation gives:

0.349 mol carnallite, 0.262 mol halite 22.11 mol bischofite.

From the computation of the precipitate (sylvite layer + carnallite layer + bischofite layer) the composition of the initial solution can be derived.

In this way all precipitates can be calculated if the composition of the initial solution and the resulting compounds are known.

β) Temperature Dependence

The strong temperature coefficient of solubility of KCl is reduced by the addition of $MgCl_2$. Fig. 10 shows, moreover, that it increases markedly with temperature and, within the limits of error, seems to be linear, whereas the dependence upon $MgCl_2$ content is not[19].

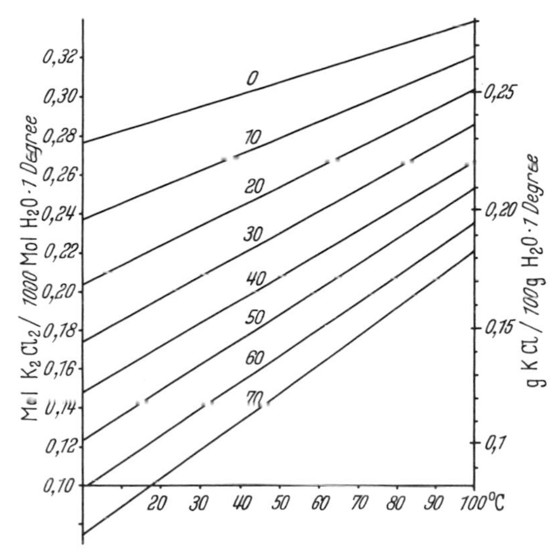

Fig. 10. Temperature coefficient of solubility of KCl at different temperatures in the system KCl–NaCl–$MgCl_2$–H_2O at KCl + NaCl saturation. The parameter is $MgCl_2$ content in mol/1000 mol H_2O

The temperature coefficient of NaCl solubility is much smaller (Fig. 11). The experimental data supporting this are not exact (relative error 10%). The essential fact is that it is negative at low $MgCl_2$ concentrations (see p. 46), while at medium and high concentrations it is positive. From about 50 mol $MgCl_2$/1000 mol H_2O it remains approximately constant and independent of temperature. The change in the sign of the temperature coefficient of NaCl solubility at about 20 mol $MgCl_2$/ /1000 mol H_2O is of great importance in the production of KCl. The $MgCl_2$ concentration should never attain this boundary value.

[19] This is not be confused with Wilson's rule (see D'ANS, 1933) whereby KCl solubility decreases more or less in proportion to the $MgCl_2$ content.

Fig. 11 is valid only with simultaneous saturation in KCl. In the KCl-free system, $NaCl–MgCl_2–H_2O$, the temperature coefficient of NaCl solubility is always positive so that NaCl crystallizes out upon cooling. Of particular interest here is the KCl-unsaturated system, which is the case of evaporating seawater (excluding $MgSO_4$) before it

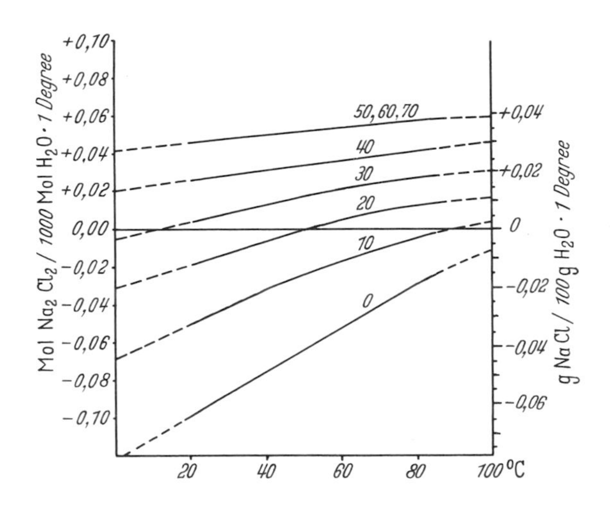

Fig. 11. Temperature coefficient of solubility of NaCl at different temperatures in the system $KCl–NaCl–MgCl_2–H_2O$ at KCl + NaCl saturation. The $MgCl_2$ content in mol/1000 mol H_2O serves as a parameter

reaches saturation in potash salts. Exact numerical values for the unsaturated KCl system are not available, but the temperature coefficients of solubility interpolated from the data available for the KCl-free and KCl-saturated solutions are a sufficiently close approximation. In Fig. 12 the temperature coefficients of NaCl solubility at three different temperatures are shown in relation to their dependence upon $MgCl_2$ concentration. NaCl solubility in a KCl-free system is indicated by light broken lines, in a KCl-saturated solution by thin lines, for KCl-unsaturated but NaCl-saturated ($MgSO_4$-free) seawater by heavy broken lines and for KCl and NaCl saturated seawater by heavy continuous lines. The interpolation of the temperature coefficient is taken as proportional to the KCl content of the solution. This can be read off Fig. 9 on the dotted line which represents the $MgCl_2/K_2Cl_2$ ratio in seawater. From this it can be observed that during the precipitation of NaCl from seawater and prior to its saturation with respect to KCl, NaCl always possesses a positive temperature coefficient of solubility. It depends but little upon the $MgCl_2$ content reached in the solution, but is clearly temperature-dependent. At higher temperatures the cooling effect is stronger. The practical application of this diagram will be made clear on p. 104.

As the carnallite isotherms are not known precisely, the exact temperature coefficient of carnallite solubility cannot be given. It is, however, certain that it is comparable with that of sylvite.

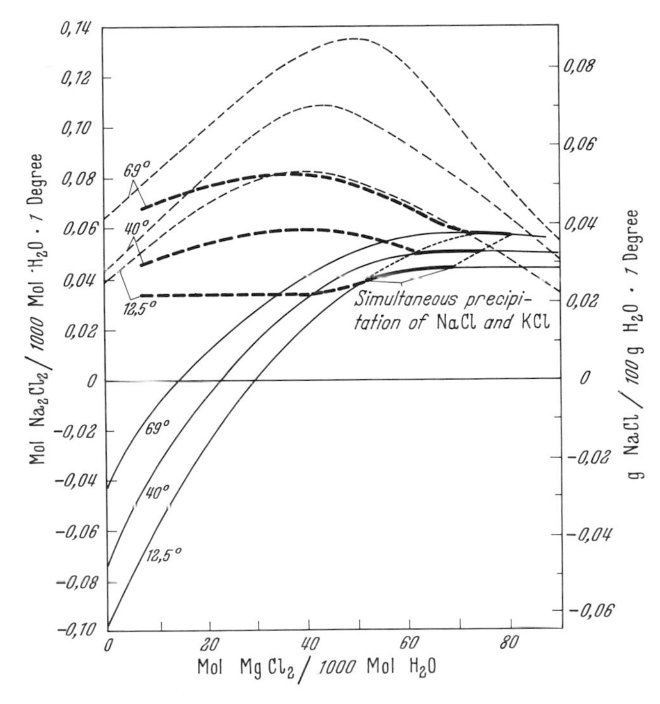

Fig. 12. Temperature coefficient of NaCl solubility at different temperatures and $MgCl_2$ concentrations. ---- system NaCl–$MgCl_2$–H_2O saturated in NaCl; —— system NaCl–KCl–$MgCl_2$–H_2O saturated in NaCl and KCl; – – – $MgSO_4$-free seawater, NaCl-saturated, KCl-unsaturated; —— the same, simultaneous precipitation of NaCl + KCl; following by simultaneous precipitation of NaCl + carnallite

γ) *Salting Out*

The solubility of KCl as well as NaCl decreases markedly with increasing $MgCl_2$ concentration. As a result, therefore, of mixing saturated NaCl and $MgCl_2$ solutions the precipitation of an appreciable proportion of the NaCl can be brought about; this is referrred to as "salting out". In a KCl-free system, in which this can be demonstrated in a test-tube and which here suffices, there are the following saturated solutions at 25° C:

$$55 \text{ mol } Na_2Cl_2/1000 \text{ mol } H_2O = 35.7 \text{ g } NaCl + 100 \text{ g } H_2O$$

$$105 \text{ mol } MgCl_2/1000 \text{ mol } H_2O = 55.5 \text{ g } MgCl_2 + 100 \text{ g } H_2O \quad .$$

Upon mixing the two in equal proportions, much less than half of the NaCl remains in solution, namely 8.3 g NaCl/100 g H_2O as calculated from the tabulated solubility data of D'ANS (1933, Table IV, 1).

Thus, merely by mixing solutions a solid phase may be precipitated, even although there has been no reaction and no exchange. This condition is hardly likely to occur in nature. There is, however, another kind of salting out which is important. When a NaCl-saturated solution comes into contact with a more readily soluble salt, the result is that the more readily soluble salt is dissolved to its saturation point, and halite separates out. The special conditions prevailing with carnallite are analogous to the dissolving of this salt in water (p. 48), with KCl precipitating along with NaCl. This is important for the formation of sylvite-halites (p. 119). Impoverishment of salts also belongs here.

III. System NaCl–KCl–Na_2SO_4–H_2O

1. Further Solid Phases

Mirabilite $Na_2SO_4 \cdot 10\,H_2O$ does not occur in typical marine salt deposits, as it forms only from SO_4-rich solutions and salt lakes etc., in particular as a winter precipitate.

Thenardite Na_2SO_4 is only found exceptionally in marine salt deposits, and then as a metamorphic product. It is the low-temperature form of Na_2SO_4 (see also glaserite). RIEDEL (1912, p. 157) reported individual grains in the barren alteration zone of the overthrust Stassfurt seam. Further data are lacking. WEBER (1931, p. 67) also referred to this occurrence without much detail. The occurrence is still not known with any certainty. The report of an outcrop with 2–7% thenardite at the Orlas-Nebra workings (Unstrutdistrict) by TINNES (1928) ought to be carefully checked.

Glaserite $K_3Na(SO_4)_2$ occurs only occasionally, as a metamorphic product[20] in salt deposits.

[20] BUDZINSKI et al. (1959, p. 62) have reported primary glaserite as small rhombohedral crystals in association with polyhalite and anhydrite, in rock salt. The primary precipitation of such a rock from seawater is not possible and not proven.

It is only a variety of aphthitalite with K: Na = 3 to 2.5 whereas aphthitalite mixed crystals contain K: Na = 5 to 1. (Etymologically: non-disappearing salt, because of its low speed of solution).

In determining solubility data, the formation of mixed crystals is ignored, a 3 : 1 ratio always being assumed. This is not inappropriate as the glaserite found in salt deposits closely approaches this ratio.

BREDIG (1942) assumed that these were merely solid solutions of the hexagonal high-temperature forms of K_2SO_4 and Na_2SO_4. This view was confirmed by HILMY (1953) who also showed that, upon crystallizing from a solution at 70° C, a solid solution may have a K/Na ratio within the range 5–1, while at room temperature only crystals with K/Na ratio of 3 are formed.

2. *Stability Conditions*

In the system Na_2SO_4–H_2O the stable phases are mirabilite (Glauber's salt) and thenardite, in addition there is a metastable heptahydrite which has not been definitely identified in nature (see however pustynite in KÜHN, 1959). The transition point mirabilite-thenardite serves as a thermometric fixed point (32.383° C). The solubility of mirabilite has an abnormally large positive temperature coefficient (Fig. 13) which is moreover markedly reduced by the addition of NaCl. Thenardite has a small negative temperature coefficient of solubility. Its solubility is also much reduced by the addition of NaCl. The mirabilite-thenardite transition point is lowered to 17.9° C in a NaCl-saturated solution. The saturation concentration of NaCl (numbers in parentheses on the isotherms of Fig. 13) is only slightly lowered by the addition of Na_2SO_4 (see also Fig. 7). It is generally true of the other systems, too, that the solubility of the more readily soluble salts is little affected by the introduction of salts with a lower solubility.

The four-component system NaCl–KCl–Na_2SO_4–H_2O will only be considered in the case of NaCl saturation so that it may be described as a three-component system (cf. p. 47). The system contains the reciprocal salt pair

$$K_2Cl_2 + Na_2SO_4 \rightleftharpoons K_2SO_4 + Ma_2Cl_2$$

(with two cations and two anions). The stability data and the course of crystallization can be particularly well represented by converting the salt concentrations into ionic percentages following JÄNECKE (1923) (see p. 30) in the present case, mol Na_2SO_4 + mol K_2Cl_2 = 100.

The invariant points can be seen in Fig. 14. The presence of KCl lowers the transition of mirabilite-thenardite to 16.3° C. The lower temperature for glaserite formation is +4.4° C. The invariant point mirabilite-sylvite-halite-hydrohalite (−2.4° C) represents a paragenetic change, a change which has little significance in the study of mineral deposits.

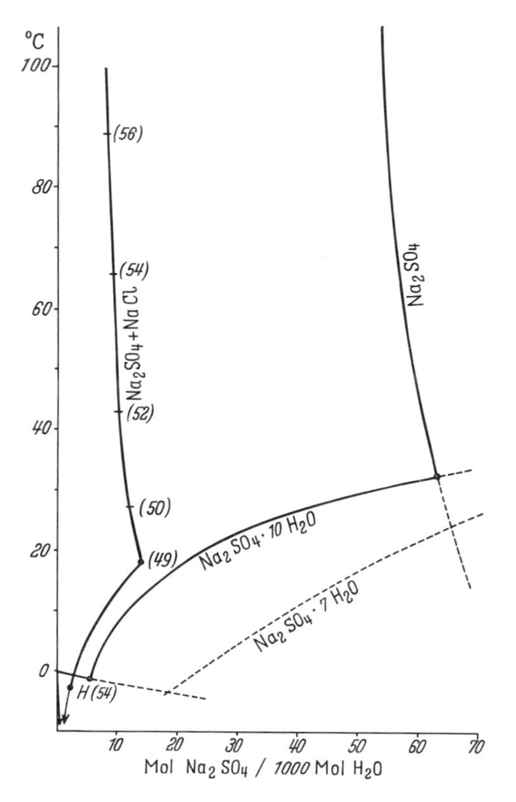

Fig. 13. The system Na_2SO_4–H_2O, right hand curve in the absence of NaCl, left hand curve at saturation with respect to NaCl (numbers in parentheses = NaCl saturation concentrations in mol/1000 mol H_2O). From D'ANS (1933). Dotted line: metastable solubilities

The point representing glaserite occurs at 42.9 ionic percent K_2 (Table 4). It is apparent from Fig. 14 that between 15° and 29° C the point will lie in the glaserite stability field. This means that within this temperature range congruent saturated solutions occur, and a solution of glaserite composition could dry up completely with the formation of glaserite. Above 29° C thenardite (+ NaCl) will precipitate first from this solution with KCl enrichment of the residual solution. Upon reaching the constant solution at the thenardite-glaserite phase boundary, further crystallization will depend upon whether thenardite is removed from contact with the solution or not. In the first case more glaserite (+ NaCl) will separate with further enrichment in KCl until the residual solution reaches the sylvite phase boundary, whereupon the solution is used up completely with the formation of glaserite + sylvite + halite. In the second case thenardite will react with the solution and glaserite will form without any change in its composition. With the assumed composition of the initial solution complete crystallization would occur at this point, but with excess thenardite all the available solution will be used before this point is reached; in contrast, with an excess of KCl, thenardite will be completely redissolved with formation of glaserite and halite, the residual solution being enriched with respect to KCl until ultimately crystallization occurs with formation of glaserite-sylvite (+ halite).

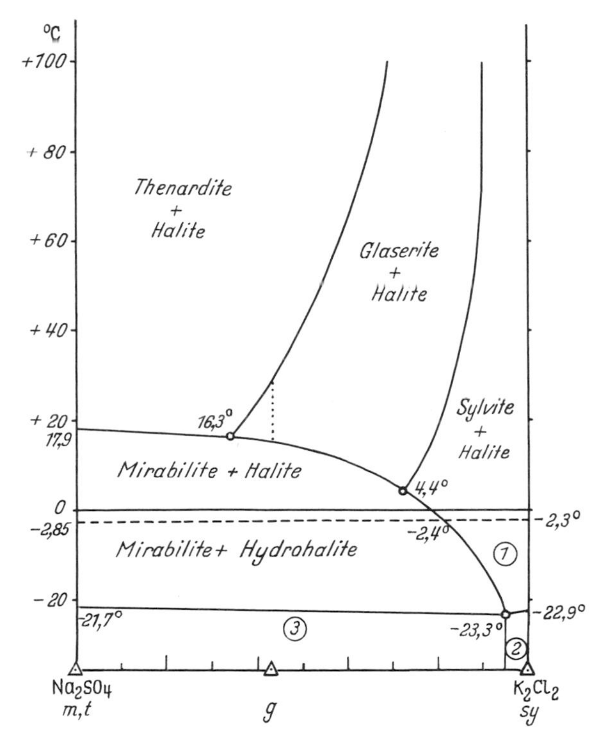

Fig. 14. The system K^+, Na^+, Cl^-, So_4^{--}, H_2O at NaCl saturation. Jänecke diagram. Solid phases abbreviated as in Table 4. Additional stability fields: *1* hydrohalite + sylvite; *2* hydrohalite + sylvite + ice + mirabilite; *3* mirabilite + hydrohalite + ice + sylvite. In *2* and *3* the first two phases named also form phenocrysts. Above the upper boundary of *2* and *3* saturated solutions are still found

The partial system described is only relevant as a boundary condition of the five-component system. It is scarcely likely to occur in nature. Only in the case of certain continental salt lakes can this system be used as a crude approximation (on the Na_2SO_4 side).

IV. System $NaCl$–Na_2SO_4–$MgCl_2$–H_2O

1. Further Solid Phases

Epsomite $MgSO_4 \cdot 7\,H_2O$ is usually found in salt deposits only as a weathering alteration product of kieserite. There is occasionally a small Fe^{II} content in epsomite. A pale green epsomite from Hallein (Salzburg) has the composition: $Mg_{0.993}Fe_{0.007}SO_4 \cdot 7\,H_2O$ (D'ANS and FREUND, 1954).

Hexahydrite $MgSO_4 \cdot 6\,H_2O$, also rare in salt deposits, readily forms through the partial dehydration of epsomite.

Pentahydrite $MgSO_4 \cdot 5\,H_2O$ is metastable. As a mineral it is reported in brine pools in the shore zone of the Gulf of Kara Bugas (SEDELNIKOV, 1959). It has some importance as a synthetic product.

Leonhardtite $MgSO_4 \cdot 4\,H_2O$ is also metastable; it occurs as a bloom on kieserite as well as on pentahydrite in brine pools on the banks of the Gulf of Kara Bugas. It may be considered as a metastable primary precipitate.

Kieserite $MgSO_4 \cdot H_2O$ is the most important Mg sulphate found in salt deposits. It contains no significant isomorphic components. Crystals are relatively rare and then mostly as intergrowths with carnallite.

Bloedite $Na_2Mg(SO_4)_2 \cdot 4\,H_2O$ is in practice found only in secondary salts (kainite caprock etc.), but it could also form as a primary precipitate from normal seawater. It is found in terrestrial salt lakes. Naturally occuring material is generally fairly pure, and mixed crystals with the analogous KMg sulphate, leonite ($K < 0.5\%$), do not occur. The Mg may be replaced to a limited extent by Fe^{2+} (up to 0.25% FeO), in which case the mineral is a dark green (sometimes regarded as a variety: simonyite of Alpine occurrences).

Loeweite $Na_{12}Mg_7(SO_4)_{13} \cdot 15\,H_2O$ is rare in salt deposits. The formula was revised by KÜHN and RITTER (1958) and confirmed by SCHNEIDER and ZEMANN (1959).

Vanthoffite $Na_6Mg(SO_4)_4$ is generally a rare metamorphic product but is occasionally a rock-forming mineral. In natural outcrop it occurs only in a massive form.

D'Ansite $Na_{21}MgCl_3(SO_4)_{10}$ is only to be expected as a metamorphic product in very rich Na_2SO_4 salts (in thenardite, glauberite and vanthoffite-halite) and until now only reported from Hall/Tyrol (GÖRGEY, 1909). The occurrence was only identified after the synthetic production of the salt (AUTENRIETH and BRAUNE, 1958; cf. STRUNZ, 1958).

2. Stability Conditions

This four-component boundary system of the five-component system, considered only in respect to NaCl saturation, has been recently re-examined by AUTENRIETH and BRAUNE (1960a, b). It contains the reciprocal salt pair

$$Na_2SO_4 + MgCl_2 \rightleftharpoons MgSO_4 + Na_2Cl_2 .$$

In addition to the solid phases cited, it contains several hydrates of $MgCl_2$ and $MgSO_4$ and several double salts. The stability data between 0–90° C, which are those of most petrogenic importance, are well seen on a Jänecke diagram (Fig. 15) in ionic percentages. Only the stable solid phases are shown. In the temperature range indicated there are also several metastable $MgSO_4$ phases:

$MgSO_4 \cdot 5\,H_2O$	pentahydrite (isotypic with copper sulphate $CuSO_4 \cdot 5\,H_2O$),
$MgSO_4 \cdot 4\,H_2O$	leonhardtite,
$MgSO_4 \cdot 2\,H_2O$	sanderite (physical and crystallographic properties not known, natural occurrence inadequately established),
$MgSO_4 \cdot {}^5\!/_4\,H_2O$	(physical and crystallographic properties and natural occurrence not known).

The first two will be considered when dealing with the five-component system.

From solutions within the stability fields halite and the appropriate solid phase always precipitate. The diagram also shows the new solid phase d'ansite, whose stability field lies between those of thenardite and vanthoffite, and in comparison with the old polytherms shows two further basic differences:

1. The lowest temperature of formation of hexahydrite is 18° C (D'ANS et al., 1960; AUTENRIETH and BRAUNE, 1960a, analogy with $ZnSO_4 \cdot 6H_2O$, D'ANS et al., 1957). It does not occur in paragenesis with bischofite as was shown on the older polytherms. The bischofite-kieserite paragenesis is stable in the whole temperature region illustrated. Only for a short temperature interval below 0° C is the bischofite-epsomite paragenesis possible, at still lower temperatures the more highly hydrated forms supervene.

2. The bloedite-kieserite paragenesis is not stable at any temperature. Below 40.3° hexahydrite intervenes, and below 30.2° epsomite. Above 40.3° loeweite comes between them and also separates hexahydrite from kieserite down to 37.6°.

The univariant phase boundary lines are not linear.

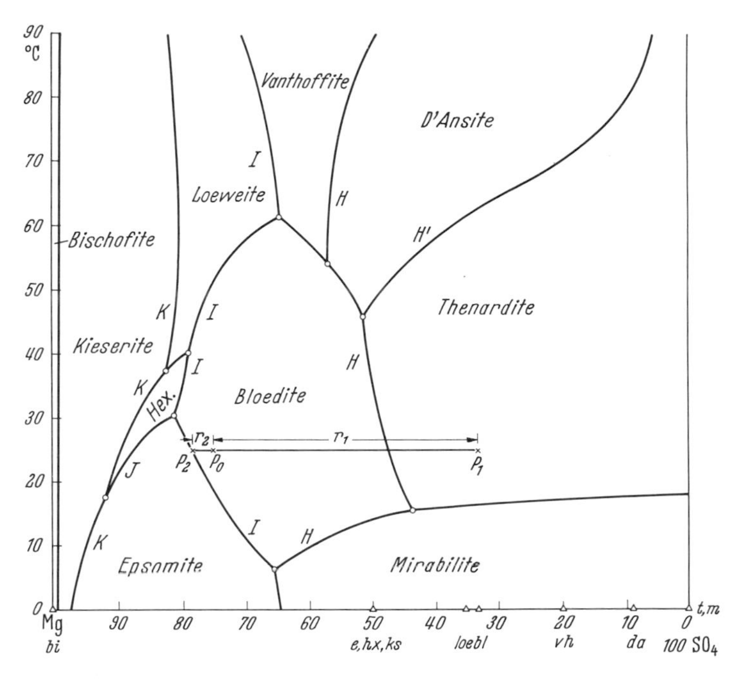

Fig. 15. Reciprocal salt pair Na^+, Mg^{++}//Cl^-, SO_4^{--} + H_2O at NaCl saturation in the temperature range 0–90° C. Jänecke diagram recalculated after AUTHENRIETH and BRAUNE (1960). Salt points from the data in Table 4. P_0–P_2 = graphical determination of precipitates according to the principle of moments

In the temperature field represented (Fig. 15) there are 9 invariant points (4 solid phases + solution + vapour). Each phase has an upper and a lower temperature limit at which it will form (Table 5). Below 0° C there are a further 10 invariant points of which three correspond to other paragenetic processes resulting from the halite-hydrohalite transition (cf. the analogous case at −2.4° C Fig. 14). There are yet other invariant points above 90° C. Loeweite disappears at about 110° C and at about 116° C bischofite alters to $MgCl_2 \cdot 4\,H_2O$. These points and others at still higher temperatures are meaningless in the study of salt deposits (see, however, JÄNECKE, 1923, and others).

The points shown on the abcissa of Fig. 15 represent the compositions of the solid phases (salt points) as per Table 4. If the points are moved up to their appropriate temperature range, it will be found that they normally lie outside their respective stability fields; therefore the solutions are incongruently saturated. Only pure $MgCl_2$–NaCl solutions, pure Na_2SO_4–NaCl solutions, and d'ansite solutions above 83° C are congruently saturated and so can crystallize out completely using all the solution. The crystallization end point of the remaining solutions[21] lies on the boundary line of the bischofite-kieserite-NaCl field. In the NaCl-free partial system $Na_2SO_4 + MgSO_4(\pm H_2SO_4) + H_2O$ double salts can form congruent saturated solutions, at least within certain temperature ranges.

Table 5. *Transition points in the system NaCl–Na_2SO_4–$MgCl_2$–H_2O at NaCl saturation (after* AUTENRIETH *and* BRAUNE, *1960 b). Upper temperature limit of a phase italicized, lowest temperature of formation underlined*

Temp. °C	Solid phases (abbreviations see Table 4				mol/1000 mol H_2O		
					$MgCl_2$	$MgSO_4$	Na_2Cl_2
6.2	m	bl	e	n	16.3	18.1	30.2
15.3[a]	m	bl	t	n	−4.5[a]	20.0	49.0
18.0	ks	hx	e	n	78.6	7.9	2.2
30.2	bl	hx	*e*	n	53.0	16.8	7.2
37.6	loe	ks	hx	n	61.3	16.7	6.4
40.3	loe	ks	*hx*	n	56.1	20.0	10.0
45.7	da	bl	t	n	1.2	17.2	47.9
54.1	da	vh	bl	n	5.4	16.3	45.6
61.4	loe	*bl*	vh	n	13.6	15.8	39.4

[a] At 15.3° C Na_2SO_4 and not $MgCl_2$ occurs. As only neutral salts will form, the Na_2SO_4 content can be expressed in purely formal terms with the help of the remaining components of the reciprocal salt pair

$$Na_2SO_4 = -MgCl_2 + MgSO_4 + Na_2Cl_2 .$$

3. Crystallization Sequence (Isothermal Evaporation)

The crystallization sequence is easy to comprehend in a qualitative sense. The point representing the composition of a given solution is determined. Halite and the solid phase in whose stability field the composition point (salt point) lies precipitate out of the solution. In consequence the composition of the solution changes linearly from the initial point and the salt point, as the salt point in general lies outside

[21] Above 83°, the crystallization end points during isothermal evaporation may also lie on the d'ansite–thenardite–NaCl boundary line.

its stability field (incongruent saturation). Upon reaching a phase boundary there are, as in previous examples, two cases to be considered, whether the solid phase already formed remains in intimate contact with the solution or not

All field boundaries with the exception of the congruently saturated bischofite-kieserite-(NaCl) boundary (and additionally the d'ansite-thenardite-(NaCl) boundary above 83°) are reaction boundaries (cf. the *E* solutions in the carnallite system p. 48, 49). Previous conclusions about constant solutions are valid here, and the calculation of precipitation can be carried out in a similar manner.

A definite advantage of the Jänecke method of representation lies in the possibility of deriving quantatively the amounts of the precipitated salts from the diagram. They are proportional to the distances between the points representing the solution composition in the diagram. As the diagrams are referred to a constant salt amount, 1 or 100%, the law of moments can be applied. For the example in Fig. 15

$$m_1 \cdot r_1 = m_2 \cdot r_2 ,$$

where r_1 is the distance from the point representing the initial solution P_0 to the point representing the composition (salt point) of solid phase P_1, r_2 is the distance from P_0 to P_2 (first constant solution), m_1 = the quantity of the solid phase P_1 formed, and m_2 = the quantity of the solution P_2 formed, both expressed in the normalized units of the Jänecke diagram. Furthermore

$$m_1 + m_2 = m_0 = \text{constant} .$$

If $m_0 = 1$, then the following can be obtained from these two relationships:

$$m_1 = \frac{r_2}{r_2 + r_1}; \quad m_2 = \frac{r_1}{r_2 + r_1}$$

For the example in Fig. 15, $P_0 = \text{Mg}\ [75],\ \text{SO}_4\ [25]$, at 25° C. $r_1 = 83.2$ mm; $r_2 = 7.4$ mm; hence $m_1 = 0.082 = 8.2\%$; $m_2 = 0.918 = 91.8\%$. (linear distances measured from the unreduced original figure with abcissa length 20 cm).

The exact calculation using the technique developed on p.45 gives

P_0		$x \cdot m_1 + y \cdot m_2$		
Mg	75	33.3	78.7	$m_1 = 8.1\%$
SO_4	25	66.7	21.3	$m_2 = 91.9\%$

To convert the normalized units to mol, which Jänecke (1923) and Borchert (1940) omitted to do, it suffices to use the ratios of the same components between their chemical formulae and the normalized units after JÄNECKE (cf. Table 4). Thus for bloedite

$$\begin{aligned}\text{1 normalized unit bloedite} &= 33.3\%\ \mathrm{Mg} + 66.7\%\ \mathrm{SO_4} + 133\%\ \mathrm{H_2O},\\ \text{1 mol bloedite} &= 1\ \mathrm{Mg} + 2\ \mathrm{SO_4} + 4\ \mathrm{H_2O},\end{aligned}$$

so that one normalized unit bloedite $= 1/3$ mol bloedite.

The amount of NaCl precipitated is obtained as the difference between the $[Na_2]$ content of the initial solution P_0 and the solution P_2. In the case of the precipitation of NaMg double salts, allowance must be made for the Na_2 abstracted from the solution. Thus

$$[\mathrm{Na_2}]_{\mathrm{NaCl\ precipitate}} = m_0\,[\mathrm{Na_2}]_{P_0} - m_2\,[\mathrm{Na_2}]_{P_2} - m_1\,[\mathrm{Na_2}]_{\mathrm{NaMg\ double\ salt}}\,.$$

It must be noted however that in the Jänecke diagram the Na_2 content of the double salt has also to be calculated from the normalized units. The conversion factor [22] for bloedite is $1/3$; for loeweite 0.3; for vant-hoffite 0.6; for d'ansite 0.955 and so on. The Na_2 content of the solution P_0 must be interpolated, as only the values along the two phase lines are known analytically. In the example just given, on the basis of $Mg + SO_4 = 100$, the amount in solution P_0 is 30, in solution P_2 is 15.2 (see point I in Table 6); hence the amount precipitated as NaCl is $30 - 0.919 \cdot 15.2 - 0.081 \cdot 1/3 = 13.3\ [Na_2]$. That is to say, of the original 30 mol $[Na_2Cl_2]$, 13.3 mol Na_2Cl_2 = 26.6 mol NaCl will be precipitated as halite and a further $\frac{8.1}{3} = 2.7$ mol as bloedite.

Using the molecular weight, the precipitate may then be converted to grams or weight percent.

All further precipitation can be calculated in a like manner. The result is, however, dependent upon whether the solid phase is removed from contact with the solution or not. An example will subsequently be described in the five-component system.

4. Temperature Dependence

The results of isothermal evaporation are in essence temperature-dependent, as can be qualitatively observed in Fig. 15. The effects of a subsequent increase in temperature on existing primary precipitates

[22] If n is the number of Na in the formula, m that of Mg (or s of SO_4), and [Mg] the Mg coordinate (or $[SO_4]$ the SO_4 coordinate) of the salt point in a Jänecke ternary diagram, as given in Table 4, then the conversion factor reads:

$$f\,[\mathrm{Na_2}] = \frac{n}{2m}\cdot\frac{[\mathrm{Mg}]}{100} = \frac{n}{2s}\cdot\frac{[\mathrm{SO_4}]}{100}\,.$$

will be considered later under thermal metamorphism. For individual solid phases the temperature gradients of solubility cannot be deduced from the Jänecke diagram. Rather one must begin (as in the carnallite system) with the projection of the isotherms on to the plane of the $MgCl_2/1000\ H_2O - MgSO_4/1000\ H_2O$ co-ordinates (see for example AUTENRIETH and BRAUNE, 1960b, Fig. 7).

In the system under consideration the relationships are highly complex and, in view of their restricted significance in the field of mineral deposits, it is sufficient to give only a brief indication of them. Most of the solid phases have negative temperature coefficients and so precipitate from saturated solutions as they grow warmer. In kieserite the temperature coefficient is always negative above 55° and between 25° and 35°, but the sign between 35° and 55° depends upon the $MgCl_2$ concentration. Further complications occur in solutions of constant composition where the precipitating phase reacts with the solution if it is not separated out. Epsomite and hexahydrite, on the other hand, have positive temperature coefficients of solubility and precipitate upon cooling.

The precipitates from a cooling solution can be calculated exactly from static solution equilibria if the composition of the initial solution and the isotherms of the appropriate temperature region are known. There are two boundary conditions depending upon whether stable or metastable solid phases are formed. The calculation is carried out in the same way as in the carnallite system so that no example need be given here.

V. The Five-Component System $NaCl–KCl–MgCl_2–Na_2SO_4–H_2O$

1. Further Solid Phases

Schoenite $K_2Mg(SO_4)_2 \cdot 6\,H_2O$ occurs only in secondary salts (as a hydration product on kainite etc.), occasionally as good crystals, but most commonly compact. It does not occur in true paragenesis with kainite, as is always stated in the older literature, but develops from kainite; its occurrence is an indication of the disappearance of kainite. The few analyses are for the most part in accord with its theoretical composition.

Leonite $K_2Mg(SO_4)_2 \cdot 4\,H_2O$ also only occurs as a secondary salt (with kainite) sometimes, though rarely, as good crystals. On account of the similarity of its formula to that of bloedite, it was formerly regarded as an isomorphic replacement of K by Na (up to $^1/_4$ of K). This assumption has not been confirmed by quantative analysis[23] nor does it seem crystochemically probable (SCHNEIDER, 1960).

Langbeinite $K_2Mg_2(SO_4)_3$ is a relatively K-rich mineral (22.7% K_2O) but because of its low speed of solubility is not of much technical importance except when it occurs as a near monomineralic deposit. Natural material corresponds closely to the theoretical composition. However, diadochic replacement of K and Mg might be expected as numerous isotypic forms have been synthetically produced (GATTOW and ZEMANN, 1958).

Kainite $K_4Mg_4Cl_4(SO_4)_4 \cdot 11\,H_2O$. The formula given is that ascertained from the analyses of KÜHN and RITTER (1958). The water of crystallization implied in the previously accepted formula $KMgClSO_4 \cdot 3\,H_2O$ has never been found even in the older analyses. In any event there is crystallochemical confirmation of the new formula.

2. Stability Conditions

As in previous cases, the chief concern is with a system saturated with respect to NaCl.

In the five-component system using the Jänecke method, only isotherms projected on to the composition plane are possible. To show the temperature dependence would require a three-dimensional model with superimposed layered isotherms. The 25° isotherm can serve as an example.

a) Stable Equilibria

Table 6 contains the equilibrium solutions at the 25° isotherm calculated in the Jänecke form from the data of D'ANS (1933) and more recent solubility data from AUTENRIETH, AUTENRIETH, and BRAUNE (1952–1960). It is illustrated on a Gibbs' ternary diagram (Fig. 16). The corners of the stability fields are indicated by conventional letters and correspond to constant solutions. From the standpoint of the phase rule they are univariant, as they are temperature-dependent, a fact that is not apparent in an isothermal section. At the corner points, NaCl and the three solid phases of the adjacent stability fields are in equilibrium

[23] Gravimetric analyses of very pure, clear leonite crystals up to 6 cm in length (collection of the Mineralogical Institute, Göttingen) from the salt mine Wittmar/Asse near Braunschweig, an occurrence which up to the present has been poorly described, gave as weight percent: SO_4^{--} 52.06; Cl^- 0.14; K^+ 20.69; Na^+ 0.45; Mg^{++} 6.73; H_2O 17.79; total 99.86 (DIERKES, 1961 unpubl.). Deducting a 0.25% NaCl impurity, the formula becomes: $K_{1.951}\,Na_{0.054}\,Mg_{1.021}\,(SO_4)_2 \cdot 4.05_4\,H_2O$. The diadochic incorporation of Na for K is sufficiently small to be neglected.

with the solution and vapour. Within the triangle the points representing the salts do not occur within their appropriate stability fields; therefore the solutions considered are incongruently saturated solutions. The Mg sulphate hydrates have the same salt point and the stability fields are so arranged that the low-hydrate solid phases lie closer to the $MgCl_2$ corner. The same is true for leonite and schoenite. In general it is true that the ordering of the fields in which the solid phases exist corresponds to their relative contact in total Mg in the solid phase.

Table 6. *Constant solutions of the five-component system at 25° C and NaCl saturation, normalized to* $K_2 + Mg + SO_4 = 100$ (calculated after AUTENRIETH, AUTENRIETH and BRAUNE, 1952 to 1960; and D'ANS, 1933; points *R* and *U*–*X* from unpublished data of AUTENRIETH and BRAUNE pers. comm. November and December 1960). Point *Y* interpolated according to note on p. 67

Point	K_2	Mg	SO_4	H_2O	Na_2	Solid phases (abbreviations see Table 4)
A	—	100	—	958	0.6	n, bi
B	100	—	—	5100	236	n, sy
C	—	—	100	7900	495	n, t
D	0.2	99.8	—	956	0.6	n, bi, c
E	7.6	92.4	—	1300	6.1	n, c, sy
F	82	—	18	4100	200	n, sy, gs
G	40.6	—	59.4	4100	240	n, gs, t
H	—	45.9	54.1	2830	141	n, t, bl
I	—	78.7	21.3	1290	15.2	n, bl, e
J	—	88.2	11.8	1090	4.15	n, e, hx
K	—	89.2	10.8	1120	3.3	n, hx, ks
L	—	99.1	0.9	936	0.7	n, ks, bi
M	22.4	55.6	22.0	1570	38.1	n, sy, gs, sh
N	20.2	58.8	21.0	1490	32.4	n, sy, sh, le
P	12.2	70.3	17.5	1230	14.5	n, sy, le, k
Q	6.9	86.9	6.2	1190	5.3	n, sy, k, c
R	1.9	91.5	6.6	1055	2.5	n, k, c, ks
S	17.4	34.4	48.2	2150	103	n, gs, t, bl
T	18.2	52.0	29.8	1560	45.1	n, gs, bl, sh
U	16.6	54.8	28.6	1510	40.2	n, bl, sh, le
V	7.3	70.4	22.3	1130	12.7	n, bl, le, c
W	7.6	71.4	21.0	1120	11.9	n, le, e, k
X	6.6	78.0	15.4	1080	6.4	n, e, hx, k
Y	3.0	88.5	8.5	1057	3.1	n, hx, ks, k
Z	0.2	98.9	0.9	942	0.6	n, ks, c, bi

In comparison with the older data the only changes at 25° C are quantitative adjustments. The parageneses have been confirmed. The bischofite field is smaller than was formerly thought, and the direction of some field boundaries has been changed, a fact of importance in quantitative consideration of the precipitates[24].

The boundaries between adjacent stability fields are either transition lines (Jänecke, 1923 and others) or crystallization paths. At transition lines (fine lines in Fig. 16), the salt points of both solid phases lie on

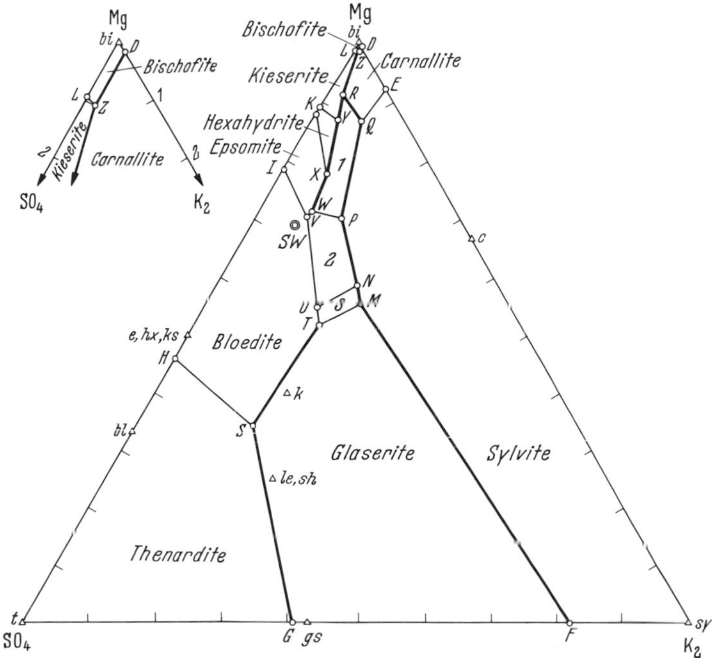

Fig. 16. 25° isotherm of the five-component system at NaCl saturation. Stable equilibria. Calculated in the Jänecke form from d'Ans (1933), Autenrieth (1955), Autenrieth and Braune (1960) with some corrections. Abbreviations of the solid phases (salt points) from Table 4. Crystallization paths, heavy lines; transition lines, fine lines. *SW* = seawater. Stability fields: *1* = kainite + NaCl; *2* = leonite + NaCl; *3* = schoenite + NaCl. Point *Y* is actually somewhat closer to point *R* (see p. 67)

[24] Even in the newer diagram, some points are still uncertain, in particular the relative position of Y_{25} in comparison with R_{25}. The following approximate values for Y_{25} were obtained by graphical interpolation between X_{25} and R_{25} in Fig. 34 and 20:

$$75.7\ MgCl_2 \quad 8.1\ MgSO_4 \quad 2.8\ K_2Cl_2 \quad 2.9\ Na_2Cl_2 \quad 1000\ H_2O$$

As the disagreement was first noticed during printing, it was not possible to correct the calculations based on the erroneous value of Y_{25} and correct Fig. 16, 20, 26, 27, and 34. This means that in those diagrams the calculated precipitation at 25° of the kieseritic halite-kainite C_3 layer is a little too large, and that for the hexahydritic halite-kainite layer C_2 is too small (see Table 10 and Figs. 26, 27). On the other hand, the relative composition of the calculated layers is very little altered. Further, it should be noted that the data for points *V* and *W* still contain small errors, as indicated by the inconsistency of the H_2O curve in Fig. 34.

the same side of the boundary. If, during isothermal precipitation, the first solid is removed upon reaching the transition line, the solution moves into the second stability field (Fig. 17) with the precipitation of the second solid phase. If the first solid phase remains in permanent contact with the solution, it must be redissolved with the precipitation of the second.

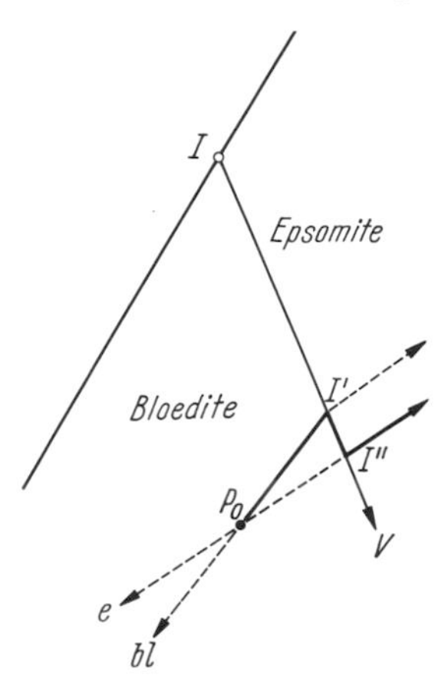

Fig. 17. Reaction of bloedite at the transition line *I–V* with the epsomite field. Schematic (after JÄNECKE, 1923)

In this event, the composition of the solution changes along the transition line. The first phase is redissolved when the solution has reached a composition which lies at the intersection of the projection of the line from the salt point of the second phase through the point of the initial solution to the transition line. There, more of the second solid phase precipitates and the composition of the solution moves into the stability field of the second phase.

Along the crystallization paths (shown as heavy lines in Fig. 16) the rays projected from the salt points on both sides intersect. The two adjacent solid phases can both precipitate simultaneously, with the composition of the solution changing along the boundary curve to a corner point. The geometric significance of this is that in such a figure the boundary lines between stability fields in isothermal figures must be linear.

Within the triangle there is only one congruent saturated solution, that represented at point *Z*, where complete evaporation with the formation of bischofite–carnallite–kieserite–NaCl occurs. At all other corner points either one or two solid phases must be redissolved, save in the cases where they have been removed from the solution. The first case involves simple incongruent saturated solutions lying upon a crystallization path. In the second case there are double incongruent saturated solutions lying at the intersection of at least two transition lines (example p. 70).

b) Metastable Equilibria

In Fig. 18 the metastable equilibria at 25° are illustrated (converted after AUTENRIETH, 1955). In comparison with the stable system, the principal difference is the absence of kainite and kieserite, with the additional occurrence of the metastable hydrates of $MgSO_4$. There also exists the metastable epsomite-sylvite paragenesis.

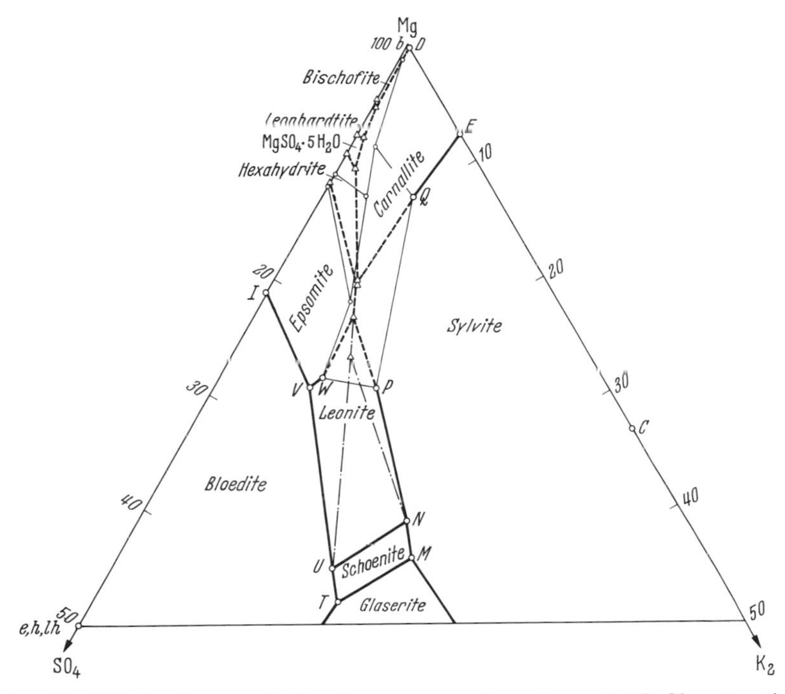

Fig. 18. 25° isotherm of the five-component system at NaCl saturation. Metastable fields indicated by broken or dot-dash lines. Stable fields indicated by continuous lines (heavy lines: solutions found in crystallization experiments, some only after seeding; fine lines: boundaries found only from solubility data)

This discovery confirms the results of crystallization experiments by the Kurnakov school dating from about 1927. The data were summarized by VALYASHKO (1958) with a detailed bibliography. The metastable equilibria are referred to in Russian literature as "solar equilibria" and the solubility diagrams as "solar diagrams" because these conditions occur during the course of evaporation of salt lakes (the "solar evaporation" of KURNAKOV).

3. Crystallization Sequence

There are various ways of ascertaining the progress of crystallization. A calculation method was described in detail when the carnallite system was considered (p. 50). Here only the Jänecke method will be described (JÄNECKE, 1923 and earlier). Further techniques have been considered exhaustively by D'ANS (1933). The construction is relatively simple and clear using the new data of AUTENRIETH (1953–1960) which can be obtained from the original publications. The same rules apply here as in partial systems. In the Jänecke triangular diagrams, too, the principle of moments is a valid means of obtaining quantative data (p. 62). If only halite and one other salt crystallize, the same construction used in the quaternary limiting case is valid. On the crystallization paths where two salts precipitate in addition to halite, the construction has to be applied twice, once to ascertain the total of both salts and once to find their relative proportions. As an example the simultaneous precipitation at 25° C of epsomite + kainite (+ halite) will be considered. The initial solution thus lies between W and X. The two salt points, kainite and epsomite, are joined by a straight line ($e-k$) and the line joining $X-W$ is prolonged to its intersection with $e-k$. The total quantity epsomite + kainite ($m_{e,k}$) may be found by the principle of moments (p. 62, 63) where r_1 is the distance from the point representing the initial solution to the line $e-k$, and r_2 the corresponding distance to the constant solution X, the two segments from the intersection points to e and k respectively giving the amounts of epsomite and kainite. If r_e and r_k are these distances, then the precipitation of epsomite is:

$$m_e = m_{e,k} \cdot \frac{r_k}{r_e + r_k}$$

and the kainite precipitate is

$$m_k = m_{e,k} \cdot \frac{r_e}{r_e + r_k}$$

in formula units. Consideration of the amount of NaCl precipitated and the calculation in moles or in weight percent follows from the technique outlined on p. 62. On p. 84 a special example, the isothermal evaporation of seawater, will be deduced.

The concepts of double incongruent saturated solutions (VAN'T HOFF, 1905, and others) can be clarified by reference to point T_{25} in Fig. 16, where a crystallization path meets two reaction lines. At this point under conditions of isothermal evaporation, the resolution of bloedite and glaserite should occur with the precipitation of schoenite. Only when the two solid phases are completely redissolved will the solution enter the schoenite field with the formation of further schoenite.

If the solution T_{25} contains more glaserite in proportion to bloedite than is necessary for the formation of schoenite, then the solution changes along the line $T-M$ after the absorption of bloedite, with resorption of the excess glaserite and formation of more schoenite. Only then can the solution enter the schoenite field. With a large excess of glaserite, the last glaserite is used up at point M with the formation of schoenite–sylvite. Conversely, with excess bloedite, after the resorption of all the glaserite, the composition of the solution changes along the line $T-U$ and, after all the bloedite is used up, enters the schoenite field; if there is a large excess of bloedite, it may reach point U where schoenite too will react with the solution to form leonite, etc. If at T_{25} there is too little solution available, then the excess of bloedite and glaserite will remain along with schoenite. All of these conclusions are contingent upon the solid phases remaining in intimate contact with the solution. In all these reactions, the solution is used up; they can only occur if water is being evaporated (a quantitative example is given on p. 93). At the invariant points the composition of the solution remains constant while its quantity diminishes.

4. Temperature Dependence

As in the limiting quaternary system (Fig. 15), temperature is the essential factor. With increasing temperature the hydrates with lower water of crystallization occur. At the invariant points in the five-component system five solid phases occur. A complicating factor is caused by the KMg sulphates which exist in a field inserted between those of sylvite and the NaMg sulphates over a definite range of temperature and concentration.

The polythermal diagrams of D'ANS (1933) form the most convenient way of following the paragenetic changes with temperature. A total of eight are needed to comprehend the whole system, four each (temperature against $MgCl_2$; $MgSO_4$, generally as the difference between total Mg and $MgCl_2$; K_2Cl_2; Na_2Cl_2) for the parageneses with sylvite and with the NaMg sulphates. The three dimensional representation of the different isotherms after JÄNECKE with temperature as an ordinate is clearer, but it gives no information about water content.

As in the present instance the paragenetic changes with temperature are the first concern, only the $MgCl_2$ polytherms will be shown (Figs. 19 and 20). The data in Fig. 19 for simultaneous saturation in KCl and NaCl (i.e. carnallite) are known in considerable detail because of their technical importance. They correspond to the polytherms of AUTENRIETH (1955, Fig. 5 No. 4) apart from a correction of the carnallite + kieserite- and the kainite + carnallite boundary. The basis

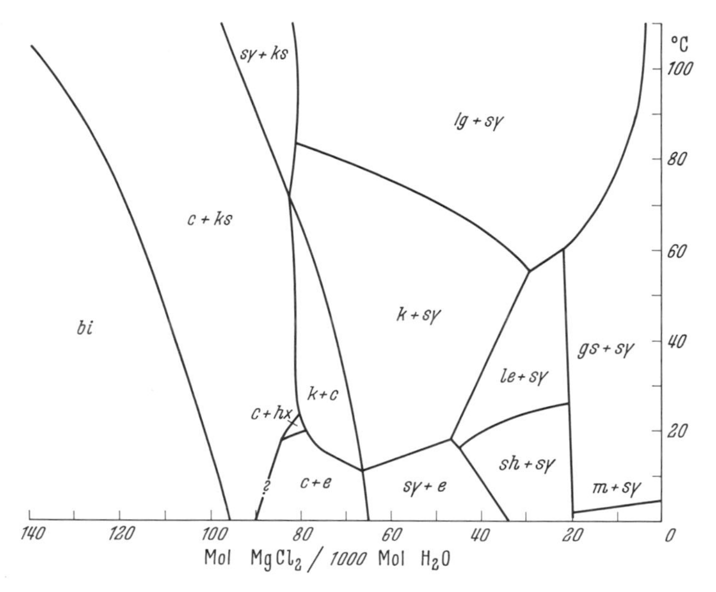

Fig. 19. $MgCl_2$ polytherm of the five-component system saturated in NaCl and KCl. Stable equilibria. After AUTENRIETH (1955). Solid phase abbreviations from Table 4

for Fig. 20 for simultaneous saturation in NaCl and NaMg or Mg sulphates is less well known because these are of less technical importance. They are, however, of very great interest for the understanding of the formation of salt deposits. There are certain essential differences in Fig. 20 with respect to the polytherms of AUTENRIETH (1955), partly resulting from conclusions drawn from new and revised data for the quaternary system Na_2SO_4–$MgCl_2$–Na_2Cl_2–H_2O (p. 59) and partly from new solubility data for points $U - X$ and R (see Figs. 16 and 21) at 25° and 35° (AUTENRIETH and BRAUNE, unpublished data made available for this monograph by the Kaliforschungsinstitut, Hannover, although the research was not completed). The important temperature range between 35° and 55° is still insufficiently investigated. Two fundamental alterations to VAN'T HOFF's data and the older polytherms of D'ANS (1933) and AUTENRIETH (1955, Fig. 6, No. 4) must be explained.

1. The hexahydrite field is smaller. It no longer occurs in paragenesis with bischofite. The lowest temperature of formation is probably 18° C. (AUTENRIETH and BRAUNE, 1960a; D'ANS et al., 1960). The point given has the same composition as that at the old invariant point at 19° C. The temperature difference, however, is not real. By extrapolation, using the data from the limiting quaternary system, the upper temperature limit is found to be 37° C (see Fig. 15). By analogy with this system, the kieserite–bloedite paragenesis was assumed not to occur, and the loeweite was interpolated between them. With the addition of KCl, the lower temperature for the formation of loeweite is lowered with respect to that of the KCl-free system; hence it must be below 37.6° C (Table 5) and extrapolation gives a value of about 35° C.

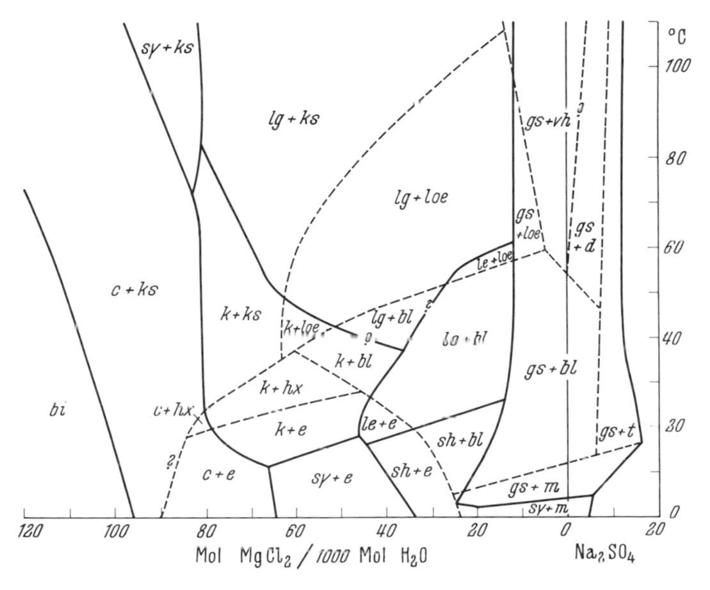

Fig. 20. $MgCl_2$ polytherm of the five-component system saturated in NaCl and in Na, NaMg, and Mg sulphates. Stable equilibria. After D'ANS (1933), AUTENRIETH (1955), with corrections in the kainite and hexahydrite fields (p. 73). Solid phase abbreviations from Table 4. Halite occurs everywhere. The hexahydrite–kieserite boundary should probably be more steeply inclined (see point Y_{25}, p. 67), and the carnallite-hexahydrite field still smaller

2. The course of kainite–leonite paragenesis with bloedite (+ NaCl) along the transition line is completely different from that of the older polytherms. It is established with the help of the newly determined point P at 18° C (AUTENRIETH, 1955, Fig. 6) and the newly determined points at 25° and 35° (see above). The most important consequences

of this boundary line are: kainite occurs in paragenesis with bloedite between about 26° and 43° C. This VAN'T HOFF (1905, p. 84) had expressly excluded, although D'ANS (1915, p. 247, note 1) early expressed doubts because the solubility data seemed to indicate the possibility of a kainite–bloedite–loeweite paragenesis. The field in which kainite can exist is

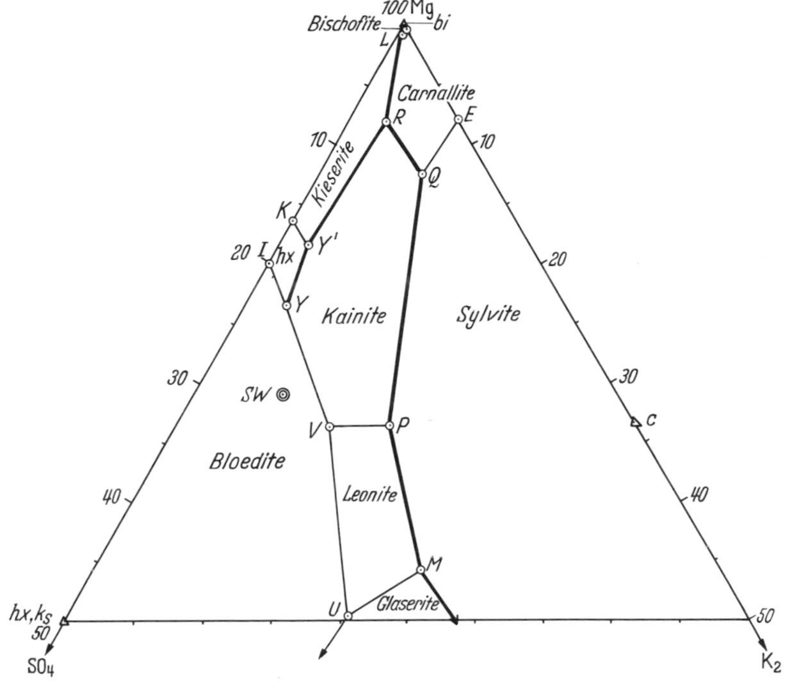

Fig. 21. 35° isotherm of the five-component system saturated in NaCl, Mg corner, stable equilibria. Crystallization paths heavy lines, fine lines are transition lines. Points *V–R* calculated from new data of AUTENRIETH and BRAUNE (pers. comm.)

larger in the present polytherm at the expense of leonite. The kainite field is now more like the kainite field of the reciprocal salt pair K_2Cl_2–$MgCl_2(-H_2O)$ than in the older figure. The lowest temperature at which langbeinite can form by extrapolation is about 37° C.

The diagram differs in another way from that of D'ANS (1933) in the absence of a schoenite–kainite paragenesis which AUTENRIETH (1955) has demonstrated in detail. It is only possible as a metastable process.

Although the present diagram is better than the figures hitherto used, it is still in need of experimental control and improvements at several points. The marked curvature of the langbeinite–kainite boundary and of the leonite–langbeinite on the NaMg sulphate side at 55° C seems questionable, and likewise the projection of the glaserite field into the epsomite field at 4° C. The limits of the d'ansite field in the five-component system have not yet been experimentally determined. It is particularly important to confirm the position of the extrapolated invariant point. The diagram should be regarded as a scheme of the parageneses, but in calculations only reliably established points should be used. It has been shown that various tensimetrically established invariant points are incorrect. Exact data can only be obtained by extrapolation between the adjacent isotherms of exactly known constant solutions (AUTENRIETH, 1955).

The polytherms shown in Fig. 19 and 20 contain 33 invariant points. They are most easily seen if the two figures are superposed. It then appears that on the sylvite side there are only three invariant points, while the remainder coincide with those on the NaMg side. The fields of the K–Mg double salts can be delimited from the two figures giving at the same time their spatial extent. At the same time it is apparent that the boundary surfaces of some stable bodies are distinctly sinuous.

The temperature coefficient of solubility of the double salts is not immediately obvious from the polytherms. For the solid phases which also occur in the limiting four-component system, the same values can be taken qualitatively as a base in the five-component system. Quantitative data are obtained from adjacent isotherms. As only occasional isotherms have been determined, additional ones must be derived from polytherms. In the temperature range above 25° C, unfortunately, the polytherms are not known in sufficient detail so that this extremely important question must remain open. The evaporation experiments with decreasing temperature of H. BORCHERT (1933, 1934) under "dynamic polythermal" conditions unfortunately provide no reliable information. They give, as a result of the strong horizontal temperature gradient (up to about $^1/_2$°/cm), stable parageneses on the warm side and metastable parageneses (e.g. kainite warm – epsomite + sylvite cool) on the cold. On account of the large temperature differences, these over-simplified experiments encompass several transition points so that nothing can be deduced about the much smaller temperature intervals occuring in nature. Nor are the experimental results directly applicable to natural data (CORRENS, 1960). The objections are not, however, directed against the great importance of temperature differences in the solution for salt precipitation.

Based upon earlier knowledge, it can be assumed that schoenite has a positive temperature coefficient of solubility, but that langbeinite on the other hand has a negative coefficient. In kainite it appears to be positive, yet it is probable that the sign may change with temperature, and possibly also with the $MgCl_2$ content of the solution.

VI. The Five-Component System with Ca and Fe Salts

1. Further Solid Phases

a) With Ca Sulphates

Glauberite $Na_2Ca(SO_4)_2$. Glauberite often occurs as idiomorphic crystals but it is not important in salts which owe their origin to water of purely marine origin. It has probably been overlooked on many occasions. The "labile salt" $2\,Na_2SO_4 \cdot CaSO_4 \cdot 2\,H_2O$ which CONLEY and BUNDY (1958) introduced following the indications of HILL and WILLS yields a powder diagram identical with that of thenardite. In spite of chemical analyses, the existence of such a composition is questionable, not least because its crystallographic and physical properties are unknown (see D'ANS, 1933, p. 207).

Syngenite $K_2Ca(SO_4)_2 \cdot H_2O$ is rare, but may occasionally be found as well-developed crystals. It is not of primary origin, but develops from $MgCl_2$-poor metamorphic solutions, perhaps containing $CaCl_2$, at low temperatures (D'ANS, 1944). Whether the compound can also occur without water of crystallization has never been clarified (see explanation to Table 3, syngenite). The compound $K_1Ca_2(SO_4)_3$ can be prepared from a melt (see D'ANS, 1933, p. 210). It is not isomorphic with langbeinite (GATTOW and ZEMANN, 1958, p. 238).

Goergeyite $K_2Ca_5(SO_4)_6 \cdot 6\,H_2O$. Goergeyite is very rare and is formed only by metamorphism from solutions with extremely low $MgCl_2$ content yet with a high KCl content.

Polyhalite $K_2MgCa_2(SO_4)_4 \cdot 2\,H_2O$. Its crystal class is not yet known with certainty[a]. According to GÖRGEY (1915) polyhalite probably crystallizes in a triclinic-pinacoidal form, but he left open the possibility that it might belong to the pedial class because of the unequal development of the crystal faces on the upper and lower surfaces. Polyhalite is by far the most important complex Ca salt in oceanic salt deposits.

Krugite was originally considered as a separate double salt, but is found as a fine-grained growth on anhydrite and polyhalite.

b) With Ca Chlorides

Tachhydrite $CaMg_2Cl_6 \cdot 12\,H_2O$ is a relatively rare salt mineral, generally not of primary origin (see bischofite p. 42), occasionally found locally enriched. Crystallographically it is not well known. The axial ratio is crudely known from cleavage rhombs to be $a/c = 1:1.768$. Fe (up to 0.05%) and Sr and Br in smaller amounts occur diadochically.

[a] But see SCHLATTI et al., reference p. 15.

Chlorocalcite $KCaCl_3$ up to the present has only been found in the Desdemona mine (near Alfeld/Leine) as the salt mineral baeumlerite. Crystallographically it has not been investigated.

c) With Fe Chlorides

Rinneite $K_3NaFeCl_6$ may be produced occasionally under extreme conditions by metamorphism.

Douglasite $KFeCl_3 \cdot 2\,H_2O$ is both chemically and crystallographically not well known. Its identity with $K_2FeCl_4 \cdot 2\,H_2O$, described by SCHABUS (1850) as monoclinic-prismatic, $a:1:c = 0.7367:1:0.5036$; $\beta = 104°\,46'$; $D = 2.162$; cleavage $(\bar{2}01)$, has been questioned by BOECKE (1911). Its natural occurrence on account of the high $FeCl_2$ concentration necessary is hardly likely.

Erythrosiderite $K_2Fe^{III}Cl_5 \cdot H_2O$ occurs only as an oxidation product on rinneite and is generally of recent origin.

In salt deposits the following could occur but have not have not yet been reported:

$FeCl_2 \cdot 2\,H_2O$ (monoclinic; $\beta = 130\tfrac{1}{2}°$; strong birefringence; extinction angle 52° against c in an obtuse angle; after BOECKE, 1911).

$FeCl_2 \cdot 4\,H_2O$ (monoclinic prismatic; $a:1:c = 1.1844:1:1.6358$; $\beta = 111°\,11'$; BOEKE, 1910b).

$FeMgCl_4 \cdot 8\,H_2O$ (triclinic; $D = 1.82$; rhombic plates; interfacial angle 82°; extinction angle 42.5°; BOEKE, 1911).

2. *Stability Conditions*

a) With Ca Sulphates

Even when the five-component system reaches saturation in its salts, the concentration in Ca is so low (of the order of 0.1 mol/1000 mol H_2O) that the solubility of the other components is not noticeably altered. There are but relatively few experimental results, so that for many questions there is insufficient quantative data available. VAN'T HOFF earlier had repeatedly remarked on the great experimental difficulties occasioned by the slow attainment of equilibrium. AUTENRIETH (1958) developed an elegant technique, the so called "coincidence method" to enable the stability fields to be defined precisely, the boundary lines being determined graphically as lines of equal $CaSO_4$ concentration in the different solid phases.

The data presently available are derived chiefly from determinations by VAN'T HOFF at 25° and 83°C and D'ANS at 55° C (Fig. 22). They should not be considered in isolation. According to AUTENRIETH, polyhalite at 25° and 35° possesses a stability field of similar extent to that existing at 55° C. There are unfortunately no new data relevant to this important matter. AUTENRIETH and BRAUNE (1959) worked only with simultaneous saturation in NaCl and KCl, under which condition the $CaSO_4$ concentration is clearly lower than that given by VAN'T HOFF'S data. It further results from this that anhydrite and polyhalite possess a negative temperature coefficient of solubility whilst syngenite on the contrary has a positive coefficient.

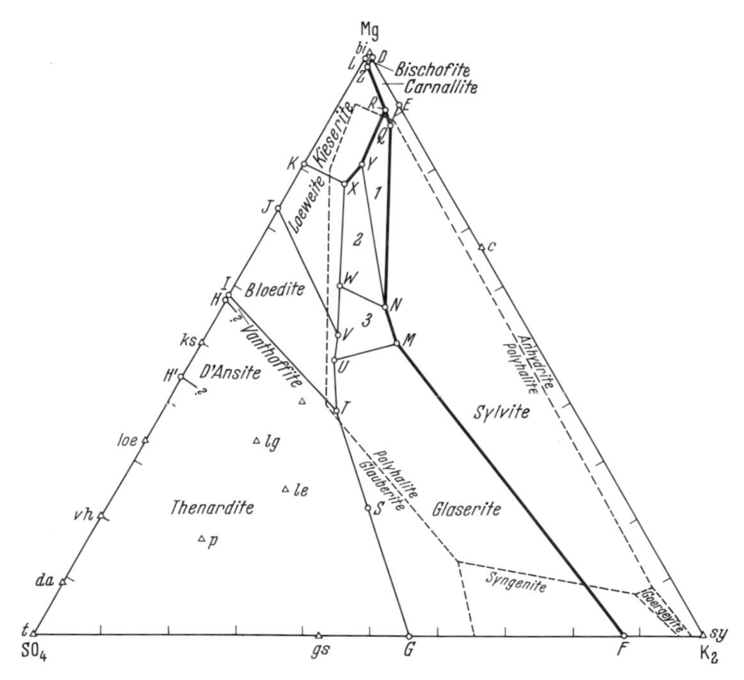

Fig. 22. 55° isotherm of the five-component system at NaCl saturation. (Limits of the d'ansite field not yet investigated. Ca-sulphate fields approximated. Points *S–Y* not known exactly.) Crystallization paths heavy lines, transition lines fine lines. *1* kainite; *2* langbeinite; *3* leonite. Salt points see Table 4 (the triangle to the left of *T* corresponds to the kainite salt point)

Polyhalite occupies a great part of the interior of the triangle. At 55° C it is stable against carnallite. Above 83° the boundary of the polyhalite-anhydrite stability fields lies between *Q* and *R* (Figs. 22, 23). Its lower temperature of formation is still not known exactly, but lies under 25° C. Its rapidity of formation at high temperatures is great[25] but at room temperature is imperceptibly small, which means that normally polyhalite would not be expected to form as a primary precipitate from evaporating seawater.

Glauberite also has a large field in which it can exist at the SO_4 corner of the Jänecke triangle. So far as is known, it reaches to the seawater point. The syngenite field decreases with increasing temperature. The goergeyite field is small, and above 30° it is inserted between the anhydrite field on the K_2–Mg side and the syngenite field. On the solubility data to hand, the formation of goergeyite is prolonged to near pure KCl–NaCl–($\pm CaCl_2$) solutions. It does not occur as a primary

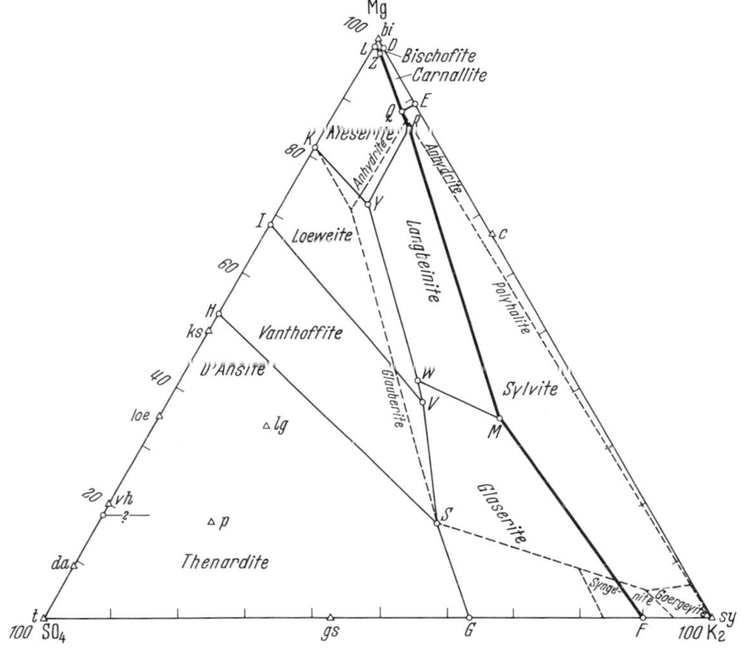

Fig. 23. 83° isotherm of the five-component system at NaCl saturation. (Limits of the d'ansite field not yet investigated. Ca-sulphate fields approximated. Points *S*–*Y* not known exactly.) Crystallization paths heavy lines, transition lines fine lines. (*Y*–*R* is also a crystallization path.)

precipitate but as a product of metamorphic change in impoverished salt rocks, that is, from solutions which result from the solution of halite and sylvite and move towards anhydrite.

[25] This has an undesirable technical consequence in the thermal-solubility working of anhydritic-kieseritic sylvite-halite, the formation of a polyhalite mud with is attendant potash loss.

b) With $CaCl_2$

With the addition of $CaCl_2$, the conditions are altered fundamentally because of its high solubility. It is not compatible with dissolved Mg sulphates, and gypsum and other Ca sulphates separate out, the solution moving nearer to the K_2-Mg side. With excess $CaCl_2$, a new system develops, one in which at high $CaCl_2$ concentrations (~ 80 mol/1000 mol H_2O) with both KCl and NaCl at lower concentrations the conditions can be approximately described by the system $CaCl_2$–$MgCl_2$–H_2O. The system at higher concentrations of $CaCl_2$ and $MgCl_2$ is well known through the works of ASSARSSON (1950–1957) and VAN'T HOFF and his co-workers (1912, p. 34, 108, 285, 303). The experimental data have been summarized by D'ANS (1961). Here the tachhydrite isotherms are of particular interest. The lower temperature of formation of tachhydrite is about 22° C, but only at the extremely high concentrations of 92.7 mol $CaCl_2$/1000 mol H_2O (~ 450 g $CaCl_2$/l). At higher temperatures the stability field of tachhydrite spreads out markedly in the direction of lower $CaCl_2$ but higher $MgCl_2$ contents.

From the isotherms D'ANS (1961) described several possibilities for tachhydrite formation. The chemical conditions for its formation can be clearly defined, yet not all the problems of tachhydrite formation have been explained. The fact that most tachhydrite occurs with carnallite and clay and not in association with bischofite indicates that KCl too must be considered. In reality carnallite and tachhydrite coexist over a wide range of concentration (Fig. 24). In normal circumstances certainly, only the $CaCl_2$-poor part of the system $MgCl_2$–$CaCl_2$–

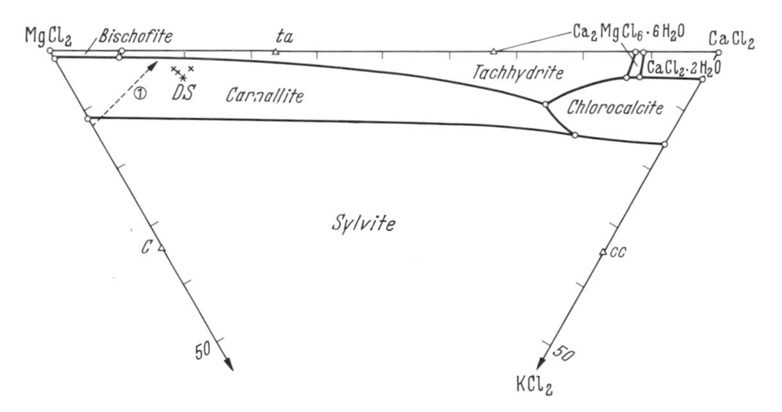

Fig. 24. System $MgCl_2$–$CaCl_2$–K_2Cl_2–H_2O at 93° C. Calculated from ASSARSSON (1957). Abbreviations of the salt points from Table 4. *DS* = Dead Sea water (asterisk = average composition after BENTOR, 1961). *1* = change in seawater composition (continuation of line *1* in Fig. 28)

$-KCl-H_2O$ need be considered. By contrast, the tachhydrite-sylvite paragenesis is probably not stable at any temperature. As carnallite as well as sylvite and halite are insoluble in tachhydrite saturated solutions, secondary tachhydrite can naturally occur in such deposits. The main problem in tachhydrite formation is undoubtedly the origin of the concentrated $CaCl_2$ solutions, for which D'ANS (1961) suggested several possible answers, none of which are completely satisfactory.

The course of crystallization is dependent upon whether the point representing the solution lies on the $MgCl_2$ or the $CaCl_2$ side of the line joining the tachhydrite and carnallite points. In the first case, crystallization leads finally to a bischofite-carnallite-tachhydrite end point, in the second to chlorocalcite-carnallite-tachhydrite. The latter is quite without practical importance. As at low temperatures the bischofite-tachhydrite-carnallite end point is displaced well towards the $CaCl_2$ corner, primary tachhydrite could not be expected to occur by isothermal evaporation. New data for the middle temperature range ($\sim 55°$ C) are, however, needed.

With isothermal crystallization of a tachhydrite-carnallite equilibrium solution at high temperatures, a mixture develops with over 90 mol percent tachhydrite (see construction in Fig. 24). Under special conditions this could occur in the pore spaces of the Salzton (p. 240).

c) With $FeCl_2$

The first researches into Fe-bearing salt systems were begun by BOEKE (1911) in connection with the discovery of rinneite. They were augmented by the solubility data of D'ANS and FREUND (1954) in the system $NaCl-KCl-MgCl_2-FeCl_2-H_2O$ with NaCl-saturation at 55° C (see Fig. 25). The field in which rinneite can exist bounds those of sylvite and carnallite. Through the addition of $MgCl_2$, the existence of rinneite at relatively low $FeCl_2$ concentrations is made possible as shown in Table 7 (after D'ANS and FREUND, 1954).

Under natural conditions, therefore, those solutions near point *Q* are of prime importance. Of the iron-bearing solid phases described up until now only rinneite is found in nature, although the discovery of others may be anticipated.

Particularly important is BOEKE's discovery that mixed crystals of Fe^{II}-containing carnallite form from $FeCl_2$-bearing solutions of the five-component system. Controlled experiments of D'ANS and FREUND (1954) have added further confirmation, and have shown that for a solution near point *Q* (Table 7) with somewhat more $MgCl_2$ and less $FeCl_2$ the

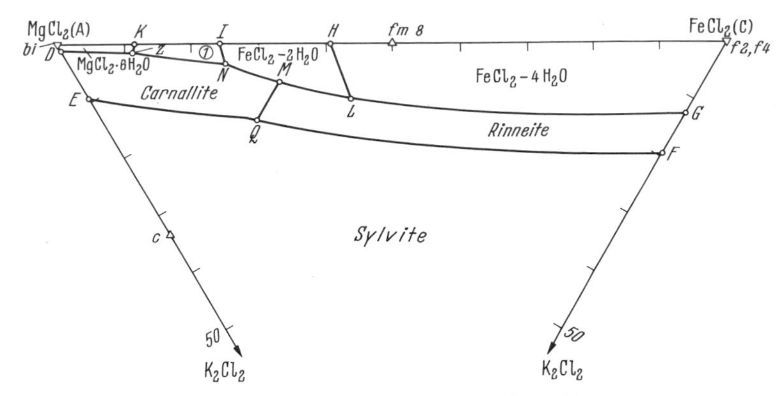

Fig. 25. 55° isotherm of the system Na^+, K^+, Mg^{++}, $Fe^{++}//Cl^-$, $+H_2O$ at NaCl saturation (after D'ANS and FREUND, 1954). (1) $FeMgCl_4 \cdot 8\,H_2O$, associated salt point "fm 8". The point representing rinneite composition at 60 K_2Cl_2 on the $FeCl_2$–K_2Cl_2 side

relationship is:

$$b = \frac{1 \text{ weight percent Fe in precipitated carnallite}}{4.2 \text{ weight percent Fe in solution}} \approx \frac{1}{4}.$$

Unfortunately there are no data available for Fe-poor solutions. By analogy with the well-known mixed crystal series with Br (see p. 131), it may be anticipated that even in solutions with a low iron content a constant distribution ratio will occur. The value so obtained should be regarded as the upper limit of iron which is taken up in carnallite.

Crystal chemistry would lead one to expect the formation of iron-bearing mixed crystals in other Mg salts, but systematic experimental data are not available.

Table 7. *Constant solutions with $FeCl_2$ at 55° C (in mol/1000 mol/H_2O)*

Solution	Na_2Cl_2	K_2Cl_2	$MgCl_2$	$FeCl_2$	Solid phases
F	10.1	25.0	—	101.9	halite, sylvite, rinneite
Q	4.4	13.1	62.8	23.3	halite, sylvite, rinneite, carnallite
M	3.9	7.9	74.9	34.8	halite, $FeCl_2 \cdot 2\,H_2O$, rinneite, carnallite
N	2.2	4.0	90.8	29.0	halite, $FeCl_2 \cdot 2\,H_2O$, carnallite $MgCl_2 \cdot FeCl_2 \cdot 8\,H_2O$
Z	1.3	2.5	106.0	14.1	halite, bischofite, $MgCl_2 \cdot FeCl_2 \cdot 8\,H_2O$, carnallite

Though the addition of $CaCl_2$ scarcely affects the limiting concentration of $FeCl_2$, the $MgCl_2$ content of the solution is lowered. The effect of increasing the SO_4 content has not been systematically investigated. Such supplementary data are needed, as in rinneite from the Hildesia mine (near Hildesheim) large kieserite crystals occur along with a little sylvite, although this is probably not a true paragenesis. (There is a sample in the mineralogical museum, Göttingen.) At higher temperatures less $FeCl_2$ is needed for the formation of rinneite (Q_{83} : 21.0 $FeCl_2$).

As in the case of $CaCl_2$, the main problem is how sufficiently high $FeCl_2$ concentrations originate.

C. Physico-chemical Models

I. Primary Precipitation

Solubility data form the basis for the calculation of primary precipitation from any given solution of the system. From these calculations different models result which depend upon the conditions selected. It seems highly improbable that the actual conditions existing in nature can be adequately represented by one of these models. The models are idealizations. In nature the systems which occur are much more complex and differ from one occurrence to the next. The conditions under which they exist are constant neither in space nor in time. This does not reduce the value of models, rather the comparison of different models with natural salt series is the only direct way of approaching the actual composition of the solutions and the conditions under which they existed, as well as the processes of secondary alteration.

Only the most important limiting cases will be considered here: stable and metastable equilibria, isothermal evaporation with or without complete reaction with constant solutions, normal and altered seawater as well as primary precipitates with a thermal gradient. At the moment this last case is still calculable only for the carnallite system without $MgSO_4$ since there are insufficient quantitative data for the calculation of the five-component system.

In all these examples progressive evaporation was considered, that is with an increase in the total salt concentration in the solution. A gradual decrease in the total salt concentration is also possible as a result of salt precipitation, every dilution phase being followed by an evaporation phase which proceeds until saturation in a solid phase is reached. An example of this "recessive salt precipitation" is presented on p. 143.

1. Static Isothermal Evaporation of Normal Seawater at 25° C

In this monograph, the term evaporation rather than vaporization is used to signify that under natural conditions evaporation occurs just so long as the atmosphere is not saturated with respect to H_2O, and so long as no liquid (vapor) phase is present.

a) Stable Equilibria without Reaction in Constant Solutions

This is the basic type. Only one example is given, as all the others can be derived in the same way.

The point representing seawater at 25° C lies in the bloedite field with the co-ordinates[26] in the Jänecke ternary diagram of Mg = 69.00, SO_4 = 24.59, K_2 = 6.41 (see Table 1). During evaporation seawater passes through the concentration stages indicated in Table 8. These can be calculated from the data of POSNJAK (1940) and USIGLIO (from VAN'T HOFF, 1905) before bloedite saturation is reached. The NaCl concentration at the beginning of bloedite precipitation was interpolated. From this point on the calculation was made from the Jänecke triangle (as explained on p. 70). The bloedite and NaCl precipitation gives a solution along the line $I - V$, the NaCl content of which must also be interpolated. Along the boundary line the bloedite formed is thought to be completely removed from the solution, so that crystallization continues with the formation of epsomite and NaCl. As there is no longer 100% of the solution the amount of epsomite (and subsequent) precipitates must be multiplied by the percentage of solution still present. The succeeding precipitates are kainite + epsomite + NaCl as now a crystallization path has been reached. Along this crystallization curve crystallization continues until residual solution Z is reached. The general division of the salt beds being formed goes back to VAN'T HOFF (1905 and earlier):

A = Precipitation before saturation with respect to salts of the five-component system.
B = Precipitation of NaMg or Mg sulphates without K-salts.
C = Precipitation of KMg-salts, without carnallite.
D = Precipitation of carnallite.
E = Terminal precipitation with bischofite.

This division has been used ever since (JÄNECKE, 1923 and earlier; BORCHERT, 1940).

The calculations for 25° C are shown in Table 9. Column 2 contains the amount of the precipitate in Jänecke's units, while column 3 gives the amount of solution remaining in terms of 100 units of solution at the onset of bloedite precipitation. Columns 4–6 give the amount of each precipitate broken down into the components of the Jänecke system (see Table 4). Column 7 contains the Na_2 precipitate calculated as described on p. 62, 63. Column 8 shows the result in mol. The last line is a check, the sum in each column being the content of the initial solution. The deviations reflect the lack of precision of the graphical method.

[26] The result differs slightly from that given by JÄNECKE (1923), but in the latter's data $CaCO_3$ was neglected.

Table 8. *Composition of seawater during isothermal evaporation at 25° C (mol/x mol H_2O)*

Initial precipitation of	Seawater	Gypsum	Halite	Bloedite	Epsomite	Kainite	Hexahydrite	Kieserite	Carnallite	Bischofite
H_2O (x)	70,700	21,100	6450	1170	1080	965	800	585	576	428
K_2	6.4	6.4	6.4	6.4	6.4	6.4	4.8	2.0	1.0	0.1
Mg	69.0	69.0	69.0	69.0	67.3	63.4	58.1	52.1	49.9	44.9
SO_4	24.6	24.6	24.6	24.6	21.0	17.1	11.6	5.7	3.6	0.4
Na_2Cl_2	303	303	303	26	11.45	9.1	4.7	1.5	1.4	0.25
$CaCO_3$	1.5	—	—	—	—	—	—	—	—	—
$CaSO_4$	11.7	11.7	1.8	—	—	—	—	—	—	—
Weight of solutions[a]	1000	322.4	121.3	24.46	21.75	19.61	16.30	12.22	11.78	9.27

[a] Weight of solutions in grams per 1000 g seawater. Corresponds to a_i in Eq. (1) p.132.

Table 9. *Static primary precipitation at 25° C in Jänecke units and in mol. For abbreviations of precipitates in cols. 2 and 8, see Table 4*

1	2	3	4	5	6	7	8
Bed	Precipitate	Residual solution	K_2	Mg	SO_4	Na_2	mol
B_1	5.3 bl	94.7	—	1.76	3.54	1.76 bl 12.8 Na_2Cl_2	1.76 bl 25.6 n
B_2	7.8 e	86.9	—	3.9	3.9	2	3.9 e 4.0 n
C_1	4.6 e	74.4	—	2.3	2.3	4.7	2.3 e +9.4 n
	7.9 k		1.58	3.16	3.16		3.16 k
C_2	0.5 hx		—	0.25	0.25	3.2	0.25 hx + 6.4 n
	14.1 k	59.8	2.82	5.64	5.64		5.64 k
C_3	0.29 ks		—	0.14	0.14	0.5	0.14 ks + 1.0 n
	4.81 k	54.7	0.96	1.92	1.92		1.92 k
D	6.45 ks		—	3.23	3.23	0.8	3.23 ks + 1.6 n
	2.85 c	45.4	0.95	1.90	—		1.9 c
E	0.86 ks		—	0.43	0.43	0.25	0.43 ks + 0.5 n
	0.27 c		0.09	0.18	—		0.18 c
	44.27 bi	0	—	44.27	—	—	44.27 bi
Check figure			6.40	69.08	24.51	26.01	

A further check may be made by recalculating the solutions at the various transition points on the crystallization curve, quoted in terms of $Mg + SO_4 + K_2 = 100$, according to the amount of solution actually present there (column 3). Table 8 gives these values. With an exact graphical method these figures would correspond to the difference between the initial constant solution and the precipitate in form of components. Within the limits of error of the graphical method, this is the case.

The primary precipitates from normal seawater at 15° and 35° C (in addition, to metastable equilibria at 15° and 25°, see below) were calculated in the same way. The results are converted to weight % within each layer and the thickness (calculated with the help of the density of the individual minerals) expressed in terms of a 100 m thick pure rock salt layer *A* (Table 10 and Fig. 26). The results were rounded off to the nearest full percent ($^1/_4$ % in layer *E*).

It is pointless to demand greater accuracy. Even if, instead of the graphic method, an exact calculation were carried out using the scheme given on p. 45, the result is uncertain, at least in the decimal places, simply because the corresponding solution equilibria themselves are uncertain. Furthermore, in some solutions the NaCl content was found by interpolation. It is therefore a mistake to consider as real the numbers resulting from multiplication with the much more exactly known molecular weights.

Calculations from more recent solution equilibrium data occasionally show appreciable deviations when compared with the older figures.

The important effects of temperature find their expression qualitatively and quantitatively in the composition and thickness of the salt layers and their sub-divisions. The potash-free layer *B* decreases in thickness at higher temperatures and changes from a pure halite-epsomite to a bloedite-halite with both forms occurring one after another at 25° C. At still higher temperatures ($\gtrsim 50°$ C) this facies is replaced by loeweite-halite (cf. Fig. 20): at the beginning of NaMg sulphate precipitation the solution contains ~ 38 mol $MgCl_2$/1000 mol H_2O but still lies within the NaMg sulphate stability field).

The carnallite-free intermediate layer *C*, within the given temperature range, is composed predominantly of kainite. At 15° C it is preceded by sylvinitic-halite-epsomite, and at 35° C it is followed by a kainite-bearing halite kieserite. Below 11° C and above 83° C kainite is entirely absent, above 50° C it is underlain by langbeinite-loeweite and at slightly higher temperatures by langbeinite-kieserite which above 83° C precedes the kieserite-sylvite paragenesis (Fig. 20). The carnallite layer *D* can contain epsomite or kieserite in stable paragenesis with carnallite and also some hexahydrite but only over a very restricted temperature range (18 to

Table 10. *Theoretical primary precipitation from normal seawater by static evaporation without reaction at the transition points. Thicknesses normalized to a 100 m-halite layer A. Composition in weight percent. Abbreviations of the solid phases as in Table 4*

Temp.	Stable equilibrium						Metastable equilibrium			
°C	15		25		35		15		25	
layer										
B_1	20 m	33 n 67 e	6.3 m	72 n 28 bl	8.7 m	74 n 26 bl	20 m	33 n 67 e	6.3 m	72 n 28 bl
B_2			4.5 m	20 n 80 e					5.8 m	24 n 76 e
C_1	3.4 m	32 n 52 e 16 sy	6.3 m	29 n 30 e 41 k	8.5 m	11 n 89 k	8.2 m	30 n 53 e 17 sy	4.5 m	33 n 47 e 20 le
C_2	5 m	33 n 67 k	5.7 m[a]	21 n 3 hx 76 k	3.2 m	14 n 1 k 85 ks			3.0 m	22 n 58 e 20 sy
C_3			1.7 m[a]	11 n 4 ks 85 k						
D_1	2.9 m	21 n 79 c	3.6 m	12 n 40 ks 48 c	4.5 m	12 n 29 ks 59 c	10.3 m	17 n 23 e 60 c	0.65 m	7 n 38 e 55 c
D_2	3.3 m	10 n 29 e 61 c					0.5 m	18 n 39 hx 43 c	14.5 m	7 n 35 hx 58 c
D_3	3.1 m	18 n 30 ks 52 c					0.8 m	17 n 37 5h 46 c	1.95 m	9 n 53 5h 38 c
D_4							0.4 m	24 n 32 lh 44 c	1.3 m	9 n 62 lh 29 c
E	33.5 m	0.5 n 1 ks 2.5 c 96 bi	38 m	0.5 n 1 ks 0.5 c 98 bi	39 m	0.5 n 0.25 ks 0.25 c 99 bi	34.2 m	0.5 n 4.5 lh 3 c 92 bi	34 m	0.5 n 5 lh 0.5 c 94 bi

[a] See p. 67

about 25°). Its thickness is particularly great at 15° C. The bischofite layer increases in thickness at higher temperatures.

In the salt sequence the amount of NaCl decreases upwards. There is, however, still nearly four times the amount in the carnallite layer than calculated by VAN'T HOFF (1905, see also BOEKE, 1908, p. 375).

The Ca-salts were neglected in the five-component system even at the beginning of the precipitation of halite layer *A*. In the model considered this is justified since at the beginning of halite precipitation

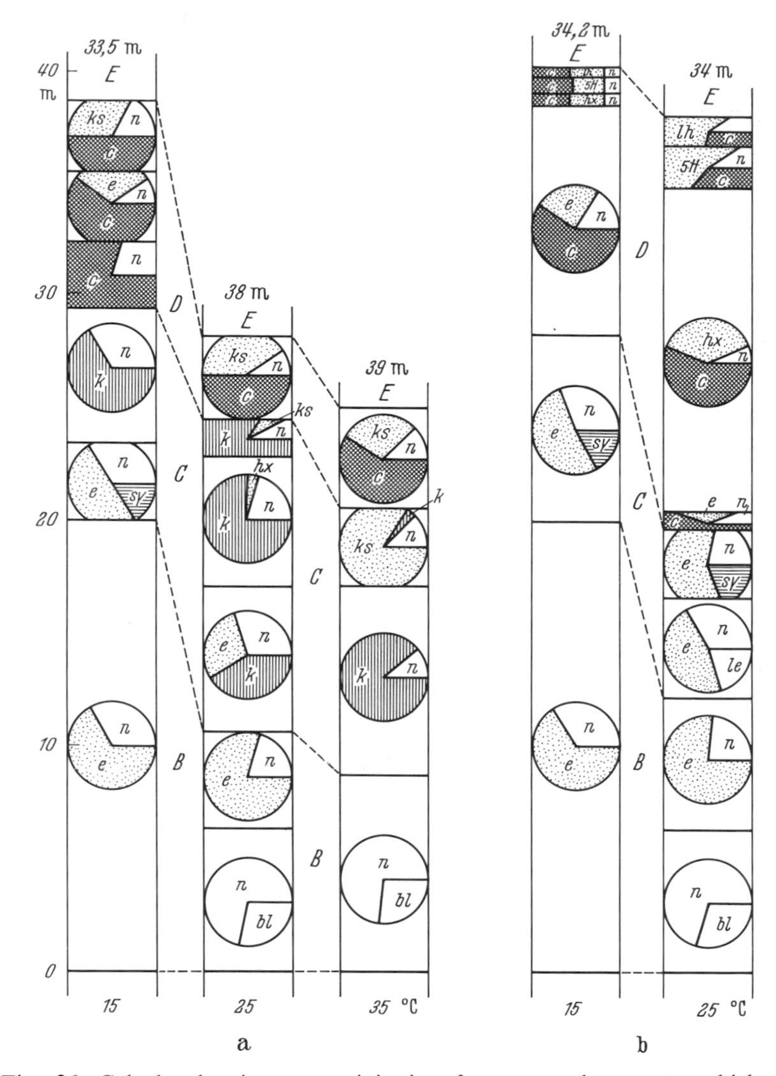

Fig. 26. Calculated primary precipitation from normal seawater; thicknesses normalized to a 100 m thick halite layer *A*. Mineral abbreviations as in Table 4. *B–E* symbols of the layers as given on p. 85. a) stable, b) metastable equilibria, both without reaction at the transition points

the $CaSO_4$ solubility is about $\lesssim 0.2$ mol/1000 mol H_2O. If this remaining $CaSO_4$ were completely precipitated in the halite layer, there would only be 0.7% anhydrite in the whole halite zone. Part of this could be altered to polyhalite, but the polyhalite content can itself be neglected.

This content is, at most, only a third of the amount assumed by JÄNECKE (1906). This means, then, that under conditions of static evaporation a nearly pure NaCl layer results; a layer which would not be altered by temperature fluctuations.

Compared to a halite layer of 100 m, the thickness of the underlying anhydrite layer is 3.0 m (as gypsum 4.8 m) and the carbonate only ~0.4 m (0.37 m).

For a salt succession to form with a 100 m halite layer from seawater of present day composition and without the addition of fresh seawater, a basin would have to be 8.5 km deep. This indicates immediately the improbability of the foregoing model.

b) Metastable Equilibria

Precipitation under stable equilibrium conditions is seldom realized in nature. It is more usual to find noteworthy supersaturation, which in evaporation experiments often leads to the precipitation of metastable phases. This is also observed in Recent salt lakes (VALYASHKO, 1958). It has been shown in the gypsum-anhydrite system that this is related to differences in nucleation energy.

Precipitation under metastable equilibria leads to quite different parageneses, as Fig. 26 illustrates. Kainite is completely absent, and in its place in layer *C* sylvite-halite-epsomite are found which are underlain by leonite-halite-epsomite at 25° C. In addition, instead of kieserite the metastable Mg sulphate hydrates are found.

Calculations have only been made for 15° C and 25° C. For 35° C the new data available are incomplete. Furthermore, the stability of the metastable forms clearly diminishes with rising temperature. At 35° C according to AUTENRIETH (1954) the rate of formation of kainite is already substantial, so that in the relatively slow precipitation of salts under natural conditions the role of metastable equilibria is diminished.

D'ANS and KÜHN (1960) stressed that the preservation of metastable primary precipitates on a geological time scale is practically excluded. It is possible to read from the 25° C isothermal diagram (Fig. 18) what happens to primary metastable precipitates once nucleation of the stable solid phase occurs. The metastable paragenesis epsomite–leonite, and the paragenesis sylvite–epsomite, which is metastable above 18° C, lie within the stability field of kainite. It is, therefore, principally a question of time before it changes to the stable paragenesis kainite + epsomite + solution (along the boundary $W-X$). The metastable paragenesis halite–hexahydrite-carnallite, which according to recent polythermal data is only stable in a narrow temperature range above 18° C in the presence of a high $MgCl_2$ concentration (Fig. 20), must change to the stable halite–hexahydrite–kainite paragenesis, and in the presence of concentrated solutions, to halite–kieserite–kainite. Whether these processes of transformation actually occur is not known. Like anhydrite, kieserite is not found as a primary precipitate under experimental conditions.

VALYASHKO (1958) who correctly emphasizes the importance of metastable primary precipitates, seems, according to his text, to consider the kieserite-sylvite paragenesis as an alteration product of the metastable primary epsomite–sylvite and hexahydrite–sylvite parageneses. This is physico-chemically incorrect. On the other hand, his Fig. 10 (VALYASHKO, 1958, p. 211) indicates that kainite can be formed diagenetically out of the metastable forms sylvite + epsomite (or hexahydrite). This change to a stable paragenesis is aided by pressure (LEPESCHKOW, 1958), particularly differential pressure.

Of the time necessary for the change from metastable to stable solid phases little is known. It is different for each solid phase. Furthermore, it depends upon several factors of which by far the most important are the concentration of the existing solutions, temperature and temperature changes. The effects of later rises in temperature will be considered in the section on thermal metamorphism. On the whole, petrographic data would indicate a diagenetic, probably an early diagenetic, alteration of the primary metastable precipitates.

In the following discussions, primary precipitation will also include the (early) diagenetic alterations of metastable to stable primary precipitates, despite the fact that they only begin with allophase recrystallization and partial removal of material; the term must not, therefore, be literally interpreted. Considered from the standpoint of conditions of formation, this expanded definition is justified. Fundamentally no changes are necessary in the parameters such as temperature, concentration etc. for the onset of stable equilibria (excluding the adjustment of the activation energies necessary for the onset of stable equilibria which in practice are usually provided by a temperature increase). In contrast to this, alterations in the parameters are a prerequisite for metamorphic changes.

This brief account shows, however, that far too little is known of the important early diagenetic changes and that these must be thoroughly investigated.

The formal distinction between the terms diagenesis and metamorphism by means of these parameters will be useful for salt deposits in which metastable primary precipitates play an important part. For sediments in general the distinction is too fine, as most of the newly formed diagenetic minerals are the result of reactions with infiltrating solutions, that is with altered concentrations (CORRENS in

BARTH-CORRENS-ESKOLA, 1939) In sediment formation, the systems involved are "open" and adaptation to changed concentrations (including hydrogen ion concentration and redox potential) must be included in the diagenetic phase for non-saline sediments.

In salt deposits, percolating secondary solutions cause such far-reaching alterations, even if the temperature is unchanged, that they are best considered under metamorphism (see p. 118).

c) Conditions in Reactions along Reaction Lines and at Transition Points

In cases where primary precipitates are not removed from the solution (by sedimentation etc.), they will be redissolved when the solution composition reaches a reaction line or a transition point, with the formation of the succeeding paragenesis (p. 67, 68). It is assumed that the solution, although becoming more concentrated, can still circulate more or less freely in the pore spaces of the previously formed precipitate. This is quite often the case, at least to a limited extent (VALYASHKO, 1958).

The extreme case (probably seldom realized) of a complete reaction of the solid phases at the corresponding points with evaporation of seawater can be seen on the 25° isotherm (Fig. 16). Only the case of stable equilibria will be considered here. The first-formed bloedite-halite upon reaching the boundary line $I-V$ reacts with the formation of epsomite. The solution I' then alters along the line $I-V$. Bloedite is used up if the solution has a composition which lies on the prolongation of the epsomite point through the seawater point (I'', Fig. 17). It then proceeds further, by the precipitation of epsomite and halite, into the epsomite field and eventually reaches the epsomite-kainite boundary, whereupon there is simultaneous precipitation of epsomite-kainite-NaCl until point X is reached. The solution there reacts with epsomite, which is completely altered to hexahydrite with the evaporation of the water of crystallization released by this change, the composition and quantity of solution remaining constant. Once the reaction is completed, the composition of the solution can proceed further in the direction Y with further precipitation of hexahydrite + kainite + NaCl. At Y hexahydrite is altered to kieserite with the solution remaining constant in quantity and composition yet with the evaporation of the water released by the change. At R the kainite in turn disappears with the formation of kieserite and carnallite. In the latter case quantitatively there results (solid phases, see Table 4):

$$100\text{ g k} + 143.7\text{ g soln. } R_{25} \rightarrow 61.26\text{ g ks} + 119.0\text{ g c} + 149\text{ g n} + 61.95\text{ g } H_2O$$

or from 1000 g seawater:

$$2.01\text{ g k} + 2.88\text{ g soln. } R_{25} \rightarrow 1.23\text{ g ks} + 2.39\text{ g c} + 0.03\text{ g n} + 1.24\text{ g } H_2O\,.$$

The kieserite and halite already within the kainite salt may remain. Part of the solution will be used up — about $1^1/_2$ times the amount of the reacting kainite, and water has to evaporate. After these changes have been concluded, the precipitation of kieserite-carnallite-halite can continue with further loss of water from the remaining solution R_{25}, with its composition altering in the direction of Z_{25}.

One important consequence of these reactions is that all the potassium is fixed as carnallite. Compared with evaporation without reaction (p. 85), more kieserite + carnallite is formed, but less bischofite. If the reaction goes to completion, there remains at the beginning of bischofite saturation only 6.42 g of solution Z per 1000 g of seawater, compared with 9.27 g in evaporation without any reaction (Table 8). Thus, with complete reaction, there will be only about two-thirds the thickness of bischofite but substantially more carnallite than shown in Fig. 26.

The important points in reactions at transition lines under conditions of constant solution composition are thus:

1. The reaction can only proceed with the evaporation of water (cf. p. 116) that is to say, by the addition of thermal energy, even in the case where the change of a solid phase is simply to one of a lower water of crystallization.

2. The composition of the reacting solution (relative to the main components) remains constant at the univariant points (eg. X, Y, R). Along the transition lines (eg. I–V) the composition of the solution alters during the reaction. Where a solid phase merely changes to a less hydrated form (eg. epsomite to hexahydrite), the quantity of the solution remains constant. In the event of another solid phase being formed, part of the solution is used up.

3. Only the solid phases stable in solution Z can remain. As a result the number of solid phases tends to a minimum. The main difference in comparison with the evaporation of seawater without reaction is in the absence of a kainite region, the greater size of the carnallite region and in the smaller size of the bischofite region.

The alteration of the solution composition (in weight %) and the amount of solution can be taken from Table 11 and Fig. 34. The values were calculated by a determinant method, and the results are given in weight %. The composition of the solution "bl" (at the beginning of bloedite precipitation), I′ and I″ were calculated by linear interpolation. A linear interpolation corresponds to the following equation (symbols for the solutions according to Fig. 16; composition calculated from Table 5, composition of seawater, see Table 1):

$$1\ \text{seawater} \rightarrow \underbrace{x\,\text{I} + y\,\text{H} + z\,\text{V}}_{\text{solution bl}} + u\,\text{NaCl} + w\,\text{H}_2\text{O}$$

	Seawater	x	y	z	u	w
$MgCl_2$	0.6281	44.5	−2.9[a]	42.6	0	0
$MgSO_4$	0.3479	16.5	19.1	19.7	0	0
KCl	0.1814	0	0	13.0	0	0
NaCl	8.5670	23.6	99.8	22.4	1	0
H_2O	1000	1000	1000	1000	0	1

$x = 0.000952$ $y = 0.00300$ $z = 0.01395$
$u = 7.933$ $w = 982.1$

[a] Na_2SO_4 written as $-MgCl_2 + MgSO_4 + 2\,NaCl$, see note to Table 5, p. 61.

where the bloedite-saturated solution is assumed to be separated into component solutions I, H, V. In an analogous manner solutions I′ and I″ are calculated as components of I + V.

I′ is obtained from

$$1\ \text{seawater} = \underbrace{x\,\text{I} + y\,\text{V}}_{\text{I}'} + z\ \text{bloedite} + u\,\text{NaCl} + w\,\text{H}_2\text{O}$$

similarly I″ is obtained from

$$1\ \text{seawater} = \underbrace{x\,\text{I} + y\,\text{V}}_{\text{I}''} + z\ \text{epsomite} + u\,\text{NaCl} + w\,\text{H}_2\text{O}$$

that is to say, all the already formed bloedite has reacted when only epsomite is left with a solution along the line I–V.

In Table 11 in the line "Weight of solution" the quantities in the residual solution, normalized to 1000 g seawater, are given; hence it is possible from this table to obtain quantitatively the precipitation (in grams) between each column. The components in each column are multiplied by the solution quantity, and the difference between the products in neighbouring columns is obtained by subtraction. These differences correspond to the precipitate plus evaporation of water.

In the lower part of Table 11 the composition of the precipitates, calculated from the minerals stable in solution *Z*, is given, along with the thickness of the salt layer normalized to a 100 m halite layer *A*. On account of the uncertainty of solution *Y*, the precipitates and changes from solutions *X* and *Y* were considered together.

In Fig. 27 the three types of primary precipitation from normal seawater by static evaporation at 25° C are contrasted. It shows that at

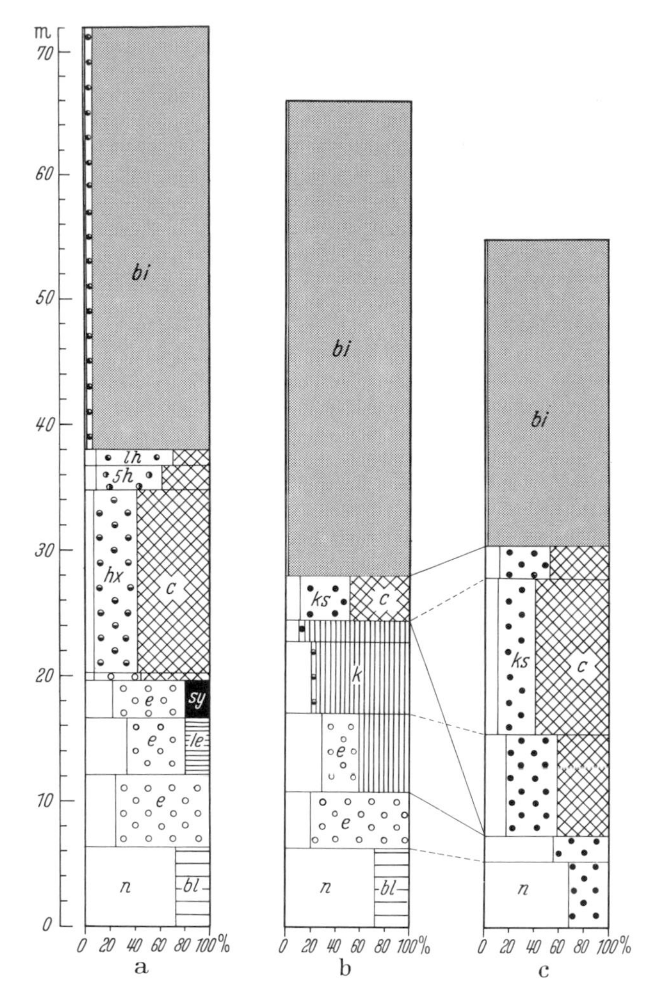

Fig. 27. Static evaporation of seawater at 25° C. Thicknesses normalized to a 100 m thick basal halite layer, a) metastable equilibria, b) stable equilibrium without reaction at the transition points; c) stable equilibria with complete reaction at the transition points. Minerals symbols as in Table 4

constant temperature, qualitatively and quantitatively different salt sequences can develop out of the same solution according to the crystallization conditions. As stated on p. 91, the main primary precipitation proceeds under metastable conditions. At the present time no more precise estimates are possible about the kind and rate of change to a stable paragenesis.

Table 11. *Composition of seawater solutions and solid precipitates from isothermal evaporation at 25° C (weight %). Complete reaction of the solid phases. Ca salts not considered*

Solution	bl	I′	I″	X″	X	Y[a]	R	Z
$MgCl_2$	12.59	15.30	15.25	16.35	20.62	26.75	28.35	34.38
$MgSO_4$	8.82	8.77	8.82	8.36	6.40	3.60	2.82	0.43
KCl	2.85	3.32	3.45	3.70	3.38	1.55	0.99	0.14
NaCl	7.81	4.94	4.92	3.96	2.57	1.25	1.04	0.28
H_2O	67.93	67.67	67.56	67.90	67.03	66.85	66.80	64.77
Weight of solution in grams normalized to 1000 g original seawater	25.45	21.81	21.03	19.61	15.55	12.0	11.30→8.41	6.42
Precipitate (weight %)								
Halite		68		55.6	17.8	11.0	11.8	0.38
Kieserite		32		44.4	40.6	30.6	41.2	0.67
Carnallite		—		—	41.6	58.4	47.0	0.70
Bischofite		—		—	—	—	—	98.25
Thickness (m) normalized to a 100 m halite layer *A*		5.27		2.05	8.10	12.5	2.54	25.39

[a] Corrected approximation see p. 67.

2. *Altered Seawater*

a) $MgSO_4$ Deficiency

The petrographic classification of salt deposits shows that the special case of $MgSO_4$-deficient parageneses is common. Such precipitates are not possible from normal seawater. The alteration of the seawater itself will be considered later (p. 246). In the calculation of models, different stages of $MgSO_4$ impoverishment will be considered. The first calculations of this kind were made by VALYASHKO (1958). In the Jänecke ternary diagram, the $MgSO_4$-deficient solutions lie on the straight line from the $MgSO_4$ point (Fig. 28) through and beyond the seawater point. In Fig. 28 four stages of $MgSO_4$ deficiency are considered:

I: up to the boundary line between epsomite–kainite,
II: up to the boundary line between kainite–sylvite,
III: up to the point [83.4] Mg, [11.4] K_2; [5.2] SO_4,
IV: up to the point [87.3] Mg, [12.7] K_2; [0] SO_4.

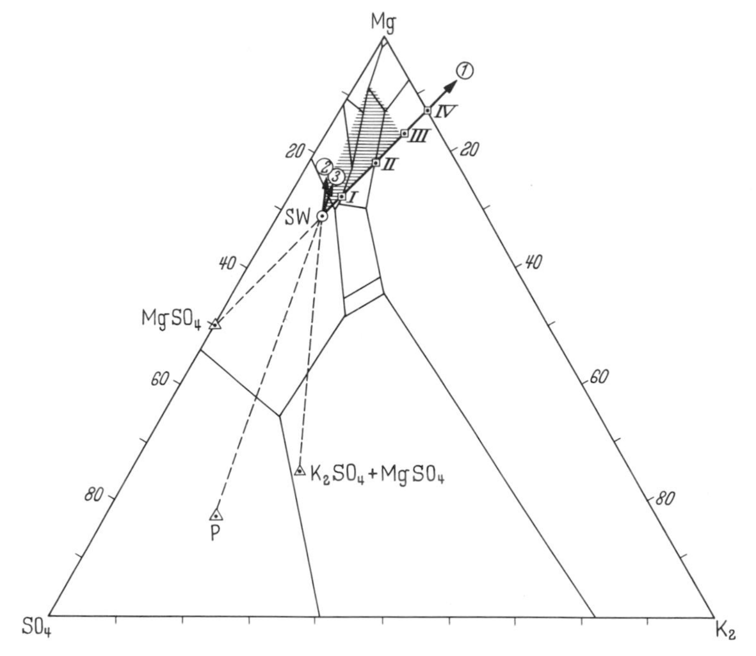

Fig. 28. Changes in seawater composition. *1* through removal of $MgSO_4$, stages *I–IV* : initial solution for the salt succession in Fig. 29; *2* through the precipitation of polyhalite by the influx of $CaSO_4$-rich solutions; *3* through polyhalite precipitation without the influx of $CaSO_4$-rich solutions. The initial solutions lie in the shaded area = a combination of *1* and *2*. Other alterations through SO_4-rich continental influx or by partial reduction of SO_4 are not considered. The stability fields of the solid phases as in Fig. 16

The result (weight %) is again normalized to a 100 m halite layer (Table 12 and Fig. 29). This, however, represents different quantities as each of the four solutions at the beginning of precipitation of their respective salts contain different amounts of halite in solution. In addition, the $CaSO_4$ precipitating in the halite layer is not included in the layer thickness.

The potash-free layer *B* is eliminated in the first stage. Layer *C* already contains in the second stage some sylvite along with the dominant kainite. In stages III and IV primary sylvite-halite develops. The carnallite layer *D* begins in stage II with a thick kainitic-halitic-carnallite over which comes a normal halitic kieseritic-carnallite as in unaltered seawater. In stage III there is a halitic carnallite at the bottom and this forms the entire carnallite layer in stage IV. Stage IV does not represent the end of the alteration, although all the $MgSO_4$ is now

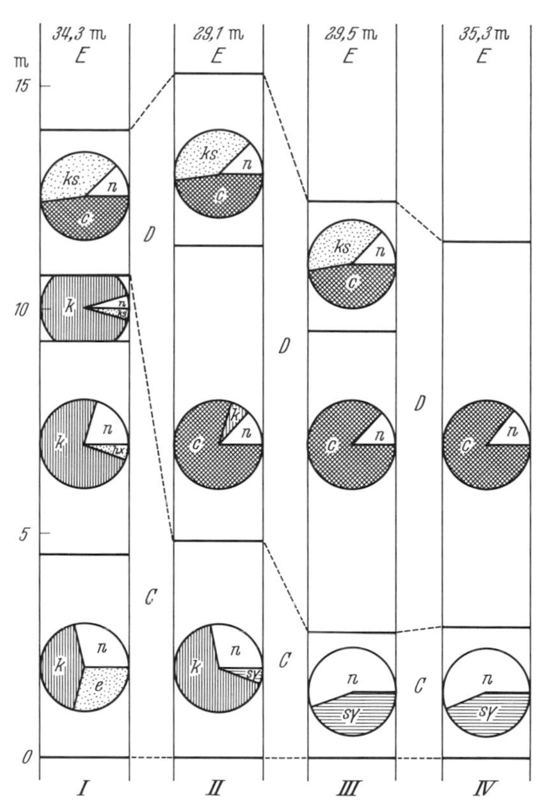

Fig. 29. Calculated primary precipitation at 25° C at different stages of $MgSO_4$ deficiency (solutions *I–IV* of Fig. 28). Stable equilibria. No reactions at transition points. Thickness normalized to a 100-m thick halite layer *A*. Abbreviations of mineral phases as in Table 4

removed from the brine.For between the system NaCl–KCl–$MgCl_2$–$MgSO_4$–H_2O and the system NaCl–KCl–$MgCl_2$–$CaCl_2$–H_2O there exist, in the case of NaCl saturation which is the only one considered here, continuous reactions so that the straight line from the $MgSO_4$ point can be extended through the K_2Cl_2–$MgCl_2$ side and into the $CaCl_2$-bearing system (Fig. 24). Incidentally the point representing the composition of Dead Sea water lies approximately in the same direction although it is not yet saturated with respect to carnallite. Since the $CaCl_2$-containing system is not well known at low and medium temperatures, no models can be calculated for these cases. Qualitatively it might be anticipated that at relatively low $CaCl_2$ concentrations carnallite would be the first potash mineral to occur.

Table 12. *Different stages of $MgSO_4$ deficiency. Primary precipitation at 25° C, stable equilibira, static evaporation. Abbreviations of solid phases as in Table 4*

	Layer	Thickness (m)[a]	Conc. factor[b]	Composition of the precipitates (in weight %)					
				n	k	MS[c]	sy	c	bi
I	C_1	4.5	74	27	42	31 e	—	—	—
	C_2	4.8		20	75	5 hx	—	—	—
	C_3	1.4		4	92	4	—	—	—
	D	3.3		12	—	40	—	48	—
	E	34.3		0.5	—	0.5	—	0.5	98.5
II			88.4						
	C	4.8		27	69	—	4	—	—
	D_1	6.6		12	6	—	—	82	—
	D_2	2.9		12	—	40	—	48	—
	E	29.1		0.3	—	0.8	—	0.7	98.2
III			89						
	C	2.8		55	—	—	45	—	—
	D_1	6.7		12	—	—	—	88	—
	D_2	2.9		12	—	40	—	48	—
	E	29.5		0.3	—	0.6	—	0.5	98.6
IV			89.4						
	C	2.9		56	—	—	44	—	
	D	8.6		14	—		—	86	—
	E	35.3		0.4	—	—	—	0.5	99.1

[a] Normalized to a 100 m halite layer A without respect to $CaSO_4$.

[b] Concentration of $MgSO_4$ deficient seawater up to the beginning of the precipitation of KMg sulphates

$$= \frac{K_2Cl_2 \text{ in solutions I–IV from Fig. 28}}{K_2Cl_2 \text{ in recent seawater}}.$$

[c] Magnesium sulphates: e = epsomite; hx = hexahydrite; no letter = kieserite.

b) Formation of Polyhalite

Besides simple deficiency in $MgSO_4$, the precipitation of polyhalite is a further process which leads to the alteration of the evaporating solution. In the static evaporation of normal seawater (without subsequent additions), this case can be neglected (p. 90). With the inflow of $CaSO_4$-bearing solutions into a solution already saturated with respect to polyhalite, the quantity of polyhalite precipitating is dependent upon the amount of $CaSO_4$ introduced. In this case it is unimportant whether the polyhalite separates immediately from the solution or whether gypsum forms first followed by anhydrite, to be replaced by polyhalite in subsequent stages of alteration.

Solutions unsaturated in salts of the five-component system have the same mixing proportions as the ions shown in the Jänecke diagram, that is to say they have the same representative points on the projection but have a higher water content than the comparable saturated solutions (cf. Table 8). The representative points thus occur above the saturation plane on the perpendicular to the point corresponding to the relative proportions of the salts[27].

As long as only $CaSO_4$ or only NaCl are precipitated, the point will remain on the perpendicular, but approach closer to the saturation plane, since the water content, related to $K_2 + Mg + SO_4 = 100$, becomes smaller. As soon as polyhalite is precipitated, the representative point is displaced.

This can be followed in the plane of projection of the Jänecke diagram (Fig. 28). With static evaporation of seawater the alteration of the solution proceeds from the polyhalite point, which lies within the thenardite field (coordinates in Table 4), but because of the low concentration of $CaSO_4$ in solution this case can be neglected. With an influx of $CaSO_4$-rich solution, on the other hand, for the most part only K_2SO_4 and $MgSO_4$ from the original solution will be used up in formation of polyhalite. (The $CaSO_4$ from the original solution can be neglected in a first approximation.) A graphic representation of the alteration of the solution must therefore begin from the $K_2SO_4 + MgSO_4$ point, which corresponds to the schoenite and leonite point (and not from the polyhalite point, as originally assumed, BRAITSCH, 1961c). On reaction with $CaSO_4$ with the formation of polyhalite, the point representing the composition of the solution moves away from the schoenite point in a straight line, before the seawater reaches saturation in salts of the five-component system.

It is not yet possible to produce quantitative models which are usable in practice, largely because of inadequate knowledge of the polyhalite and glauberite solubilities in the relevant temperature and concentration ranges. Furthermore, many different cases would have to

[27] The D'ANS figure (1915, p. 234, Footnote 1; 1933, p. 28) is a central projection in which the point representing the solution is found by vector addition along the edges of a trigonal pyramid. Here the origin of the co-ordinate system corresponds to pure water and the edges to K_2Cl_2, $MgCl_2$, Na_2SO_4; the unsaturated solution lies along the radius vector from the origin to the representative point on the saturation plane, that is within the pyramid. In contrast the Jänecke figure is a parallel projection with the points representing pure water and NaCl saturation without the components of the five-component system at infinity. Solutions unsaturated in the salts of the five-component system lie above the plane of projection at a distance determined by the factor for converting the salt content per 1000 mol H_2O to actual water content, normalized to $K_2 + Mg + SO_4 = 100$ (p. 30).

be considered depending upon the point at which the salts of the quinary system reached saturation. Only the limiting cases can be considered here, and then only when the boundaries of the polyhalite field are known.

For these reasons the alteration of brine solutions by the formation of polyhalite will be considered qualitatively. In Fig. 28 the shaded area indicates the region in which occur those altered brines likely to be found in nature. The major alteration corresponds to the impoverishment in $MgSO_4$ (line 1) already described. The alterations which lie within the shaded area are caused by the precipitation of glauberite and particularly of polyhalite, since polyhalite precipitation is still possible even from partially $MgSO_4$-deficient solutions. The boundary of the shaded area on the $MgSO_4$ side is unfortunately not well defined. It is obtained from the field in which polyhalite can exist, which is in turn temperature-dependent. The limit used is derived from the old and somewhat inexact data of VAN'T HOFF and D'ANS (p. 78) for 25° and 55° C. From this it can be seen that a large part of the kainite field (cf. Fig. 16) is covered. (That part of the polyhalite field below line 1 and left of 2 is not relevant in this respect, cf. Fig. 22.) Although according to the new data for 35° C the kainite field begins with a distinctly lower K_2 content than suggested by the older results, it is not known whether in precisely this important temperature region the polyhalite boundary lies within the kainite field or not. A further uncertainty is introduced by the positioning of the glauberite-polyhalite stability boundary, which according to present data should lie close to the seawater point. In seawater without an initial deficiency in $MgSO_4$, the first alteration in brine composition is caused by the formation of glauberite. The latter because of its low importance can be neglected. The change can be described by the ray from the SO_4 corner through the seawater point.

If the initial solution within the shaded area is known or assumed, then the salt succession which develops from it can be found either graphically or by calculation. In every case polyhalite precipitation proceeds at the expense of the kainite layer. This is the most important qualitative result: the thickness of the kainite layer is reduced. The case where the kainite layer is totally missing is even possible, but only above 35° C. In such a hypothetical, extreme example the following "primary" sequence occurs: anhydritic halite, polyhalitic halite, kieseritic halite or halite-kieserite, halitic kieserite-carnallite. The polyhalite content in this extreme event would exceed 10%, with the polyhalite region comparable in thickness with the other layers.

In general, however, the kainite layer will not be completely lacking, but will occur, reduced in thickness, above the polyhalite layer. In addition, in instances of partial $MgSO_4$ deficiency, that is to say from solutions close in composition to II in Fig. 28, followed by polyhalite formation caused by the influx of $CaSO_4$-bearing solution, carnallite could precipitate directly above the polyhalite region. There appears as yet to be no known example of this special case.

Qualitatively it is understandable that large amounts of polyhalite can be formed by the influx of $CaSO_4$-rich solutions into brines already saturated with respect to polyhalite or oversaturated in $MgSO_4$ or K_2SO_4 (in contrast to polyhalite as the solid phase). The influx of fresh brine or seawater concentrated to the point of gypsum precipitation can provide the $CaSO_4$ needed for polyhalite formation; for example, from $8^1/_2$ m seawater (cf. p. 91) 3 mm of anhydrite could be formed. If there were complete reaction, 14 mm of polyhalite could result from the anhydrite. The amount of polyhalite formed is controlled only by the excess of $MgSO_4$ and K_2SO_4 present in pre-existing brine. Since the later influxing, less concentrated brines overlie the denser, more concentrated brines in all known oceanographic occurrences, gypsum can be precipitated by evaporation of the surface layers and later react with the underlying solution.

Even when the precipitation occurs directly from a polyhalite saturated solution, it cannot be assumed that stable equilibrium exists at the time of primary precipitation. Rather, as indicated on p. 91, stable paragenesis is attained only by (early) diagenetic alteration of the metastable primary precipitate, in this case, gypsum. According to petrographic observations (see p. 161) anhydrite occurs as an intermediate phase in polyhalitization.

In addition to the changes in solution composition treated here, there are other quite different changes which result from the addition of water from continental areas. These influxes may have quite different compositions, usually characterized by a higher SO_4 content (as for example in the Caspian Sea). Their effects are easily identified in salt sequences through the interbedding of appreciable amounts of bloedite, glauberite or even thenardite or other sodium sulphate or carbonate minerals. These, however, lie outside of the scope of the problems considered here. It is nevertheless clear that from the composition of the solution can be deduced the salt sequence itself, insofar as the associated salts named were the result of primary precipitation.

3. *Primary Precipitation with a Temperature Gradient*

The effects of temperature variations on primary precipitation from $MgSO_4$-free, altered seawater can be calculated. The composition of such a brine at various concentration stages can be read from the dotted line in Fig. 9. The origin of this line corresponds to the $MgCl_2$ and K_2Cl_2 content at the beginning of NaCl precipitation. KCl-saturation is reached at 20° C with 55 mol $MgCl_2$/1000 mol H_2O. From Fig. 9 the influence of temperature variation can be quantitatively deduced. According to Fig. 10, the temperature coefficient of KCl solubility at 20° C and 55 mol $MgCl_2$/1000 mol H_2O amounts to 0.11_3 g KCl/ 100 g H_2O per degree. The appropriate saturated solution contains, by calculation from the salt concentrations in Fig. 9, in units of g/100 g H_2O (for factors see note 11, p. 28, 42.04 g salt + 100 g water; its density according to the determinations of FEIT and PRZIBYLLA (in BOEKE, 1910c) is 1.255 gcm^{-3}, so that the 100 g H_2O is contained in $\frac{142\text{ g}}{1.255\text{ gcm}^{-3}} = 113\text{ cm}^3$ solution. In this particular example, the temperature coefficient is actually $\frac{0.1\text{ g KCl}}{100\text{ cm}^3\text{ solution per degree}}$. As the precipitated sylvite has a density of 2, from a 10 m deep saturated solution cooling from 20–19° C, a uniform sylvite layer 5 mm thick results.

In a similar way the NaCl precipitated by cooling can be found from Fig. 11. From the same calculated 10 m brine concentrate cooling from 20 to 19° C, 0.28 g NaCl/cm^2 would be precipitated. The total precipitate obtained by cooling consequently consists of:

$$\frac{6.3\text{ mm salt }(78\%\text{ KCl} + 22\%\text{ NaCl})}{10\text{ m deep solution} \cdot 1\text{ degree cooling}}.$$

The result is practically independent of the temperature of evaporation as check calculations at various temperatures show.

The NaCl precipitated by cooling before saturation in KCl is reached can be determined with the help of Fig. 12, but as the curves are interpolated the value is only approximate. Both the composition and density of the brine itself from the beginning of NaCl precipitation up to KCl saturation must also be found by interpolation (cf. Fig. 36). These interpolations can be made with reasonable accuracy in $MgSO_4$-free systems. The total salt content of the solution rises from 35 g/100 g H_2O at the beginning of halite precipitation to more than 40 g/100 g H_2O at the beginning of KCl precipitation (see above). However, the volume of the brine containing 100 g H_2O scarcely alters because of the concomitant increase in density. At 20° C the temperature coefficient of NaCl-solubility is constant over the whole interval of NaCl precipitation (Fig. 12). It is approximately:

$$\frac{0.024\text{ g NaCl}}{100\text{ g H}_2\text{O per degree}} = \frac{0.0215\text{ g NaCl}}{100\text{ cm}^3\text{ solution per degree}}$$

or calculated in terms of salt thickness

$$\frac{1 \text{ mm NaCl}}{10 \text{ m deep solution} \cdot 1 \text{ degree cooling}}.$$

At higher temperatures the precipitate thickness is correspondingly greater since the cooling effect increases with mounting temperature.

The cooling effect can also be determined for carnallite. In Fig. 9 the line must be drawn from the carnallite point (83.33 mol K_2Cl_2, 166.7 mol $MgCl_2$, 1000 mol H_2O) through the point representing the composition of the initial solution. The solution which results from the precipitation of carnallite has a composition which lies at the intersection of this line with the lower isotherm. At the boundary between the carnallite and sylvite fields two possible cases may be distinguished:

a) No sylvite is present. Then a simple calculation similar to that for sylvite gives the precipitation (reversible!) on cooling at 20° C

$$\frac{2 \text{ cm salt } (98.7\% \text{ carnallite} + 1.3\% \text{ NaCl})}{10 \text{ m deep solution} \cdot 1 \text{ degree cooling}}.$$

b) Some sylvite is present. In such an event, the brine composition cannot be far from the curve $E-E$, and it will react with the sylvite present with the precipitation of carnallite. This is the most important possibility for the secondary carnallitization of sylvite (cf. p. 128).

These examples should underline the importance of temperature variations. They are also probably important in causing the precipitation of salts from the warmer surface water layer which, having reached saturation through evaporation, as a result of ever-present water circulation, subsequently sinks to the cooler, deeper parts of the basin (Figs. 46, 47). Those salts with a positive temperature coefficient of solubility, for example, the chlorides in seawater, precipitate out on cooling. As the current is usually directed towards the deeper parts of the basin, this mechanism can explain thickness differences in the primary chloride precipitates within the salt basin. It must be emphasized that the cooling effect is reversible. The salts precipitated can only persist if, on reheating, water is immediately evaporated.

The sinking of the warmer surface waters into the deeper and cooler parts of the basin is based upon the well-known densities of KCl and NaCl saturated solutions at varying $MgCl_2$ concentrations (measurements of Feit and Przibylla in Boeke, 1910, p. 300) and hence in the presently considered case. From these data it is clear that at higher temperatures KCl and NaCl saturated solutions (with the same $MgCl_2$ content) always have higher densities. For example, a solution derived from $MgSO_4$-free seawater upon reaching saturation in KCl and NaCl at 40° C (62.5 mol $MgCl_2$/1000 H_2O) has a density of 1.268_5, but at 25° C it is 1.265 g · cm^3. Thus the warmer brine will sink and cool by heat energy exchange with the lower layers with the contemporaneous precipitation of some KCl and NaCl (Figs. 10, 11). The same holds true at other temperatures and other temperature differences, although before KCl saturation is reached only NaCl will precipitate (Fig. 12). The important point is that temperature differences between the surface and lower water layers bring about circulation.

No quantitative data have yet been obtained for the $MgSO_4$-bearing system. Certainly similar conditions to the $MgSO_4$-free system must exist during the precipitation of carnallite or NaCl, however, the behaviour of the Mg sulphates and double salts is different and not yet well known.

4. *The Effect of Pressure*

The cases considered here no longer belong to primary precipitation but to diagenetic alteration which leads into metamorphism.

Under hydrostatic conditions the pressure on all phases is the same and equilibrium is scarcely affected. The solubility of NaCl is only slightly increased under pressure (p. 44). In natural salt deposits the pressure corresponding to maximum solubility will never be reached (because before this the salts would because of their buoyancy flow plastically out of the deeper crustal regions). When, however, there is a sudden pressure release in closed salt beds overlain by a more or less gas-tight beds (so that there can be no further evaporation), as may happen through the sudden opening of fissures, pore solutions from the adjacent rocks can penetrate and deposit some halite in the tension fractures. The amount of this salt is negligible, e.g. for a pressure decrease from 100 to 1 atmosphere at 25° C, only 0.74 g NaCl will be deposited from 1000 g of solution. Additional precipitation caused by adiabatic cooling can be neglected in the case where the pore solutions remain in an enclosed space, for the cooling effect is reversible. For all practical purposes, then, the effects of pressure decrease can be neglected.

Usually the pore solution pressure is less than the pressure on the solid phase. This is the case of differential pressure which was already considered in the discussion of the pressure dependence of the gypsum anhydrite transition. For diagenetic recrystallization, the reduction of pore spaces and the infilling of cavities, this type of pressure difference is not very important. Much more important is the inhomogeneity of pressure distribution on the solid phases. It is found not only experimentally (particularly RUSSELL, 1953; CORRENS and STEINBORN, 1939) but also petrographically (although in sandstones: HEALD, 1956; v. ENGELHARDT, 1960, p. 21–23) and thermodynamically (especially in MACDONALD, 1957) that crystals have greater solubilities at points of greatest stress. Such stresses occur at the points of contact of crystals, be they angular, round or any other shape. Solution in the pore fluids (which are saturated under hydrostatic pressure) occurs at these points. The resultant oversaturation leads to precipitation of the solid phase at points of minimal pressure whether on free crystals growing in cavities or at points of minimum pressure on the stressed crystals. By this mechanism, with a given amount of solution, and without temperature variations or evaporation, continuous recrystallization can occur. The process is only ended by the equalization of the pressure differences and the minimum deformation energy is distributed in the rock. This type of recrystallization leads to a flattening of the crystals perpendicular to the principal stress axis, and to an intertonguing but not to a reorientation of the crystals, i.e. no structural realignment. Such a structural realignment can occur only in anisotropic crystals by preferential growth of those favorably oriented with respect to the maximum stress directions.

It is worth remarking that this process of pressure solution according to TURNER and VERHOOGEN (1960 p. 476) is not identical with Riecke's principle, since under homogeneous stress conditions no differential solubility is found at different parts of the crystal.

A special case, commonly observed in salt rocks and Salzton is the fibrous crystal growth in joints or fissures. Probably here too there is an influx of material from inhomogeneously stressed crystals in the adjacent rock. The growth process according to MÜGGE (1928) is mainly influenced by the velocity with which the fissure opens and the velocity of crystal formation.

More research still needs to be done on the details of growth and the mechanism of solution under non-hydrostatic conditions. Further, a distinction must be made between plastic flow and the structural alterations brought about by solution and reprecipitation. As indicated on p. 41 the plastic deformation of salt begins at relatively low pressures, so that in many instances both effects may be operative. This means that after the diagenetic closing of pore spaces only small pressure differences can exist between pore solutions and the enclosing salts (KÜHN, 1955a, BAAR, 1958), and that pore solutions can only enter newly developed, tectonic openings to a limited extent. Finally, in all the processes in which solutions participate, the forces of cohesion must be considered. With the decrease of pore space these forces become increasingly important (v. ENGELHARDT, 1960).

II. Thermal Metamorphism

1. Generalities and Calculation

The relatively restricted stability field of the hydrated salt minerals means that thermal metamorphic changes of primary precipitates can occur at relatively low temperatures (below 100° C). Such changes had long been considered, see, for example, ARRHENIUS (1912), and especially RINNE (1913, 1920) and others, yet it was JÄNECKE (1923 and older works) who first deduced the physico-chemical basis of the thermal metamorphism of salt hydrates.

Discussion can be restricted to salt parageneses in the presence of an excess of NaCl. It will be further assumed that the salts are dry (see p. 117) with an impervious overlying layer so that no water can be lost by evaporation. Only stable equilibria are considered; because of the short duration of metastable equilibria over geological time, the latter can be safely excluded (BAAR, 1960) in the highly reactive salt parageneses. In addition only stable primary precipitation is taken into account since the metastable are changed to stable at a very early stage (p. 92). The negligible pressure dependence of the transition points (pp. 44, 129) will not be considered, although this may cause temperature errors of a few degrees (less than 10° C). Over the operative pressure range no vapor phase occurs so the transition points are still univariant (p. 28, 29). With these restrictions, thermal metamorphism occurs at the transition points. From the phase rule the number of participating phases can be deduced, although it does not give how many and which phases are newly formed or disappear. In the NaCl-saturated five-component system there are five solid phases, one of which is halite, at the transition points, as well as the equilibrium solution. In the case of the kainite transition point there are, for example the three following cases:

Nature of transition point	Disappearing phases	Newly formed phases	Example (cf. Table 13)
Upper temperature of formation	1 (+NaCl)	3	83° kainite
Alteration of paragenesis	2 (+NaCl)	2	72°
Lower temperature of formation	3 (+NaCl)	1	11° kainite

The scheme above is generally valid only in relation to the sum of the phases; in particular cases at the upper temperature of formation two phases may disappear and two new phases appear and so on. In the quaternary partial system without K, where there are three solid phases in addition to NaCl, two phases may disappear at the upper temperature limit for the formation of hexahydrite, whilst only one new phase appears (see Table 13).

Calculation

In the schematic simplification it is common practice to represent the possible reactions in the form of simple equations. This procedure is good for teaching purposes as long as it uses only the compositions belonging to the relevant transition points. It is possible, for example, to represent the kainite transition at 72° C (in an NaCl-free system at 76° C) by the following equation (ESKOLA in BARTH, CORRENS, ESKOLA, 1939, p. 332; with the new kainite formula):

$$2 \text{ kainite} + \text{carnallite} \rightleftharpoons 2 \text{ kieserite} + 3 \text{ sylvite} + \text{solution}$$

$$2(KCl \cdot MgSO_4 \cdot 2.75\, H_2O) + KCl \cdot MgCl_2 \cdot 6\, H_2O \rightleftharpoons 2\,(MgSO_4 \cdot H_2O) + 3\, KCl + MgCl_2 + 9.5\, H_2O\,.$$

But the formula for kainite alteration at 83° C according to ESKOLA (p. 322)

$$2 \text{ kainite} + \text{kieserite} \rightleftharpoons \text{langbeinite} + \text{solution}$$

$$2(KCl \cdot MgSO_4 \cdot 2.75\, H_2O) + MgSO_4 \cdot H_2O \rightleftharpoons K_2SO_4 \cdot 2\, MgSO_4 + MgCl_2 + 6.5\, H_2O$$

is wrong even qualitatively and in contradiction to the phase rule, as it does not contain all the phases belonging to the transition point. It would be better described as below:

$$4 \text{ kainite} \rightleftharpoons \text{langbeinite} + \text{kieserite} + 2 \text{ sylvite} + \text{solution}$$

$$4(KCl \cdot MgSO_4 \cdot 2.75\, H_2O) \rightleftharpoons K_2SO_4 \cdot 2\, MgSO_4 + MgSO_4 \cdot H_2O + 2\, KCl + MgCl_2 + 10\, H_2O\,.$$

Although these and similar formulae are stoichiometrically correct, they have only a qualitative value because the solution from which they form in reality contains the other components. Worse still, they are not unique, since there are several alternative modes of formulation according to the solution assumed. A qualitatively and quantitatively correct and unique formulation can only be obtained by introducing the composition of the equilibrium solution at the transition point. Only then is the system univariant (neglecting pressure effects), in other words uniquely defined with respect to the relative amounts of the participating phases. In this case there can be no simple numerical ratio of phases, so that no readily understood equation for the reaction is to be expected.

In the following only the cases of practical interest will be considered with contemporaneous NaCl saturation.

JÄNECKE (1923 and earlier) considered thermal metamorphism graphically. In the five-component system this involves a rather complex spatial construction. For this reason calculation using the determinant method described on p. 45 was preferred, especially as it can be applied directly to moles or grams. The result is more exact than that obtained graphically and much easier to test. As a single example, the equation for 72° C will be demonstrated. The solid phases participating in the reaction at the transition point are obtained from Figs. 19, 20, the composition of the equilibrium solution from D'ANS' tables (1933) or from the newer solubility data of AUTHENRIETH and co-workers (1952–1960). In this example the solution data for solution R_{72} was taken from AUTHENRIETH (1955). Since there are five unknowns in the equation, all the components in the quinary system must be known.

In principle, it is unimportant in what sequence the phases are ordered or which phases are brought to the right-hand side of the equation (or to the left) for the calculation. In the result the sign indicates which phase was used up or newly formed. The result is not affected by the order of the lines and columns. The following equation[28] can be so used:

1 kainite		72° C	x carn. +	y kieser. +	z sylvite +	u NaCl +	v sol. R_{72}
$MgCl_2$	0		1	0	0	0	83
$MgSO_4$	1		0	1	0	0	3
K_2Cl_2	0.5		0.5	0	0.5	0	10
Na_2Cl_2	0		0	0	0	0.5	3.5
H_2O	2.75		6	1	0	0	1000

from which $x=-0.2911$; $y=0.9895$; $z=1.2209$; $u=-0.02455$; $v=0.003507$ (solution R_{72}), that is to say, the carnallite is used up in the reaction, also some halite, although it only goes back into

[28] The order used follows the most practical and simplest method of calculation. Most of the solutions given in Tables 13, 14, and 16 were obtained using a standard program on an IBM 650 which allowed diagonally only components that differed from zero. For an expansion of the standard program thanks are due to Professor BAUR. In calculations without a computer, equations were usually solved by lower determinants, with the equilibrium solution brought to the right hand side of the equation. This was done particularly with simple combinations, that is, when in two or more columns there is only one coefficient not equal to zero.

solution, and kieserite, sylvite and equilibrium solution R_{72} are formed. The transition point corresponds to a change in the paragenesis.

The x, y, z etc. are not molar proportions so that their sum never equals unity. On the other hand, the components are in molar proportions. Thus, introducing the values found for x, y, z, ... on each side of the equation gives the same values in each row. It is essential to make this check.

2. *Examples*

Table 13 contains the results for some transition points after correction for density and normalization (sum of the reacting components = 100 g). The first four relate to the quaternary limiting system Na_2SO_4–$MgCl_2$–NaCl–H_2O [28a]. Due to overlapping of the polytherms, the metamorphic products of this system are quite different from those

Table 13. *Thermal metamorphism of salt hydrates (in grams: sum of each side of the equation = 100 g). Abbreviations from Table 4. Solutions in parentheses from literatures sources*

Temperature °C	Phases disappearing	New phases formed
30.2	94.3 e + 5.7 n	64.8 hx + 14.3 bl + 20.9 ($I_{30.2}$)
37.6	85.3 hx + 14.7 n	37.3 loe + 13.2 ks + 49.5 ($K_{37.6}$)
40.3	44.1 hx + 46.7 bl + 92 n	67.7 loe + 32.3 ($I_{40.3}$)
61.4	96.0 bl + 4.0 n	66.7 loe + 16.8 vh + 16.5 ($I_{61.4}$)
11	31.2 c + 58.3 e + 9.9 sy + 0.60 n	55.7 k + 44.3 (Q_{11})
72	74.8 k + 24.8 c + 0.4 n	41.9 ks + 27.8 sy + 30.3 (R_{72})
83	99.5 k + 0.5 n	32.0 lg + 17.2 sy + 23.9 ks + 26.9 (R_{83})
167.5	100 c	20.1 sy + 79.9 ($H_{167.5}$)

calculated by JÄNECKE. A possible primary paragenesis epsomite-NaCl at 30.2° C gives bloedite-hexahydrite-halite, but this is scarcely likely to occur in salt deposits because of $MgSO_4$ deficiency. It changes at 40.3° C to loeweite-halite. Only the excess hexahydrite alters to kieserite-loeweite-halite. The resultant rock is thus a loeweite-halite with accessory kieserite, not a pure kieserite-halite.

[28a] For loeweite and kainite in Table 13 the new formulae of KÜHN and RITTER (1958) were used.

Unfortunately the alteration of the primary hexahydrite-kainite cannot be calculated, as the older polytherms are incorrect (see p. 73) and the extrapolated polytherms are far too uncertain for quantitative calculations. Qualitatively from the extrapolated isotherms in Fig. 20 the change from hexahydrite to loeweite + kieserite occurs at about 35° C, whilst kainite remains unaltered. The paragenesis loeweite–kieserite–kainite–NaCl is stable up to about 54° C, at which temperature loeweite + kieserite + NaCl + langbeinite forms (if the langbeinite-kieserite boundary is correctly extrapolated, see p. 74, 75). The primary precipitate of halite-epsomite-kainite (see Fig. 26) changes at about 27° C to halite-bloedite-hexahydrite-kainite. At about 37° C the hexahydrite alters to loeweite and at 47° C the kainite + bloedite goes to langbeinite + loeweite.

It is still uncertain how important the incongruent alteration of carnallite at 167° C is in the thermal metamorphism of salt deposits (Table 13). The resultant, and poorly known, solution H consists of 8.4% KCl; 42.9% $MgCl_2$; 48.7% H_2O (according to VAN'T HOFF from D'ANS, 1933, p. 83). The upper temperature limit of formation of carnallite, 152° C (D'ANS, 1933, p. 78, 1961) probably does not have to be considered in thermal metamorphic changes. The resultant equilibrium solution at this temperature, J, contains more $MgCl_2$ than the carnallite. In the absence of bischofite it cannot be formed from the alteration of carnallite since, besides separation of KCl, additional $MgCl_2$ or additional evaporation is required. In reality the thermal alteration of carnallite may be much more complex, there is possibly a metastable intermediate stage (D'ANS, 1961) with the formation of $KMgCl_3 \cdot 2\,H_2O$ (TITTEL, 1958). In the temperature range above 100° C there is still insufficient experimental detail, and the changes given in Table 13 can only be regarded as a crude approximation.

The decomposition of bischofite at 116° C, treated in detail by JÄNECKE (1923 and others), will not be examined in detail here. Should bischofite play a major role as a primary precipitate (see p. 42), which was certainly not the case in the German Zechstein then its importance in deeply buried salt beds is as a source of $MgCl_2$-metamorphic solutions which react with kainite, langbeinite, etc. with the formation of kieserite-sylvite or of carnallite (from sylvite). This secondary enrichment will be described on p. 124 using as an example the influence of solution Q which has similar effects. Secondary carnallitization in such cases will be described on p. 128. Carnallite also reacts with solutions resulting from the decomposition of bischofite to form a thin sylvite reaction rim, both polythermally and isothermally. Since it is of little importance the need to give an example does not arise.

3. *Kainite Alteration at 72° and 83° C*

Of particular interest here are the two transition points at 72° and 83° C. The old polytherms are probably correct so that quantitative results are possible (Table 13). The alterations at 72° have already been described. At 83° the upper stability limit of kainite is reached. It forms predominantly langbeinite besides some sylvite and kieserite and an appreciable quantity of liquid.

When some of the reaction products (see Table 13) are present before the alteration of kainite, these are preserved. Of the four phases in the right side (column Table 13), only three can be present with kainite in stable equilibrium, thus

k + lg + ks + sy

or

k + lg + ks + solution $R_{83}(-R_{72})$ and other combinations.

All thermal metamorphic changes are reversible so long as the solution produced is not removed (squeezed out). The changes move from left to right upon heating. The solution which results can alter

1. by changes in temperature
2. when it encounters salts which are products of a different evaporation stage (D'ANS and KÜHN, 1960).

This second case is especially important in the thermal metamorphism of kainite, as a carnallite rock occurs above possible primary kainite if reaction with solution *R* does not occur. This complication will be treated in the section on thermal metamorphism.

4. *Reactions in the Solid State*

Thermal metamorphic changes are most easily understood with the help of solution equilibria. About the kinetics of the process still nothing is known, and this has led to doubts about the process (LEONHARDT and BERDESINSKI, 1949/50). The counter arguments, however, only show that there is an appreciable retardation in the establishment of equilibrium, as, for example, in occurrence of kainite in a metastable region. As in nature the condition of a dry salt mixture does not occur (cf. KÜHN, 1952), it can be assumed that during metamorphism the whole would adjust to an equilibrium state in the course of time. Solutions already present, for example solution inclusions, favor the reactions. True they alter the mass balance slightly, but do not, however, alter the equilibrium position. With salt deposits, significant solid state reactions can certainly be excluded since solutions always form under thermal metamorphic conditions. In the laboratory experiments of LEONHARDT and BERDESINSKI (1949/50) on the other hand, salts were mixed which do not occur in a stable paragenesis so that the formation of a stable compound on the grain boundaries would be expected here. However, it is impossible that, when there are layers of salts belonging to different parageneses, growth should continue after the formation of a thin boundary layer of the stable compound (reaction rim) if there is no solution present.

5. Geothermal Metamorphism as Defined by Borchert[a]

BORCHERT (1959, p. 55) observed that upon heating a mixture composed of damp layers of epsomite and halite the chief alteration products at 80° C were loeweite and halite. This could not be explained by the thermal metamorphism as described by JÄNECKE (1923) according to whom kieserite and halite should develop. JÄNECKE'S statement, however, was based upon incorrect older polytherms and the new polythermal data give as main products loeweite-halite in this example. The data in Table 13 show that there exists at 15° C a possible halite–epsomite primary paragenesis (Fig. 26), which at 30° gives a bloeditic halite-hexahydrite. Upon further heating to 40.3° C (with prior removal of the solution formed) a kieseritic halite-loeweite is formed as a result of a two part reaction: 1. the alteration of bloedite + hexahydrite; 2. the alteration of the excess hexahydrite to loeweite and kieserite. The second part of the reaction can, to a good approximation, be calculated from the alteration at 37.6° C for alteration is already beginning at this temperature and because the equilibrium solution $K_{40.3}$ compares closely with equilibrium solution $K_{37.6}$. The complete reaction is shown in Fig. 30 in terms of the initial thickness given in Fig. 26. It can also be seen by comparison with Fig. 26 that, as a result of thermal metamorphism, a layer B develops out of layer B_1 (15° C) which is similar to the primary precipitate at higher temperatures.

BORCHERT concluded from this experiment that a stable paragenesis was not reached but rather an intermediate step. This raises the question of the kinetics of the alteration, a question which up to the present time still cannot be completely answered for salt metamorphism. Yet it can be shown by quantitative calculation that BORCHERT'S "geothermal metamorphic (Stufenmetamorphose) hypothesis", which was the first attempt to solve this question, is not correct.

[a] See footnote 12, p. 32.

According to BORCHERT (1940, 1959), the first metamorphic product must be that salt in whose stability field the salt point of the initial mineral lies. In the case of bloedite this is the field of thenardite (cf. Fig. 15, 16). The solution formed simultaneously should, in the continuous presence of excess NaCl, lead to the formation of additional thenardite with the consequent enrichment of the residual liquor in

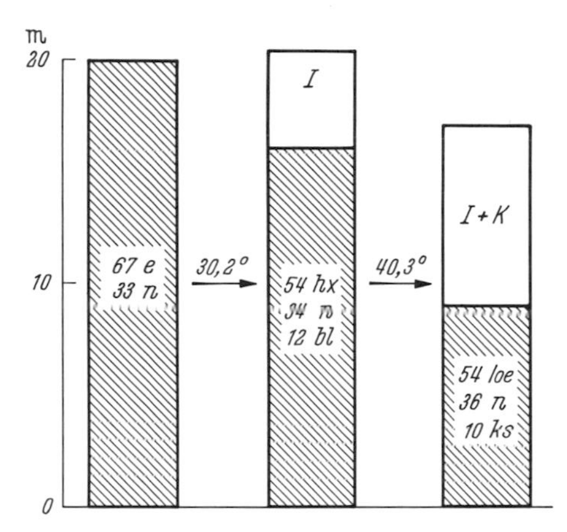

Fig. 30. Theoretical thermal metamorphism of epsomite-halite (layer B_1 at 15° C, Fig. 26). Composition in weight %, abbreviations of solid phases as in Table 4, solutions as in Table 5

$MgCl_2$. Upon reaching the margin of the stability field, the thenardite formed first reacts with the solution to give vanthoffite (according to the newer polytherms of D'ANS, cf. Fig. 15), as was shown on p. 93, 94 for the reaction of the primary precipitate with constant solution. At each phase boundary the composition of the solution remains constant whilst within its stability field it alters rapidly. BORCHERT terms this "dynamische Laugenentwicklung", the pause at the phase boundary being the "step". The development may proceed further in the direction of solution Z.

As BORCHERT (1940, p. 66) excludes the congruent melting of salt hydrates, it is impossible to understand why the first metamorphic product to occur should be the salt in whose stability field salt point of the decomposing salt hydrate lies. For only in congruent melting can the salt point and melt occur at the same point on the diagram. Furthermore, it is not possible through a pure melting process without the addition of water to derive an equilibrium solution of the five-component system with the composition of the melting hydrate. The salt points of the salt hydrates lie below the saturation plane because of their much lower water content (Fig. 31). They fall in the stability fields only as a result of projection (p. 32). Even if intermediate steps occur in the reaction, of which there is still no proof, they cannot be described with the help of solution equilibria, as the solutions of the intermediate stages would also contain less water than the equilibrium solutions.

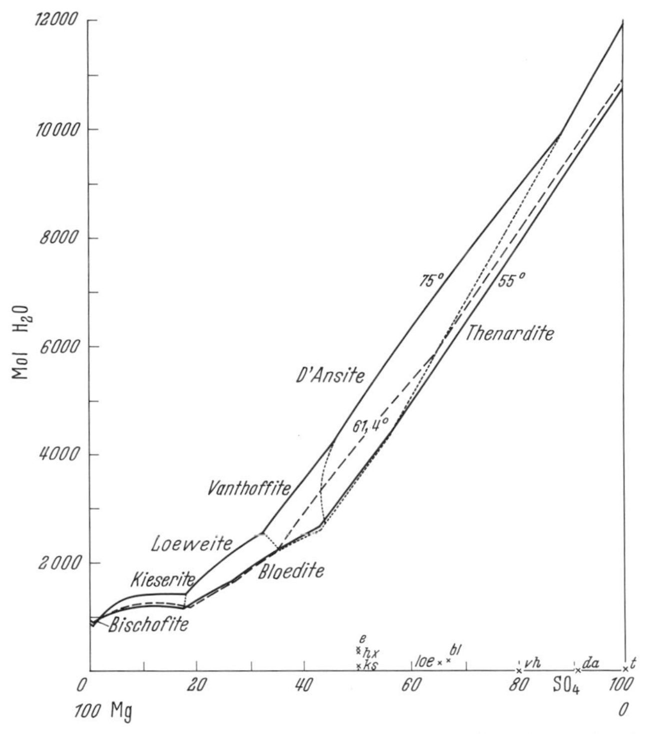

Fig. 31. Water content of saturated solutions and solid phases in the system Na^+, $Mg^{++}//Cl^-$, $SO_4^{--} + H_2O$ at different temperatures. Ionic percentages after Jänecke, calculated from Autenrieth and Braune (1960). The dotted curves delimits the stability fields of the various solid phases of Fig. 15

Should the intermediate steps occur, they cannot in this case be compared with the reaction in constant solutions (p. 93). This reaction is only possible if water can be evaporated. Consider as an example solution I_{25} in which the reaction is from bloedite to epsomite (Fig. 15) which by calculation (determinant method) gives the following results:

$$1\,[I_{25}] + 44.5\ \text{mol bl} \rightleftharpoons 105.5\ \text{mol e} + 112.6\ \text{mol n} + 439.5\ \text{mol}\ H_2O\,.$$

In this case solution I_{25} is completely used up with the precipitation of epsomite, halite and water without changing its composition. If less bloedite is added, then correspondingly less solution is used, that is to say, the amount of solution used and the amount of water evaporated are proportional to the amount of bloedite reacting. When no water can evaporated, the bloedite will remain, as it is stable in solution I_{25}. The same is true by analogy for all other alterations in constant solutions at reaction boundaries. From this it follows that it is impossible to derive a $MgCl_2$ concentration higher than the equilibrium solution at the transition point by incongruent melting processes.

In every chemical reaction, the left and right sides of the equation must balance, even if the details of the reaction process in some of the equations are not known exactly. In BORCHERT's geothermal metamorphism the two sides of the equation do not agree, as the amounts of water on each side are different.

The application of solution equilibria to thermal metamorphism is only possible in the manner introduced by JÄNECKE (1923 and earlier). Spontaneous nucleation of the newly formed phases, perhaps already in the decomposing salt hydrate, must be assumed. The water of crystallization released becomes saturated in the ions present so that the equilibrium solution at the transition point and the new solid phases develop at once. Therefore there must be incongruent melting. The solution formed in this case is much more dilute than that obtained from melting salt hydrates. However, intensive experimental and theoretical research is necessary into thermal metamorphism, for an explanation of the actual reaction process is urgently required. With the formation or disappearance of two or three solid phases, a delay in the establishment of equilibrium or perhaps even intermediate stages have to be considered, but these cannot be predicted. Presumably these alterations are only possible in existing solution residuals. Since the liquid inclusions present adjust to equilibrium with the solid phases present with changing temperature, when the transition temperature is reached, they will already have the composition of the equilibrium solution. This naturally strongly favors the initiation of the reaction.

Apart from aiding alterations with changing temperatures, the solutions present influence the quantitative composition of the solid phases. Because the composition of the solution compared with the solid phase is temperature-dependent, some phases will dissolve and others will form. The preference for transition points in the calculation is only justified by the fact that at such points large amounts of solution can develop. In general it can be assumed that this will only be of limited duration in salt rocks. As soon as the liquid is squeezed out, the alterations and adjustments of the solid phases are drastically reduced.

All thermal metamorphic changes are reversible so long as the liquid remains. When it is squeezed out, high-temperature parageneses remain on cooling. For changes in the reverse direction, solution is just as necessary as the solid phase.

III. Solution Metamorphism

1. Isothermal Solution Metamorphism

On account of the great solubility of salt minerals, salt parageneses are very sensitive to subsequent penetrating unsaturated solutions. Very important in this context is the origin of incongruent saturated solutions brought about by the precipitation of new solid phases. The results depend as much on the temperature as on the composition of the percolating solutions and the existing salts. If stable equilibrium can be assumed, the changes can easily be calculated from the solution equilibria, as first D'ANS and KÜHN (1940, p. 79) and later D'ANS (1947) have shown. To a great extent the processes have been proved and explained by BAAR (1952) in detailed calculation. In a qualitative form, models were presented a long time ago by TSCHERMAK (1871), NAUMANN (1913) and RÓZSA (1915, and others). The dissolving of salts in various solutions was considered theoretically by VAN'T HOFF (1905, p. 81) and in greater detail by JÄNECKE (1923, p. 31). BORCHERT (1959) termed solution metamorphism "retrograde metamorphism" (Rückläufige Stufenmetamorphose). As it is quite distinct from thermal metamorphism and as the geothermal metamorphism hypothesis is unsound, the term will not be used.

a) Incongruent Carnallite Alteration

In a theoretical discussion it is only necessary to consider the most important limiting cases. The calculation here too is particularly clear when made according to the method of determinants. It must be applied to the solid phases developing from the constant solution formed and to the composition of the percolating solution. The equilibrium solution formed is placed on the right-hand side of the equation. Negative signs indicate that that the solid phases on the left-hand side disappear. That is to say, they reappear in a new form on the right-hand side together with the equilibrium solution. Consider as an example the equation for a halitic kieserite-carnallite percolated by a saturated solution of NaCl at 83°. A general formulation is obtained by taking 1 mol of the solid phase and the composition of the solution normalized to 1000 mol H_2O. The resulting equilibrium solution is solution Q_{83}, for the dissolving of carnallite ceases when the carnallite-saturated solution reaches its minimum $MgCl_2$ concentration. The relationship can be expressed as follows:

	x carn. +	y kieserite +	z sylvite +	u NaCl +	v NaCl sol. →	sol. Q_{83}
$MgCl_2$	1	0	0	0	0	87.4
$MgSO_4$	0	1	0	0	0	2.6
K_2Cl_2	0.5	0	0.5	0	0	9.9
Na_2Cl_2	0	0	0	0.5	58.9	3.8
H_2O	6	1	0	0	1000	1000

The result is

$x = +87.4, y = +2.6, z = -67.6, u = -48.12, v = +0.473$

(check as p. 110, 111),

or, calculated in grams normalized to 100 g carnallite:

$$100 \text{ g c} + 1.48 \text{ g ks} + 48.5 \text{ g NaCl solution} \xrightarrow{83^\circ}$$
$$\xrightarrow{83^\circ} 20.75 \text{ g sy} + 11.58 \text{ g n} + 117.65 \text{ g solution } Q_{83}\,.$$

The excess kieserite and the halite in the initial rock remain. The halite in the equation is precipitated (salted out) of the percolating NaCl solution. In order to apply the reaction to particular examples, the results must be calculated from the actual carnallite content. In the case of the Stassfurt carnallite (from Table 24, col. I–O), one obtains:

$$52.1 \text{ g c} + 0.77 \text{ g ks} + 25.26 \text{ g NaCl solution} \rightarrow$$
$$10.8 \text{ g sy} + 6.03 \text{ n} + 61.3 \text{ g solution } Q_{83}\,.$$

There were, however, originally present:

$$52.1 \text{ g c} + 16.3 \text{ g ks} + 28.8 \text{ g n} + 2.8 \text{ g clay and anhydrite}.$$

The calculated sylvite-kieserite-halite originating is thus

$$(16.3 - 0.77) \text{ g ks} + (28.6 + 6.03) \text{ g n} + 10.8 \text{ g sy} + 2.8 \text{ g clay and anhydrite}$$

or

29.1% kieserite, 20.2% sylvite, 45.5% halite, 5.2% clay and anhydrite.

In comparison with the observed layer (Table 24), the calculated kieserite is too great and the amount of halite estimated is too small.

The alteration of carnallite has so far been considered as a *single* process. This is something of an idealization. In fact, until saturation in sylvite is reached, carnallite dissolves congruently. Until that point is reached only NaCl will be precipitated from the infiltrating solution, so that at first it is impoverished. As soon as KCl saturation is reached, all further KCl from the alteration of carnallite is precipitated as sylvite so that at this stage more sylvite separates than would correspond to the amount calculated. Examples of this kind have already been calculated by KÜHN (1955) and BRAITSCH (1960).

Because the actual geological process is unknown, only the broad principles will be explained without resort to particular examples. During the first stage of the congruent solution of carnallite, the equilibrium solution which arises corresponds to the molar ratio $KCl/MgCl_2$ in carnallite. Its composition at the transition temperature can be found from Fig. 9 where it is represented by the intersection of the isotherm with the ray inclined at 45°. The corresponding NaCl content can be obtained from D'ANS' tables. To a very good approximation it is also calculable from the 20° NaCl isotherm (Fig. 9) bearing in mind the temperature coefficient of solubility of NaCl (Fig. 11).
For example at 80° C:

$$43.6 \text{ mol } MgCl_2; \quad 21.8 \text{ mol } K_2Cl_2; \quad (14.7 + 60 \cdot 0.04) \text{ mol } Na_2Cl_2; \quad 1000 \text{ mol } H_2O\,.$$

Thus on the right-hand side of the scheme given on p. 119 the following can be inserted:

$$43.6 \text{ mol } MgCl_2; \quad 0\ MgSO_4; \quad 21.8\ K_2Cl_2; \quad 17.1\ NaCl; \quad 1000\ H_2O$$

wherc to a first approximation the $MgSO_4$ content can be taken as constant. There are no changes on the left-hand side.

In the second stage of the process, solution Q_{83} is formed while on the left-hand side the NaCl solution of the first stage is replaced by a KCl-saturated solution.

The incongruent alteration of carnallite is probably the most important process in the alteration of potash salts. For that reason it has to be investigated under a variety of conditions. The results normalized to 100 g carnallite are given in Table 14. With regard to exactness the comments on p. 85 re. Table 9 are valid. The application of the alteration equations to naturally occurring examples is explained on p. 118, 119.

In $MgSO_4$-bearing systems the kieserite-sylvite paragenesis occurs above 72° C. Below this temperature kainite forms in its place. If in this case there is more carnallite present than is required for the alteration of the total kieserite (Table 14, Nos. 6–9, 13, 14), the excess carnallite gives rise to sylvite and solution Q (Table 14, Nos. 10, 11, 15, 16). This has the result that some kainite is dissolved, as solution Q below 72° C contains more $MgSO_4$ than solution R. If more kieserite was present

Table 14. *Incongruent alteration of carnallite in various solutions (abbreviations of solid phases as in Table 14)*

No.	Temp. °C	Percolating solution g	Solid phases, disappearing g	Solid phases, appearing g	Solution produced g
1	83	827.2 R_{83}	100 c	0.8 ks + 32.3 sy + 1.1 n	893.0 Q_{83}
2		106.1 Y_{83}	100 c	4.4 ks + 23.5 sy + 5.8 n	172.4 Q_{83}
3		48.5 sat. NaCl soln.	100 c + 1.5 ks	11.6 n + 20.8 sy	117.6 Q_{83}
4		51.9 sat. NaCl soln.	100 c	12.0 n + 19.8 sy	120.1 E_{83}
5		60.4 sat. NaCl–KCl soln.	100 c	8.3 n + 31.6 sy	120.5 E_{83}
6	55	536.1 Y_{55}	100 c + 38.6 ks	103.1 k + 1.5 n	570.1 R_{55}
7		117.8 W_{55}	100 c + 44.5 ks	93.8 k + 8.6 n	159.9 R_{55}
8		69.0 sat. NaCl soln.	100 c + 43.7 ks	73.9 k + 17.3 n	121.5 R_{55}
9		50.4 water	100 c + 43.7 ks + 1.2 n	73.9 k	121.4 R_{55}
10		62.0 sat. NaCl soln.	100 c + 3.8 k	21.9 sy + 14.4 n	129.5 Q_{55}
11		45.3 water	100 c + 3.8 k + 2.3 n	21.9 sy	129.5 Q_{55}
12		62.0 sat. NaCl soln.	100 c	14.3 n + 20.9 sy	126.8 E_{55}
13	25	70.8 sat. NaCl soln.	100 c + 51.5 ks	84.0 k + 17.5 n	120.8 R_{25}
14		52.1 water	100 c + 51.5 ks + 1.2 n	84.0 k	120.8 R_{25}
15		74.4 sat. NaCl soln.	100 c + 6.6 k	24.3 sy + 17.1 n	139.6 Q_{55}
16		54.7 water	100 c + 6.6 k + 2.6 n	24.3 sy	139.6 Q_{55}
17		71.6 sat. NaCl soln.	100 c	16.2 n + 22.4 sy	133.0 E_{25}

than indicated by the equations, this will remain along with the newly formed kainite.

In the $MgSO_4$-free system sylvite forms at all temperatures (Table 14, Nos. 4, 12, 14). At lower temperatures a greater volume of solution is required for the alteration of carnallite, and the resulting sylvite-halite will be richer in KCl. This is brought about by the lower solubility of KCl at low temperatures.

Up to the present, percolating solutions saturated with NaCl, and water, have been considered (as in the formation of salt cap at low temperatures). Reactions do not however cease with the sylvite-kieserite paragenesis and, as further alterations can lead to $MgSO_4$-bearing KCl-free parageneses as will be discussed in the next section, more or less $MgSO_4$-saturated percolating solutions can develop. Of the many possibilities only a few are presented in Table 14 (1, 2, 6, 7).

It is apparent that of the many possible solutions which can result through these alterations most do not need to be considered, although example 1 in Table 14 has a certain practical importance, as will be shown on p. 125.

It is clear that with the help of the reaction equations in Table 14 the reverse process can also be performed, and from observations on a metamorphic paragenesis we can arrive at the composition of the original carnallitic rock.

b) Impoverishment

The alterations do not end with the disappearance of carnallite, since the sylvite and kainite can also be further altered by dilute solutions. In $MgSO_4$-free sylvite-halite the product of improverishment is halite. The calculation is done as previously with a KCl-NaCl saturated solution resulting. Here, too, from the NaCl-saturated infiltering solution NaCl is precipitated. Original clay (and anhydrite) impurities can be considered as constant to a first approximation, although their relative amount increases as the alteration proceeds (BRAITSCH, 1960; HOFFMANN, 1961). Some examples, normalized to 100 g of dissolved sylvite are given in Table 15 without respect to their derivation. The $CaSO_4$ present in the percolating solution will be precipitated with the alteration of carnallite. There thus develops, in addition to the anhydrite already present, a secondary anhydrite generation (see p. 187). The halite already present persists. At lower temperatures much greater amounts of NaCl solution are necessary.

Table 15. *Sylvite dissolution in NaCl solutions; $MgSO_4$-free parageneses*

Temp. °C	Percolating saturated NaCl solution g	Disappearing solid phases g	Appearing solid phases g	Resultant KCl–NaCl sat. soln. g
83°	445.4	100 sy	34.1 n	511.3
55°	574.4	100 sy	35.9 n	638.5
25°	834.2	100 sy	37.4 n	896.8

In kieserite-sylvite-halite further alteration is much more complex and very different rocks may develop according to the nature of the percolating solution and the temperature. These will not be systematically discussed here, but a few examples will be presented. A consideration of the 83° isotherm (Fig. 23) may help understanding of the process.

In the $CaSO_4$-free system the first stage in impoverishment at 83° C is the formation of langbeinite at the expense of kieserite and sylvite (Table 16). Further development depends upon whether kieserite or sylvite remains with the langbeinite. From kieserite + langbeinite, loeweite is the next to form (Table 16, No. 3). The further alterations of

Table 16. *Alteration of kieserite-sylvite-halite, langbeinite-halite and kieserite-halite by solutions at 83° C (no. 11 at 55° C) (abbreviations of solid phases after Table A)*

No.	Penetrating solution g	Solid phases disappearing g	Solid phases appearing g	Resultant solution g
1	130.4 Y_{83}	94.3 ks + 38.8 sy	100 lg + 7.6 n	155.9 R_{83}
2	55.4 NaCl solution	101.3 ks + 41.2 sy	100 lg + 13.8 n	84.0 R_{83}
3	646.3 NaCl solution	100.0 lg + 337.1 ks + 29.6 n	431.6 loe	681.5 Y_{83}
4	663.0 saturated NaCl solution	100.0 lg + 150.9 sy	75.9 gs + 60.7 n	777.2 M_{83}
5	885.6 Q_{83}	100.0 lg + 1.8 n	98.6 ks + 27.9 sy	860.8 R_{83}
6	—	100.0 lg + 99.2 c + 0.73 n	99.5 ks + 60.0 sy	40.4 R_{83}
6a	—	32.0 lg + 31.7 c + 0.23 n	31.8 ks + 19.2 sy	12.9 R_{83}
7	303.1 I_{83}	100.0 ks + 11.6 n	109.4 loe	305.3 K_{83}
8	198.1 H_{83}	100.0 ks + 6.2 n	105.4 loe	198.9 K_{83}
9	169.7 saturated NaCl solution	100.0 ks + 3.5 n	96.3 loe	176.9 K_{83}
10	178.6 saturated NaCl solution	100.0 loe + 1.1 n	76.73 vh	202.9 I_{83}
11	30.2 saturated NaCl solution	100.0 loe	108.8 bl + 3.7 n	17.7 I_{55}

langbeinite and loeweite are quite complex (see JÄNECKE[29], 1923) and although they may be calculated in the same way they are without great practical importance. As a further example the alteration of langbeinite-sylvite may be considered, which as the first step produces the glaserite-sylvite paragenesis (Table 16, No. 4).

[29] JÄNECKE'S figure (1923, p. 31) is not quite correct. At point *V* (1923, p. 34) after the resolution of all vanthoffite between points *V* and *W* with the precipitation of some loeweite, the resolution of a certain amount of glaserite was assumed. The opposite is the case. With the resolution of all vanthoffite at point *V* some excess loeweite is formed, which is redissolved between *V* and *W*. As *V* – *W* is a reaction line between the loeweite and glaserite fields, the glaserite precipitates alone as soon as the loeweite is used up.

In the presence of anhydrite in the kieserite-sylvite-halite, polyhalite occurs as the first alteration product, before even langbeinite is formed. As, however, insufficient solubility data are available, no calculation is possible.

The final result of all these and further paragenic alterations is an impoverishment consisting of a halitic rock.

The alteration of original kieserite-halite, for which examples are given in Table 16, has a certain practical significance. Each of the solutions formed by an alteration process is capable of reacting with precipitates representing a different evaporation stage, with the formation of new salts. Only a few of the possible cases have been considered. At higher and medium temperatures loeweite is the first alteration product of kieserite. From this vanhoffite develops above 61.4° C and bloedite below 61.4° C. As VAN'T HOFF (1905) has already stressed, in general, even in secondary alteration, only normal parageneses occur which can be taken from the solubility diagrams. Qualitatively these alterations can be demonstrated in Figs. 15 and 20. From this it becomes obvious that thenardite is just as unlikely to be a metamorphic product in pure oceanic salt deposits as it is to be a primary precipitate. From vanthoffite, d'ansite is the first to form and this mineral has not yet been found in the German salt deposits. Only from d'ansite or from bloedite can thenardite form through further penetration of NaCl solution.

c) Enrichment

One consequence of the great variety of solutions resulting out of solution metamorphism is that they can also lead to KCl enrichment and other alterations. The most important case extensively considered is that of KCl enrichment by the alteration of carnallite. Other examples should also be closely examined in this context. If, for example, solutions saturated in KCl–NaCl as a result of dissolving sylvite (Table 15) should encounter unaltered carnallite, the latter would also be broken down and much more sylvite would be precipitated (Table 14, No. 5). The great volume of solution Q derived from the alteration of carnallite may enrich previously existing langbeinite and other rocks (Table 16, No. 5).

Particularly interesting is the paragenesis langbeinite + kieserite + sylvite + solution R_{83} (p. 113) which results from the thermal metamorphism of kainite at 83° C. As in the absence of early diagenetic alterations (see p. 93, 94) carnallite rocks should occur above primary kainite rocks, further reactions must result from the production of solution R_{83} in which carnallite is not stable. As long as langbeinite is present, the solution composition cannot shift from point R. It can, however, dissolve carnallite and react with the langbeinite present forming kieserite + sylvite with the production of more solution R_{83} (Table 16, Nos. 6, 6a). After the resorption of langbeinite, more carnallite dissolves according to the manner of Table 14, No. 1. The net result of the thermal metamorphism of kainite with the subsequent alteration of carnallite, corrected to the initial amounts in grams assumed in Table 13, is as follows:

$$\begin{aligned} &99.5 \text{ kainite} + 36.6 \text{ carnallite} + \ \ 0.7 \text{ halite} \\ \rightleftharpoons &37.9 \text{ sylvite} \ + 55.8 \text{ kieserite} \ + 43.0 \text{ solution } Q_{83} \end{aligned}$$

or in cubic centimetres:

$$\begin{aligned} &46.3 \text{ kainite} + 22.8 \text{ carnallite} + \ \ 0.3 \text{ halite} \\ &19 \quad \text{sylvite} \ + 21.7 \text{ kieserite} \ + 35 \text{ solution } Q_{83}\,. \end{aligned}$$

The primary halite and kieserite (in kainitic and carnallitic rocks) already present will persist, and these are not included in the above enumeration.

There is only a small amount of solution. Even if the solution is not squeezed out, the volume on the right-hand side of the equation is only a little greater, while in most other solution metamorphic processes the volume effects are much larger. The volume of the newly formed phases is always much lower, so that where the squeezing out of the solution is a possibility the right-hand side of the equation is favored. As might be expected, the combination of solution and thermal metamorphism produces the same result as the thermal metamorphism of the primary kainite-carnallite paragenesis (see $MgSO_4$ deficiency, stage II, p. 98) at 72° C (see Table 13).

d) The Reactions of $CaCl_2$-bearing Solutions

Solutions containing $CaCl_2$ play an important part among the solutions circulating in rocks (v. ENGELHARDT, 1960). Their importance in metamorphism has been stressed by various authors (p. 187), and they must equally be considered here. To calculate their effect it is convenient to express the $CaCl_2$ content in terms of the reciprocal salt pair $CaCl_2 + MgSO_4 \rightleftharpoons CaSO_4 + MgCl_2$ and written in the form $CaSO_4 + MgCl_2 – MgSO_4$[30]. Infiltrating solutions containing $CaCl_2$, because of the insolubility of anhydrite will exchange with the Mg-sulphates present and precipitate anhydrite. Since naturally occurring

[30] Conversion factor: 1 g $CaCl_2 \rightarrow 1.2267\ CaSO_4 + 0.8580$ g $MgCl_2 - 1.0847$ g $MgSO_4$ (or -1.25 g $MgSO_4 \cdot H_2O$).

$CaCl_2$-bearing solutions always contain some KCl and $MgCl_2$, they cannot alter the same amount of carnallite as an NaCl–$CaCl_2$ solution. Unfortunately, insufficient solubility data are known for these complex solutions so that the NaCl saturation concentration cannot be exactly determined. In the calculation, therefore, analytically determined $CaCl_2$-containing solutions are inserted [HERRMANN (1961a) to simplify the calculation, neglected the KCl and $MgCl_2$ in the solution, and used the figure from D'ANS (1933), for the NaCl saturation concentration referred to the $CaCl_2$ present in NaCl–$CaCl_2$-bearing solutions]. The NaCl content of the solution used in the following calculations is probably too small as the solutions were analyzed at room temperature and the amount of NaCl precipitated upon cooling to room temperature is not known. However, an exact correction is not possible and none is made. The calculated NaCl in the Hartsalz (potash salt rock) formed is thus too low. Here too it is possible to calculate in a general way how much $MgSO_4$ (per 100 g of dissolved carnallite) can be converted to $CaSO_4$. As an example of an infiltrating solution the solution from the Langensalza bore (HERRMANN, 1961b) can be used. Calculated in mol/1000 mol H_2O, it has the composition given in column x, with $CaCl_2$ converted to $CaSO_4 + MgCl_2 – MgSO_4$ (see above).

The resultant Q solutions can be assumed to be $CaSO_4$-free without introducing significant errors.

	Infiltrating solution x	Kieserite y	Sylvite z	Halite u	Anhydrite v	Carnallite w	Resultant solution Q_{83}
$MgCl_2$	29.17	0	0	0	0	1	87.4
$MgSO_4$	−24.50	1	0	0	0	0	2.6
KCl	7.19	0	1	0	0	1	19.8
NaCl	54.10	0	0	1	0	0	7.6
$CaSO_4$	24.55	0	0	0	1	0	0
H_2O	1000	1	0	0	0	6	1000

The result is:

$x = 0.5568,\quad y = 16.24,\quad z = -55.36,\quad u = -22.52,\quad v = -13.67,\quad w = 71.16$

or calculated in grams and normalized to 100 g carnallite:

$$70.1\text{ g } CaCl_2 \text{ solution (with } 10.93\%\ CaCl_2) + 100\text{ g carnallite} + 11.4\text{ g kieserite} \xrightarrow{83^\circ} 20.9\text{ g sylvite} + 6.7\text{ g halite} + 9.4\text{ g anhydrite} + 133.5\text{ g } Q_{83}\,.$$

To this, on the right-hand side, must be added the amounts of halite, anhydrite, clay and excess kieserite already present in the carnallite.

The calculation based on the Langensalza solution is an example of particularly high $CaCl_2$ content. Yet the amount of sylvite formed is not very different from that resulting from the alteration with an NaCl solution (Table 14, No. 3). As instead of kieserite almost equal amounts of anhydrite are formed, the sylvite content is relatively unaffected. The amount of anhydrite formed, however, is rather high. In order to apply the alteration equations to concrete examples, the results have to be calculated from the actual amount of carnallite present (example on p. 119).

With percolating solutions having a lower $CaCl_2$ concentration, the sylvite-halite relationship is not appreciably altered, only the amounts of kieserite used and anhydrite precipitated are correspondingly lower. During formation of Hartsalz, the kieserite and anhydrite occur together. For example, from the percolating solution III from Sollstedt-Craja (17.22% NaCl; 6% KCl; 3.16% $MgCl_2$; 0.063% $CaSO_4$; 2.88% $CaCl_2$; HERRMANN, 1961b) the reaction gives:

$$57.9 \text{ g solution III} + 100 \text{ g carnallite} + 3.8 \text{ g kieserite} \rightarrow 23.7 \text{ g sylvite} + 8.3 \text{ g halite} + 2.1 \text{ g anhydrite} + 129.4 \text{ g solution } Q_{83}.$$

To distinguish the process of formation of kieseritic-anhydritic Hartsalz, some additional criteria are therefore necessary.

In connection with these questions, the alteration of primary kieserite-halite (that is to say, the kieseritic transition layer between halite and the potash layer) by $CaCl_2$ solutions is very important for, in this case too, anhydrite would result. It would be especially interesting to know the amount of $CaCl_2$-solution needed to completely alter a thick kieserite-halite. As the KCl content of the solution is unaltered in this case, and the NaCl is also little affected, the kieserite can be simply approximated by equivalent amounts of $CaSO_4$. Calculation shows that 732 g of the $CaCl_2$-bearing solution from the Langensalza bore can alter 100 g kieserite to 98 g anhydrite (+ 734 g solution without $CaCl_2$ and $MgSO_4$ but with about 11% $MgCl_2$). Although the $CaCl_2$ concentration in the Langensalza bore solution is unusually high, large volumes are needed for the reaction. The resultant solution is able to dissolve many times as much carnallite in the overlying layer (for every 100 g of kieserite altered about 900–1000 g carnallite are dissolved, compare the kieserite carnallite ratio in the calculation on p. 126). Under these conditions no carnallite would be present and most of the sylvite would be dissolved.

e) The Formation of Kainite from Kieserite and Sylvite

In Section a) the formation of kainite by the incongruent alteration of carnallite at temperatures below 72° C was considered. In addition, not infrequently kainite may form from kieserite + sylvite. There are two limiting conditions:

a) When kieserite is present in excess. After reaction of all the sylvite with kieserite to give kainite, some kieserite remains and by reaction with further NaCl solution is converted to loeweite or bloedite. With this reaction kainite is also gradually dissolved. After the resorption of kainite, leonite and finally glaserite form (Fig. 20). The same sequence can be derived from kieseritic langbeinite.

b) When sylvite is present in excess, and it can be preserved together with kainite. By further reaction with NaCl solution, sylvite and kainite are likewise redissolved with the formation of leonite and glaserite, but no bloedite. Below 26° C leonite is replaced by schoenite.

The first condition was discussed in more detail by KÜHN (1957). It can be stated in general that the solution processes correspond to a reversal of the direction of crystallization with complete reaction of the solid phases along reaction lines and at transition points.

2. *Polythermal Solution Metamorphism*

All the preceding cases are valid at constant temperatures. In comparison, polythermal solution metamorphism plays a subordinate role. This will be examined here quantitatively only for the $MgSO_4$-free system, and it is sufficient to consider secondary carnallitization. Assuming a solution E_{100} (Fig. 9) produced by the incongruent alteration of carnallite at 100° C, injected into a sylvite rock and there cooled to 80° C, the reaction gives 11.3 mol carnallite – 6.64 mol sylvite (the negative sign indicates the mineral is used up) and 0.932 solution E_{80}. Recalculated in terms of weight, the reaction gives from 495 g sylvite 3130 g carnallite, that is six times the amount.

	$E_{100°}$	→	sylvite	+ carnallite	+ $E_{80°}$
$MgCl_2$	89.5		0	1	83.9
K_2Cl_2	24.4		0.5	0.5	21.2
H_2O	1000		0	6	1000

There occurs with this such a great volume increase that, if there is a rock overburden, the process can have no appreciable geological significance (BAAR, 1960). For example, the carnallitization of a sylvite-halite with 40% KCl and 60% NaCl gives a 3.3 times volume increase.

Moreover, it is known from solubility data (VAN'T HOFF, 1905) that sylvite at the transition point is rapidly crusted over with carnallite and hence cut off from contact with the solution. This always happens in natural salt precipitation, so that on these grounds, apart from thin reaction crusts, no secondary carnallitization is possible.

Carnallite can, of course, occur in joints through the cooling of a hot E or Q solution. This cooling effect has already been described on p. 104. By further extrapolation of isotherms and the curve $E-E$ in Fig. 9, it is found that a high $MgCl_2$ content close to point D (Z of the $MgSO_4$-bearing system) can arise by incongruent alteration of carnallite at about 140° C. Following this the precipitation of small amounts of bischofite would be possible.

IV. Dynamic Metamorphism

The plasticity of salt minerals is much greater than that of silicates so that they are deformed at much lower stresses (BORCHERT, 1959). The transition points are little affected by pressure so that to a first approximation the pressure dependence of the equilibria can be neglected. VAN'T HOFF showed that the temperature at which tachhydrite is formed increases by only 0.016° C per atmosphere, in other words by 1° C for each 240 m of overburden (cf. p. 108). Dynamometamorphism is concerned mainly with isochemical and isophase recrystallization. The special case of isothermal lateral secretion where pressure is inhomogeneously distributed has already been considered (p. 106).

The deformation of salts has been extensively observed and also experimentally investigated. One consequence of their low specific gravity is that they possess a certain bouyancy when buried in the lower part of crust. This buoyancy is without doubt a most important energy source in salt tectonics (ARRHENIUS, 1912). For this reason too, salt deposits are not normally found at depths greater than 3000–5000 m (TRUSHEIM, 1957, 1960). As the maximum depth, the figure of 10,000 m, the depth of the "mother layer" of the Gulf coast diapirs in the Gulf of Mexico given by MURRAY (1962), can be assumed. At a depth of 5000 m the pressure is of the order of 1300 atmospheres which is already sufficient to cause a shift in the transition points but in general does not produce new parageneses.

In the present work the extensive field of dynamometamorphism is not considered in detail, despite its fundamental importance for the form of present-day salt deposits. Many changes brought about by solutions were also not possible without the effect of pressure (the forcing in of solutions, opening of circulation channels, etc.). As the parageneses are qualitatively unchanged and the purpose of this work is to examine the physico-chemical conditions of formation of salt minerals, this restriction appears justified.

In comparing calculated and observed salt sequences, however, the alteration of the mass relationship of the minerals by tectonic demixing (p. 193, 194) is not very well known as a source of error. The original thickness and the original composition are altered beneath the sedimentary overburden at the beginning of plastic flow. The conditions observed in salt diapirs are therefore not quantitatively comparable with the calculated models. In this area a lot of research remains to be done and the attempt should be made to follow the deformation process quantitatively by means of physical models and experimental data.

V. The Behaviour of Minor Components

The minor (and trace) components of seawater can be separated into four categories according to their behaviour during the main crystallization process, although many elements can occur in two or more groups[31]:

a) Elements diadochically included in the main precipitates (masked in the case of Br, Rb[32], Cs, Tl, captured in the case of Pb etc. in the alkali chlorides).

b) Precipitated as distinct mineral phases upon reaching their solubility product. To this group belong B, F, P, etc.

c) Adsorbed on the major components and especially on impurities such as the clay minerals. Many of the heavy metals, Pb, Cu, etc., belong in this category. This group may also comprise the heavy metals included diadochically in the impurities and authigenic new phases; pyrite and haematite are the most important carrier substances.

---

[31] See also discussion HERRMANN in „Mineralsalze ozeanischen Ursprungs", Freiberger Forschungsheft A 123, Akademie-Verlag, Berlin, p. 165 (1958).

[32] For the Rb-contents of potash minerals see, for instance: KÜHN, R.: Rubidium als geochemisches Leitelement bei der lagerstättenkundlichen Charakterisierung von Carnalliten und natürlichen Salzlösungen. N. Jb. Miner. Mh, 107–115 (1963).

BRAITSCH, O.: Bromine and rubidium as indicators of environment during sylvite and carnallite deposition of the Upper Rhine Valley evaporites. Second Symp. on Salt, Northern Ohio Geol. Soc. 293–301 (1966).

KÜHN, R.: Geochemistry of the German potash deposits. The Geological Society of America, Special Paper **88**, 427–504 (1968).

d) Enrichment of the residual solution. Among the elements occuring here are Li and I for which there is no suitable host in the precipitation of the main phases and in the impurities. Many of the elements occurring in the other groups may be found here too.

A systematic treatment of these groups is not envisaged, as only for bromine and strontium are sufficient data available for the calculation of models which is the main concern here. For the later application to questions of genetic interest, however, it seems practical to consider certain individual components even if the theoretical basis is uncertain.

1. Bromine

a) Theoretical Bromine Distribution

Bromine occurs diadochically almost exclusively for Cl in chlorides, although most of it remains in the residual solution. Bromates and perbromates have not yet been recorded. Sulphates and Cl-free borates have so far proved to be free of bromine, although traces might be expected in liquid inclusions. Traces may also be expected in clay residues (BEHNE, 1953; CORRENS, 1956) and have been demonstrated by VALYASHKO (1956).

BOEKE (1908) investigated the behaviour of bromine during the crystallization of salt solutions and this work has formed the basis for all subsequent theoretical considerations (cf. KÜHN, 1955b; VALYASHKO, 1956). Since the time of D'ANS and KÜHN (1940) the investigation of bromine has been a standard technique in the examination of salt deposits, especially since BAAR (1952) demonstrated its value in the determination of stratigraphic horizons. Bromine is also a useful genetic indicator (bromine test) of unusual conditions of formation, even if, unfortunately, attemps have been made to prove erroneous conclusions by bromine analysis.

BOEKE (1908) used the following simple differential equation for the distribution of bromine between solution and crystalline precipitate:

$$a \cdot p = (a + da)(p + dp) - q \cdot b \cdot p \cdot da$$

where a = amount of the solution (in grams),

p = bromine content of the solution (in weight %),

da = total salt precipitate + water vapour (change in the amount of solution)

dp = increase in the bromine content of the solution,
q = proportion of the Br-bearing salts in the total precipitate

$$\text{that is: } = \frac{\text{g chloride}}{\text{g total precipitate}}$$

(N.B. KÜHN, 1955, uses q with a different meaning)

$$b = \frac{\text{weight \% Br}_{(\text{minerals})}}{\text{weight \% Br}_{(\text{solutions})}}$$ = fraction of bromine present in the solution included in the chloride = the partition coefficient.

[It must be noted that this definition is not identical with the Berthelot-Nernst partition coefficient where the bromine content in the denominator is related to the amount of salt in solution. In practical applications the definition used here which covers a directly determined quantity seems more suitable. The same physical laws apply to both definitions. VALYASHKO (1956) used the Berthelot-Nernst definition.]

The first term on the right-hand side of BOEKE's equation describes the change in the amount of solution and its bromine content, and the second the bromine content of the precipitate. Transformed this gives:

$$a\,dp = -(1-qb)\,p\,da\,.$$

From this it follows that at different evaporation stages ($i = 0, 1, 2, \ldots$) and after eliminating the integration constant using the initial values under the assumption that q and b remain constant

$$\ln \frac{p_i}{p_{i-1}} = -(1-qb)\ln \frac{a_i}{a_{i-1}}\,, \tag{1}$$

a_i is the amount of seawater left out of 1000 g at concentration stage i. This amount is obtained from the calculation of the primary precipitate (last row, Table 8) and following from this (after re-calculation in g) the value for q. These are considered as constant, although strictly this applies only to crystallization from constant solutions. From the evaporation of seawater no constant solution can result. However, even in the carnallite system the effect of marked variations in q on the bromine content of the solution is quite small, as can be shown by a check calculation from formula (1) for ten intermediate stages. q may therefore be considered as constant.

The assumption that b is constant is not obvious and certainly not fulfilled in all cases of mixed crystal formation. Besides a linear dependence, there is in the case of trace elements also the possibility of a logarithmic or other dependence (HAHN, 1936, p. 70).

The partition coefficient b must be determined from crystallization experiments. At high bromine concentrations in the solution, the only ones investigated (BOEKE, 1908), b depends upon the bromine content of the solution itself, and upon the other components of the solution,

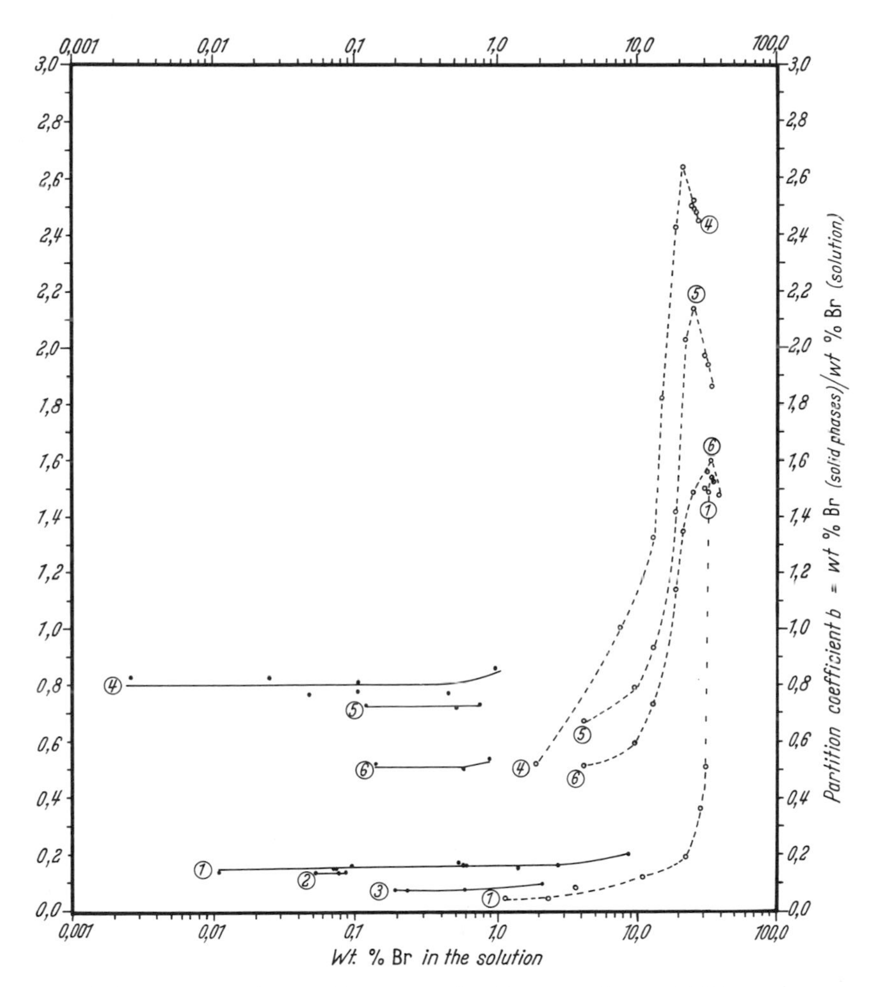

Fig. 32. Bromine partition coefficients. Broken lines after BOEKE (1908), continuous lines after HERRMANN (from BRAITSCH and HERRMANN, 1962, 1963). *1–3* solid phase Na (Cl, Br). System $Na^{+}//Cl^{-}$, $Br^{-} + H_2O$ ($\pm$ $MgCl_2$); *1* without $MgCl_2$; *2* $MgCl_2$ concentration as at the beginning of NaCl saturation of seawater; *3* $MgCl_2$ concentration at the beginning of carnallite saturation; *4, 5* solid phase K (Cl, Br). System $K^{+}//Cl^{-}$, $Br^{-} + H_2O$ ($\pm$ $MgCl_2$); *4* without $MgCl_2$; *5* immediately before carnallite saturation; *6* solid phase KMg $(Cl, Br)_3 \cdot 6\,H_2O$ system *5* immediately after carnallite saturation

but on the other hand it depends very little on temperature. To further delimit the area over which BOEKE's Eq. (1) is valid, further experiments at low bromine concentrations were necessary, since the extrapolated values from BOEKE's experiments are too uncertain. Furthermore the new determinations of BLOCH and SCHNERB (1953/1954) gave somewhat different values. For this reason the partition coefficient in this interesting concentration region has been investigated by HERRMANN (BRAITSCH and HERRMANN, 1963). The results quoted for the most important chlorides are shown in Fig. 32. With $<1\%$ Br in the solution the partition coefficient is practically constant. The natural salt solutions occur only in this concentration region. BOEKE's Eq. (1) is thus applicable to the bromine distribution in salt deposits. It indicates:

1. NaCl absorbs the least amount of bromine, but distinctly more than was to be expected from BOEKE's investigations (in an $MgCl_2$-free system).

2. KCl takes up the most bromine, even more than BOEKE found.

3. In solutions of equal bromine concentration, carnallite takes up less bromine than sylvite, also as reported in the experimental data of BOEKE (1908) and by BLOCH and SCHNERB (1953/1954). The higher value used by D'ANS and KÜHN (1940) and KÜHN (1955) does not correspond to the experimental data.

4. The partition coefficient b for NaCl and KCl is reduced by the addition of $MgCl_2$ to the solution. BOEKE concluded the contrary for KCl, although his analytical results also showed a decrease in b. BOEKE's assertion is true of the ratio $\frac{\text{no. of bromine ions in the mineral}}{\text{no. of halogen ions in the solution}}$ that is to say, the mol % of halogens, but his application of this to natural occurences indicates that he meant the weight %. In addition, of course, BOEKE's Eq. (1) contains the weight % ratio.

As in natural salt deposits $MgCl_2$-rich solutions usually have to be considered, it is only necessary to be concerned with the partition coefficient for these solutions (Table 17).

For sylvite consideration need only be given to the value in the region of the sylvite-carnallite transition point, since the $MgCl_2$ concentration itself increases relatively little during the precipitation of sylvite. For carnallite too, at present, only the value near this point is known. This is, however, of great importance for understanding of deposits, as under normal geological conditions the point at which bischofite forms is not reached. Precipitation of NaCl on the contrary can occur over a wide range of different $MgCl_2$ concentrations (cf. Fig. 9) in which the bromine partition coefficient b steadily declines

Table 17. *Bromine partition coefficients during the precipitation of salts from an $MgCl_2$-bearing or $MgCl_2$-rich solution (after* HERRMANN, *published in* BRAITSCH *and* HERRMANN, *1963)*

Mineral	Partition coefficient $b = \frac{\text{weight \% Br (mineral)}}{\text{weight \% Br (solution)}}$			Relative Br content simultaneous precipitation
	25° C[a]	55° C	83° C	25°–83° C
Halite				
at the beginning of NaCl precipitation	0.14 ± 0.01			
at the beginning of Na–Mg sulphate precipitation	0.073 ± 0.004			
at the beginning of carnallite precipitation	0.073 ± 0.004	0.078 ± 0.004	0.079 ± 0.004	1
Sylvite (from $MgCl_2$-rich solution)	0.73 ± 0.04	0.77 ± 0.04	0.83 ± 0.04	10 ± 1
Carnallite	0.52 ± 0.03	0.56 ± 0.03	0.59 ± 0.03	7 ± 1
Kainite	0.25 (0.22)			$3^1/_2 \pm 1$
Bischofite	0.66 ± 0.03			

[a] KÜHN and RITTER [in KÜHN: Geochemistry of the German Potash Deposits, The Geological Society of America, Inc., Special Paper **88**, 427–504 (1968)] determined the partition coefficients for halite and carnallite at 30° C. The results agree with the determinations listed in this Table 17.

from the beginning of NaCl precipitation. Since the magnitude of b is small, it has little effect on the bromine content of the solution in BOEKE's Eq. (1), and with an average value ($b = 0.1$ during halite precipitation) a good approximation is obtained. In contrast, in calculating the bromine in NaCl crystals (using the defined equation for the partition coefficient b) the value for the appropriate concentration stage must be applied each time.

The value for kainite was not determined by BOEKE (1908). It was estimated on the assumption that the KCl in kainite takes up the same amount of bromine as that in sylvite. The value in brackets is valid for $MgCl_2$-rich solutions and is probably to be preferred for natural kainite. Calculations from the crystallization experiments of D'ANS and KÜHN (1940, p. 77) give approximate values (the Br content of the dissolved chloride not being known): 0.27 (for ~0.36% Br/solution; 0.096% Br/kainite) and 0.21 (for 3.54% Br/solution; 0.737% Br/kainite) which on average corresponds to the estimated value used. D'ANS and KÜHN (1940) and KÜHN (1955) had derived the value as 0.4, but this is not in agreement with the experiments. These authors obtained their value by indirect conversion of the molar ratio (Br/halogens) into units by weight (Br/halogens + non-halogens), using a wrong equivalent for Br/NaCl.

Table 18 gives the values calculated by means of Eq. (1) for various primary precipitates. If more than one chloride is formed, the sum of the products $q \cdot b$ for all chlorides must be inserted in the brackets. The table shows the Br content of the solution on the appearance of a new chloride and the Br content at each change of paragenesis.

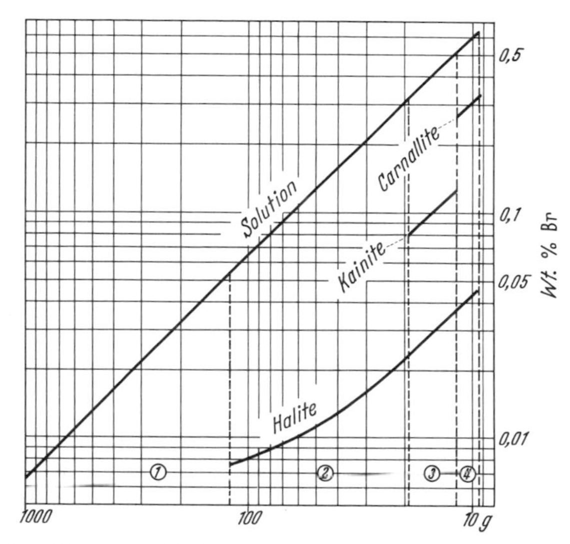

Fig. 33. Calculated bromine distribution during static evaporation of seawater at 25° C. Abcissa = amount of solution in grams normalized to 1000 g seawater. *1* pre-concentration up to NaCl saturation; *2* precipitation of halite (with less than 24.5 g solution, together with bloedite or epsomite, see Table 8); *3* precipitation of halite, kainite, magnesium sulphate; *4* precipitation of halite, carnallite, kieserite. From 9.27 g of solution: bischofite precipitation

Fig. 33 gives the results for static evaporation of seawater. The other models can be presented similarly. The Br contents for any required concentration can be read off directly. As the Br contents of the minerals are proportional to the Br content of the solution (factor b), the observed Br contents of the minerals provide a measure both of the concentration reached in the solution and of the concentration of its major components.

This can be clarified by a further model in which there is complete reaction of the initially formed solid phases in concentrating solutions (see p. 93, 94, and Fig. 34). Fig. 34 shows only the composition of the

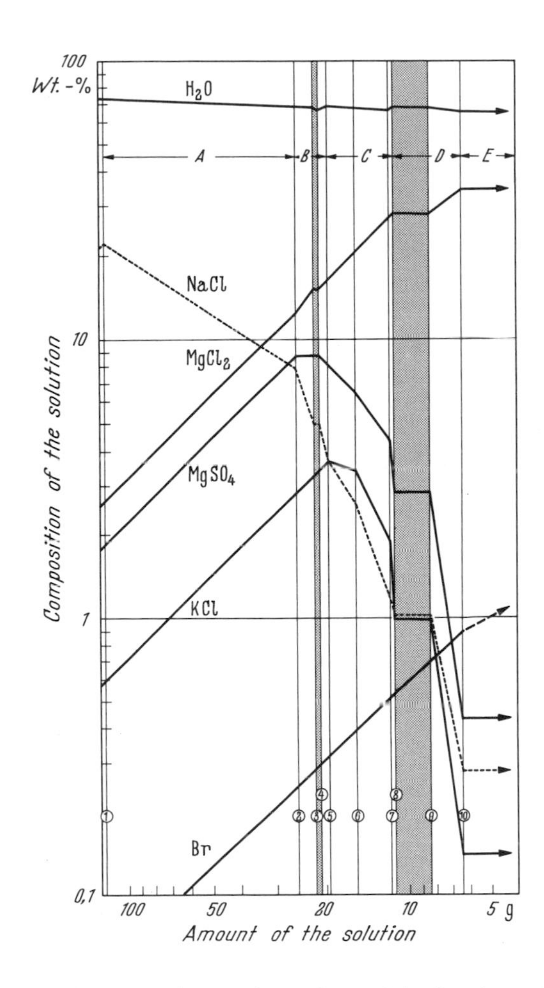

Fig. 34. Change in composition during salt precipitation from the beginning of halite precipitation at 25° C (stable equilibria). Complete reaction of solid phases at the transition points. Abcissa = amount of solution in grams normalized from 1000 g of original seawater. *1* beginning of salt precipitation; *2* beginning of bloedite precipitation; *3* solution *I'* (Table 11); *4* solution *I''* (between *3* and *4* reaction of newly formed bloedite to epsomite); *5* solution *X''*; *6* solution *X*; *7* solution *Y* (composition still incorrect, see p. 65, corrected values see Table 11); *8–9* solution *R*, alteration of newly formed kainite to carnallite and kieserite in which solution is used up; *10* solution *Z*. Precipitates: *A* = NaCl; *B* = NaCl + NaMg or Mg sulphates, no potash salts; *C* = NaCl + kainite + Mg sulphate; *D* = NaCl + carnallite + kieserite; *E* = bischofite + carnallite + kieserite + NaCl. The bromine content in the precipitated chlorides is found by multiplying the bromine content of the solution by the partition coefficient *b* (Table 17)

solution at each evaporation stage after NaCl saturation is reached. The bromine content of the solution is obtained from the bromine content of the halite (with weight % Br/solution $= 1/b \cdot$ weight % Br/NaCl) and from this also the stage of evaporation; it is further possible to read from the abcissa of the same diagram the content of other components in the solution ($CaSO_4$ is excluded since the experimental data are still lacking).

The diagram gives a complete picture of the alteration of the solution during salt deposition. The BOEKE equation is equally applicable to the principal components. As soon as one component is precipitated, the slope of its concentration line is altered. In the course of bloedite precipitation Na_2SO_4 is also withdrawn from the solution. This is not represented on the figure, and must be expressed through the reciprocal salt pair $Na_2SO_4 + MgCl_2 \rightleftharpoons MgSO_4 + Na_2Cl_2$. This causes an apparent increase in the $MgCl_2$ concentration of the solution during the precipitation of bloedite. During the reaction of bloedite with the solution to form epsomite the Na_2SO_4 withdrawn is again released, so that after complete

Table 18. *The calculated bromine distribution between solution and solid phases (letters in column q = solid phases after Table 4)*

a) Static evaporation of seawater at 25° C to bischofite saturation: without reaction of the solid phases at transition points

i	Solution	q	b	1000 a_i g	p_i % Br/ solution	% Br/ NaCl	% Br/ kainite	% Br/ carnallite
0	Seawater	0	0	1000	0.0065			
1	NaCl sat.	n 0.254	0.14	121	0.0538	0.075	—	
2	*W'*			19.6	0.316	0.023	0.07	
		n 0.0942	0.073					
		k 0.2512	0.22					
3	*R*			11.8	0.508	0.037	0.11	0.26
		n 0.0257	0.073					
		c 0.1465	0.52					
4	*Z*			9.3	0.632	0.046	—	0.33

b) As in a) but with complete reaction of the solid phases at the transition points

0–2 as under a)

i	Solution	q	b	1000 a_i g	p_i % Br/ solution	% Br/ NaCl	% Br/ kainite	% Br/ carnallite
3	R_1		0.073	11.3	0.527	0.038	—	0.27_4
		n(0.056)	0.52					
		c 0.218						
	R_2			8.41	0.687	0.050_2	—	0.35_7
		n 0.0257	0.073					
		c 0.1465	0.52					
4	*Z*			6.42	0.888	0.065	—	0.46_2

Table 18 (continued)

c) Complete $MgSO_4$ deficiency, static evaporation at different temperatures

i	Solution becoming sat. at	q	b	$1000\,a_i$ g	p_i % Br/ solution	% Br/ NaCl	% Br/ KCl	% Br/ carnallite
0° C								
1	Halite			120.4	0.0539	0.0073	—	—
		n 0.251	0.13_5					
2	Sylvite			18.4	0.339	0.023_7	0.237	—
		n 0.161	0.07					
		sy 0.079	0.7					
3	Carnallite			12.7	0.478	0.033_5	0.335	0.229
		n 0.054	0.07					
		c 0.215	0.48					
4	Bischofite			8.4	0.692	0.048_5	—	0.332
25° C								
1	Halite			119.4	0.0544	0.007_6	—	—
		n 0.253	0.14					
2	Sylvite			15.6	0.396	0.028_9	0.289	—
		n 0.130	0.073					
		sy 0.102	0.73					
3	Carnallite			12.5	0.485	0.035_4	0.354	0.252
		n 0.05	0.073					
		c 0.31	0.52					
4	Bischofite			7.62	0.735	0.053_6	—	0.382
50° C								
1	Halite			115.8	0.0562	0.008	—	—
		n 0.257	0.15_5					
2	Sylvite			13.2	0.465	0.035_8	0.358	—
		n 0.113	0.077					
		sy 0.166	0.77					
3	Carnallite			11.8	0.515	0.039_7	0.397	0.288
		n 0.043	0.077					
		c 0.358	0.56					
4	Bischofite			7.02	0.775	0.059_7	—	0.434

reaction the further concentration of $MgCl_2$ lies on a linear elongation of the $MgCl_2$ line during NaCl precipitation.

The case where complete reaction occurs of the kainite precipitated between solution R and X'' is very important in considerations of the distribution of bromine. As was indicated on p. 93, 94, during the alteration to kieserite + carnallite the composition of solution R with respect to the major components remains constant while its volume is greatly reduced. As only a small part of the bromine present in the solution is taken up in the carnallite, the bromine content of the solution further increases, and hence the amount of bromine in the precipitate increases proportionally. In contrast to the model where there is no reaction of the initially precipitated kainite, there is a distinctly higher level of bromine in the precipitated carnallite and halite (cf. Table 18).

This situation requires further investigation. There is first of all the question of whether the precipitated NaCl is also recrystallized during the reaction so that bromine reaches equilibrium at any given time. According to the experiments of D'ANS and KÜHN (see KÜHN, 1955a), this process has a certain measurable significance (the so-called Br/Cl exchange effect). It could only occur during the true recrystallization. It is, however, for the present not known what fraction of the precipitated halite recrystallizes.

As the first limiting condition the total absence of NaCl recrystallization can be considered. From the beginning to the end of the alteration of kainite, the bromine content in the precipitated carnallite rises from about 0.27% Br/carnallite to about 0.36% Br/carnallite (Table 18, b). If the last precipitated kainite, which remains in any case in contact with the solution, reacts first, the normal paragenetic bromine ratio of 0.038% Br/NaCl and 0.27% Br/carnallite is obtained at this interface. If now the reaction continues downwards so that the first kainite precipitated is the last to react, without recrystallization of the initially precipitated halite, then at the base of the carnallite layer present after the reaction, 0.023% Br/NaCl and 0.36% Br/carnallite is found, that is, a non-paragenetic bromine ratio of 1 : 16. This produces a decrease in the bromine content of the carnallite from bottom to top, whereas it increases in the halite. The most marked deviations from the paragenetic bromine ratio occur in the basal layers.

The second limiting condition assumes the complete recrystallization of NaCl, which gives only the paragenetic bromine ratios. If the reaction proceeds from the top downwards, then there are again high Br contents at the base of the carnallite layer (see solution R_2 in Table 18, b), decreasing upwards to the value appropriate to solution R_1. If after complete alteration the carnallite precipitation goes further, then again the normal upward bromine increase – from R_2 to Z in the example considered – continues with a clear, if only apparent, variability towards the end of the reaction, because the last salts formed by the reaction are spatially separated from the salts produced by further evaporation.

Obviously the actual process will lie between the two limiting conditions, nor is a process operating from above downwards or in the reverse direction usually to be reckoned with, yet the theoretical distinction between the limiting conditions is important. Above all else it shows that non-paragenetic bromine ratios can also result from non-metamorphic reactions. Thus important genetic information can be obtained from the careful analysis of the bromine content.

Finally attention ought to be called to the fact that the preceding analysis is also qualitatively valid for the reaction of metastable primary precipitates at reaction points. True, it is not yet known if, when and whether in each individual case metastable epsomite + sylvite or hexahydrite + sylvite etc. first alter to kainite or are altered directly to carnallite + Mg-sulphate. So long as adjustments in the concentration of the pore solution with the evaporating mother liquor are possible, the new Br partition equilibrium will be established in the solid phases, and the final compositional values of the stable parageneses listed in Table 18b may be realized.

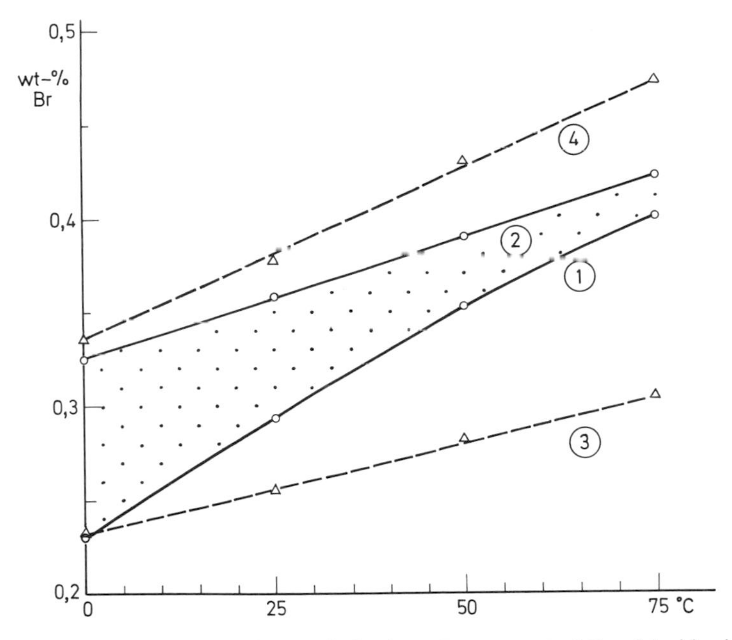

Fig. 35. Temperature dependence of the bromine content of the chlorides in an $MgSO_4$-free seawater system. Weight percent Br/KCl in the shaded area; *1* at the beginning *2* at the end of sylvite precipitation (at the same time 10 times the amount Br/NaCl during sylvite precipitation). Weight % Br/carnallite between *3* and *4*; *3* at the beginning, *4* at the end of carnallite precipitation. Curve 1 is suitable as a geological thermometer for primary sylvite-halite. (Newly calculated fig. from BRAITSCH and HERRMANN; Konzentrations-, Dichte- und Temperaturverteilung in der unteroligozänen Salzlagune des Oberrheins. Geologische Rundschau **54**, 344–356, 1964)

The temperature dependence of the bromine partition coefficient *b* has recently been experimentally re-examined by HERRMANN (see Table 17). As BOEKE (1908) had already shown, the bromine content increases but little with rising temperature; despite the low temperature dependence of *b*, there is a strong temperature dependence of the bromine content of the primary precipitated sylvite and carnallite because of the strong positive temperature coefficients of sylvite and carnallite solubility. Sylvite precipitation begins at lower temperatures and at lower solution concentrations, that is, at lower bromine concentrations, so that there is a lower bromine content in the first sylvite and the paragenetically simultaneously formed halite (Fig. 35)[33]. The bromine content of carnallite from solutions which have reached the beginning of bischofite saturation is also strongly temperature-dependent. In contrast, the bromine content of halite at the beginning of NaCl saturation is practically independent of temperature (Table 18).

Naturally the bromine in the five-component system is also temperature-dependent. In particular the first sylvite formed from the low-temperature sylvite-epsomite paragenesis must possess a low bromine content. Models have not been calculated for this case, however. It probably has no practical importance, as primary sylvites can only persist in an $MgSO_4$-free paragenesis; in an $MgSO_4$ system upon subsequent temperature increase the sylvite goes to kainite. Such kainite must have a lower bromine content than primary precipitated kainite.

b) The Effect of Influxes on Bromine Distribution

Fig. 33 is only valid for the static evaporation of seawater in the absence of influxes. Nevertheless periodic and continual influx may be treated in the same manner. In this case the initial solution considered also includes all the influxing seawater. Of course, no regular logarithmic bromine growth curve is to be expected within the salt precipitation

[33] The applicability of the bromine method for the temperature determination in primary potash deposits of the $MgSO_4$-free type was tested in the Oligocene salts of the upper Rhine valley [see: BRAITSCH and HERRMANN: Zur Geochemie des Broms in salinaren Sedimenten. Teil II: Die Bildungstemperaturen primärer Sylvin- und Carnallitgesteine. Geochimica et Cosmochimica Acta **28**, 1081–1109 (1964). — BRAITSCH and HERRMANN: Konzentrations-, Dichte- und Temperaturverteilung in der unteroligozänen Salzlagune des Oberrheins. Geologische Rundschau **54**, 344–356 (1964)]. Three sylvite beds immediately below the exploited part of the lower potash seam proved to be suitable for temperature estimation. The average values found range from about 10° C (or a few degrees less) to $51 \pm 5°$ C. Distinct gradients in temperature and density were found, both in space and geological time. Temperature and density of the brine increase from bed to bed and within the same bed towards the deepest part of the individual basins.

sequence. The ideal case on which Fig. 33 was based, that the influxes are uniform and always occur at the same time intervals with constant evaporation, is highly improbable. The individual concentration stages need not occur with the same speed. Rather to the contrary, dilution phases can intervene yet nevertheless chloride precipitation may continue. The condition necessary for this is that the influxes exceed evaporation losses yet occur periodically so that in the intervening time between fresh influxes chloride saturation is again reached.

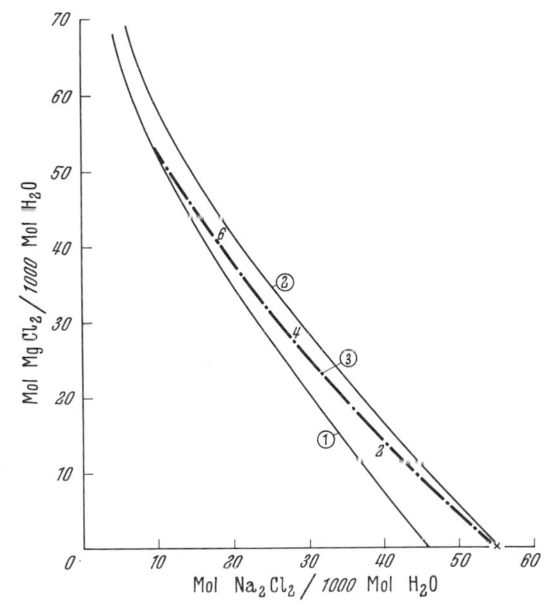

Fig. 36. NaCl saturation concentration at 25° C for a varying $MgCl_2$ content. *1* system $NaCl–KCl–MgCl_2–H_2O$ saturated in NaCl and KCl; *2* system $NaCl–MgCl_2–H_2O$; *3* $MgSO_4$-free seawater before saturation in KCl, interpolated linearly between *1* and *2*. Numbers on curve 3 = mol K_2Cl_2/1000 mol H_2O. *1* and *2* from D'ANS (1933)

Because of the importance of such a case, and because of the contrasting conclusions of D'ANS and KÜHN (1940–1960), it will be quantitatively formulated.

Assume a sylvite-saturated solution to be present initially. Without restricting the general validity, let it be supposed to be SO_4-free solution (Fig. 9) which at 20° C contains 54 $MgCl_2$, 7.7 K_2Cl_2, 9 Na_2Cl_2 per 1000 mol H_2O. It contains 1.28 mol Br per 1000 mol H_2O calculated from the K_2 enrichment in Table 1 where all the bromine diadochically included in the halite precipitate is neglected. Now fresh seawater influxes: as a result the solution is no longer saturated and the KCl and NaCl concentrations cannot be as in Fig. 9. With continuous evaporation, after gypsum, halite saturation will again be reached. The NaCl concentrations can be interpolated with sufficient accuracy between the NaCl curve of Fig. 9 and the NaCl concentration in $MgCl_2$-bearing solutions (Fig. 36). The necessary amount of influxing seawater for each KCl-unsaturated but NaCl-saturated solution can then be readily calculated, from which naturally an exactly determined amount evaporates leading to the precipitation of $CaSO_4$. The precipitation of NaCl should begin again, for example, at 50 mol $MgCl_2$/1000 mol H_2O. The required amount of seawater is then (seawater, see Table 1 with:

$$MgCl_2 = Mg^{++} + Ca^{++} + Sr^{++} - SO_4{}^{--} - \tfrac{1}{2} HCO_3^-$$

that is to say, seawater totally deficient in $MgSO_4$):

	x initial solution	$+y$ seawater	NaCl saturated solution
$MgCl_2$	54	0.63	50
Na_2Cl_2	9	4.28	11.5

$x = 0.917$, $y = 0.759$.

Before the beginning of NaCl precipitation $917 + 759 - 1000 = 676$ mol H_2O must evaporate. The bromine content of the solution at the beginning of NaCl precipitation amounts to $0.917 \cdot 1.28 + 0.759 \cdot 0.015 = 1.18_3$ mol Br/1000 mol H_2O. The bromine content of the solution falls relative to that of the KCl-saturated solution, so that the bromine in the halite precipitate also decreases. The result is scarcely altered by the introduction of the full Mg value (0.976 instead of 0.63) so that it is not dependent upon the manner of calculating the $MgSO_4$ deficit. If there is a further influx before evaporation has proceeded as far as sylvite precipitation, the bromine content of the solution is again diminished, so that over a long interval the step-like progress of NaCl precipitation can be followed in a reversed, recessive direction. The result is generally valid, for at the beginning of NaCl precipitation at low $MgCl_2$ concentrations x decreases (= a less concentrated solution) and y increases (= more seawater); as far as bromine is concerned, x is the decisive factor. This remains valid when the initial solution is not saturated with respect to sylvite or if it is already saturated with respect to carnallite or bischofite (recessive potash precipitation is thus possible even if less likely). The relationship also applies to preconcentrated solutions; they must only be more dilute than the existing solution. Obviously in such cases a smaller influx is necessary and less water has to be evaporated. The same arguments apply to $MgSO_4$-containing solutions.

The argument up to the present has concerned periodic influxes separated by an interval during which evaporation occurs. The amount of influx can be made arbitrarily small, and the evaporation interval correspondingly short, so that in principle the limiting condition of continuous influx with continuous evaporation is possible and for a given ratio between influx and evaporation salt precipitation will occur despite the decreasing total concentration of the solution. The duration of the recessive salt precipitation will be determined by the initial concentration (that is to say, the degree of concentration reached) and the influx/evaporation ratio.

What has been demonstrated here for recessive salt precipitation is also applicable within a single bed in a progressive evaporation sequence where each bed corresponds to an influx. At the beginning of the formation of each bed the bromine in the NaCl should be less than at the end of the formation of the immediately preceding bed. If evaporation is greater than influx, then the bromine content at the end of the formation of each bed is greater than that at the end of each preceding bed. Each bed corresponds to a progressive succession even when the total succession is recessive.

The calculations were made upon the assumption of complete mixing of the influx and the solution already present. In nature, however, more or less extensive layering occurs. As a result the diminution of the bromine content of the first precipitates is magnified.

Influxes of preconcentrated solutions have many times been considered (e.g. KÜHN, 1953, and others). These cases may also be deduced from Figs. 33 and 34. Nevertheless, this is only possible if assumptions are made about the concentration of the influx. The diagrams indicate clearly that certain well-defined limits cannot be passed. At the beginning of carnallite saturation the bromine content of the solution is about 0.5%. The assumption of a preconcentrated influx with 1.78% Br (KÜHN, 1953, p. 651) thus is not in agreement with natural salt solutions.

c) Partition of Bromine in Thermal Metamorphism

The Boeke differential equation (p. 131, 132) is also applicable to metamorphic processes. The relations during thermal metamorphism are particularly simple. With the establishment of equilibrium the appropriate constant solution for the transition temperature is developed, there is thus no change in the solution composition or its bromine content. From that it follows that for the upper limit of formation of kainite (Table 13) there is the simple relation:

$$a_k \cdot p_k + a_n \cdot p_n = a_{\text{soln}} \cdot p_{\text{soln}} + a_{sy} \cdot p_{sy} \tag{2}$$

(with the solid phase symbols from Table 4 as indices of the existing amounts a, and the bromine content p). Because

$$p_{sy} = b_{sy} \cdot p_{\text{soln}}$$

it follows

$$p_{\text{soln}} = \frac{a_k \cdot p_k + a_n \cdot p_n}{a_{\text{soln}} + a_{sy} \cdot b_{sy}}$$

With the average bromine contents of the initial minerals from Fig. 33 and $b_{sy} \approx 0.8$ (from Table 17) and the amounts of the participating phases (from Table 13) in the reaction, there results for the thermal metamorphism of kainite at 83° C

$$p_{\text{soln}} = \frac{99.51\ \text{g}_k \cdot 0.10\,\%_{\text{Br}} + 0.49\ \text{g}_n \cdot 0.03\,\%_{\text{Br}}}{26.81\ \text{g}_{\text{soln}} + 17.16\ \text{g}_{sy} \cdot 0.8} = 0.25\,\%\ \text{Br}\,.$$

The sylvite contains on the average

$$p_{sy} = 0.8 \cdot 0.25\,\%\ \text{Br} = 0.20\,\%\ \text{Br/KCl}\,.$$

The halite not participating in the reaction retains its original Br content. If the reaction were possible without the mixing of solutions from various primary beds (because of intervening rock salt layers), then an increase in the bromine content of developing sylvite could occur between different layers (from 0.16 – 0.24% Br/KCl).

d) Bromine Distribution in Solution Metamorphism

In solution metamorphism consideration must be given to the bromine content of the infiltrating solution. If in the reaction the amounts of solution and salt are known (e.g. Table 14), the average bromine content of the salts which are forming can also be calculated to a first approximation from Eq. (2) by multiplying the amounts of salts by the appropriate bromine content. For example, the incongruent solution of carnallite in a saturated halite solution (amount $a_{\text{NaCl-soln}}$) results in a carnallite saturated solution (E or Q) with

$$p_{\text{soln}\,Q} = \frac{a_{\text{NaCl-soln}} \cdot p_{\text{NaCl-soln}} + a_c \cdot p_c}{a_n \cdot b_n + a_{sy} \cdot b_{sy} + a_{\text{soln}\,Q}} \tag{3}$$

(a = amount in grams, b = partition coefficient, p = % Br), from which by use of the partition coefficients the bromine contents of the precipitated halite and sylvite can be calculated.

The result is somewhat temperature-dependent, since the amount a depends upon temperature. The result is furthermore strongly dependent upon the bromine content of the penetrating NaCl solution. For the $MgSO_4$-free system, using the data in Table 14, Nos. 17, 12 and 4, and the above equation, several models can be calculated, in which a ratio of 0.3% Br/carnallite is always assumed for the altering carnallite and the temperature dependence of the partition coefficient b is taken from Table 17 (Table 19).

Table 19. *The bromine content in the incongruent solution of carnallite at various temperatures and bromine content of the percolating solutions*

$T°$ C	$p_{NaCl\text{-}soln.}$ weight %	$p_{soln.\,Q}$	p_n	p_{sy}
25	0	0.20	0.015	0.146
	0.1	0.246	0.018	0.180
	0.2	0.295	0.021	0.205
	0.3	0.346	0.025	0.252
55	0	0.208	0.016	0.160
	0.1	0.251	0.019	0.193
	0.2	0.294	0.023	0.226
	0.3	0.338	0.026	0.260
83	0	0.219	0.018	0.175
	0.1	0.257	0.021	0.205
	0.2	0.295	0.024	0.236
	0.3	0.333	0.027	0.266

The p_n values refer only to the bromine content of the halite precipitated from the metamorphosing solution. As the carnallite already contains halite with a higher bromine content (~0.04% Br/NaCl, Fig. 33) and this is preserved during the alteration, its effects must be taken into account. The average value of % Br/NaCl in the sylvitic potash rock (Hartsalz) is thus naturally higher and the ratio between this average value and the bromine in sylvite no longer corresponds to the paragenetic 1 : 10 ratio. However, since the original halite can recrystallize in the metamorphic solution, a paragenetic ratio may occur even here. The absolute values of the bromine in the sylvite (and halite) are smaller than in primary sylvite.

A more exact consideration has to take into account the alteration of the solution during the alteration of the carnallite. This can be done by means of a modified differential equation which formally corresponds to BOEKE'S. The final result is, however, relatively unaffected so that the much more extensive calculation is not warranted (cf. KÜHN, 1955).

In support of this statement an example of KÜHN'S (BAAR and KÜHN, 1962, p. 303) was calculated with the simplified formula (3) using KÜHN'S assumptions (and his smaller partition coefficients $b_{sy} = 0.67$; $b_n = 0.067$). This gave 0.179% Br/KCl and 0.034% Br/NaCl, while KÜHN'S detailed calculation gave 0.173% Br/KCl and 0.033% Br/NaCl. The difference between the two results lies within the limits of analytical error. Even a slide rule gives sufficient accuracy for these calculations.

In general the chlorides which are formed as a result of metamorphic processes have a lower bromine content than primary precipitates. The metamorphic solutions owe their bromine content to a large extent to Br-poor primary precipitates (in comparison with primary solutions because of partition coefficient b). Should, however, the infiltrating solution already possess an appreciable bromine content, then the metamorphic products may possess a Br content similar to those of primary precipitates.

2. Strontium

a) Diadochically in Ca-Sulphates

BOEKE'S Eq. (1) (p. 131, 132) is applicable to many enrichment processes of trace elements in the solution and their diadochic inclusion in precipitates of the major components. In the case of strontium $p_0 = 0.008\%$ (Table 1), q is in the gypsum region

$$\frac{1360 \text{ g } CaSO_4}{246{,}000 \text{ g } H_2O + 1360 \text{ g } CaSO_4} = 0.005,$$

in higher regions q tends to zero. The distribution equilibrium of carbonate precipitation has been investigated (OXBURGH et al., 1959). The specific data have not yet been published so that they cannot be converted into the form used here. Nevertheless it was found, as might have been expected from data on natural occurrences, that aragonite accepts about ten times more Sr^{++} from the solution than does calcite.

Since $SrSO_4$ is less soluble than $CaSO_4$, much Sr would be expected in the precipitating $CaSO_4$. Its inclusion is, however, hindered since it is isotypic neither with gypsum nor anhydrite. The product $q \cdot b$ probably should not be neglected during gypsum precipitation. For $b = \frac{\% \text{ Sr in mineral}}{\% \text{ Sr in solution}} = 10$, from Eq. (1) of BOEKE, we obtain values of 0.0022% Sr in solution at the beginning of gypsum precipitation and 0.0065% Sr in solution at the beginning of halite precipitation with in both cases about ten times that amount in the precipitating crystals. The figure of ten, however, may be the lower limit for the partition coefficient and a number around 50 may be more correct[34]. During halite precipitation, the precipitated portion q of the $CaSO_4$ falls to 0.002 of the total precipitate, so that BOEKE's equation can with good approximation be simplified to:

$$\ln \frac{p_i}{p_{i-1}} = \ln \frac{a_i}{a_{i-1}}. \tag{1a}$$

Using only this formula, we get the following: Sr in solution at the beginning of gypsum precipitation 0.0022%; at the beginning of halite precipitation 0.0067%; at the beginning of polyhalite precipitation 0.012% (with the values for the beginning of polyhalite precipitation uncertain).

This simplified Eq. (1a) is valid for all other trace elements present in the solution, so long as no independent minerals are formed and no anomalously large values are reached for the partition coefficient. Thus at every evaporation stage the concentration of the solution can be found by referring to Fig. 33 for the initial concentration in seawater of the appropriate element and constructing a ray through that point inclined at $+45°$.

b) Sr Minerals

When the solubility product of celestite is reached, in addition to diadochically included Sr in Ca sulphates (and chlorides) free celestite also occurs. From this point onward the Sr concentration of the solution can no longer be found from BOEKE's Eq. (1) or (1a), as it corresponds the saturation concentration of $SrSO_4$. Strontium borates (p. 23) are so rare that there is no need to consider them here.

The solubility data for $SrSO_4$ contain some contradictions. Fig. 37 shows the determinations of BUDZINSKI (pers. comm.[35]) and MÜLLER (1960). At low chloride concentrations the results are in approximate

[34] The partition coefficient between Sr in a solution with 1–100 ppm Sr and the mineral gypsum during gypsum precipitation has been experimentally determined, at 25° C $b = 48$, for anhydrite with a Sr content of 0.5–12 ppm in the solution $b = 296$ was considered (USDOWSKI, 1967). USDOWSKI, E.: Der Einbau von Sr in Gips und Anhydrit. Annual meeting of the Deutsche Mineralogische Gesellschaft, Berlin, October 1967. — USDOWSKI, E.: unpublished manuscript, 1967.

[35] The author acknowledges with thanks BUDZINSKI's courtesy in making available unpublished solubility data.

agreement. As might be expected, the molar solubility increase is about the same for KCl and NaCl. At high salt concentrations there are deviations, and here the values of BUDZINSKI should be preferred for it seems improbable that the behaviour of $SrSO_4$ is fundamentally different from that of $CaSO_4$ (Fig. 5). The results of the old determinations of VIRCK (1862) are certainly incorrect. MÜLLER and PUCHELT (1961)

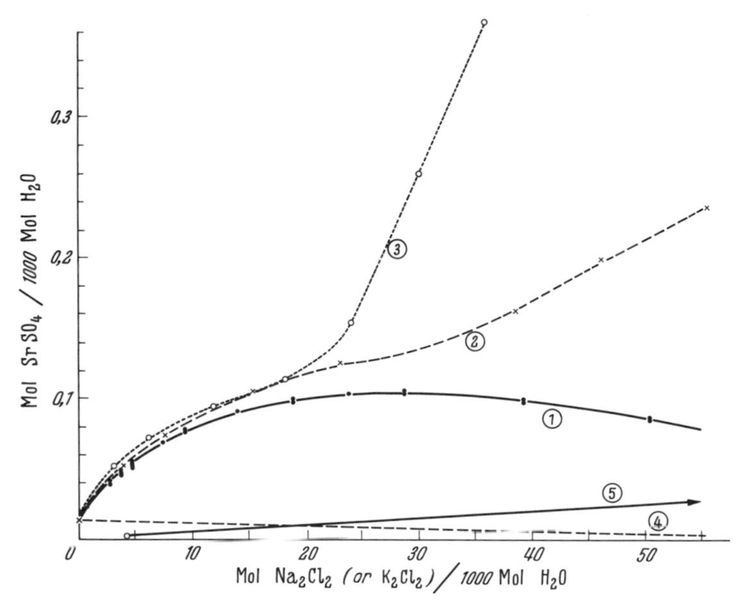

Fig. 37. Solubility of $SrSO_4$ in brines at 25° C. *1* in NaCl-solution after BUDZINSKI (pers. comm., gravimetric Sr determination); *2* in NaCl solution after MÜLLER (1960); *3* in KCl-solution after MÜLLER (1960); *4* in seawater after MÜLLER and PUCHELT (1961); *5* increase in $SrSO_4$ concentration during evaporation of normal seawater. The conversion of the seawater data from MÜLLER and MÜLLER and PUCHELT into units mol/1000 mol H_2O is not reliable, as the concentration data in the original work were not clear and were different on the abcissa and ordinate. The data for seawater concentration (curve 4) are especially crude. Only curve 1 (and 5) appear reliable

also investigated $SrSO_4$ solubility in natural and artificial seawater. They found a decrease of the same order in $SrSO_4$ solubility with increasing seawater concentration. This result also seems uncertain since at low salt concentrations a solubility maximum is first to be expected (at one or twofold seawater concentration). Qualitatively, however, it is certainly correct that there is a distinct reduction in solubility, in contrast to SO_4-free solutions, at higher salt concentrations.

MÜLLER and PUCHELT (1961)[36] believed they could assume that primary celestite precipitated in the transition region between carbonate and sulphate precipitation (at about 15 mol Na_2Cl_2/1000 mol H_2O, Fig. 5). This would require, from Fig. 37, a solubility decrease of less than 1/10 in comparison with the chloride solution, a value which seems somewhat high when compared with $CaSO_4$ (from Fig. 5). Furthermore this would lead to contradictions with natural observations (p. 207). Thus it seems more correct to assume that celestite precipitates from normal seawater at the earliest at the beginning of halite precipitation, and at the latest at the beginning of polyhalite precipitation[37]. As furthermore $SrSO_4$ solubility must further decrease with increasing evaporation, the quantitative ratio of celestite to halite must increase with progressive evaporation. Because the Sr content of the solution cannot increase after celestite saturation is reached, isomorphous Sr in anhydrite should decrease slightly, or at best remain constant, from the beginning of celestite precipitation. On the other hand, an alteration in the partition coefficient b (p. 132) with increasing salt concentration could produce the opposite behaviour.

There are at present no solubility data for $SrSO_4$ in $MgSO_4$-poor solutions, nor unfortunately has the temperature coefficient of solubility of $SrSO_4$ been determined experimentally. By analogy with anhydrite (Fig. 4), a negative temperature coefficient may be anticipated. In view of the regional distribution of celestite in the Stassfurt rock salt (p. 211, 212) this seems to be a point of great importance.

c) The Sr/Ca Ratio

In normal seawater the ratio $\frac{\% \text{Sr}}{\% \text{Ca}} \cdot 1000 \approx 20$ (Table 1). During the static evaporation of normal seawater the ratio $\frac{\% \text{Sr}}{\% \text{Ca}} \cdot 1000$ in the solution increases from the beginning of precipitation of Ca compounds (the Sr in the solution is enriched). At the onset of gypsum precipitation it has the value 22 (neglecting the very small amount of Sr diodochically

[36] See also G. MÜLLER: Zur Geochemie des Strontiums in ozeanen Evaporiten unter besonderer Berücksichtigung der sedimentären Coelestinlagerstätte von Hemmelte-West (Süd-Oldenburg). Geologie, Beiheft **35**, 1–90 (1962).

[37] HERRMANN (1961a) assumed that the precipitation of celestite began only at the beginning of carnallite precipitation, based upon Virck's incorrect (too high) solubility data. According to the determinations of USDOWSKI (1967), celestite precipitates from normal seawater begin at the earliest at the end of gypsum precipitation (see also footnote p. 149).

included in the carbonates; cf. HERRMANN, 1961a), and at the beginning of NaCl precipitation it has the value 44. Within the halite and potash salt regions exact data are not possible as the solubility data for $CaSO_4$ are only approximately known. In addition celestite now begins to precipitate so that further enrichment of Sr is limited by the solubility product of $SrSO_4$. However, a constant value of the Sr/Ca ratio in a solution undergoing further evaporation is not to be expected for the solubilities of $CaSO_4$ and $SrSO_4$ are affected to different degrees by the presence of foreign ions.

D. Natural Salt Sequences and Physico-chemical Models

I. The Main Components

The calculation of the different models was carried out with due regard to natural occurrences. It is not to be expected, however, that these models will be closely approximated in Nature. Nevertheless, despite their manifold differences, natural salt deposits can be reduced to a few types. The classification given here is based upon purely petrographic and quantative chemical data. Other classifications, based upon geological criteria (cf. LOTZE, 1957), will not be discussed here.

In establishing a model, the distinction between primary precipitates and altered salts does not present any difficulty. This is not so with fossil salt deposits, whose history has to be deciphered. They are scarcely ever found unaltered. Furthermore, different alteration processes may succeed one another or operate simultaneously to give the deposit its form. It is the goal therefore of special geological, petrographic and geochemical research to determine the processes of formation and to deduce each phase of the reaction for each deposit separately. In the following examples, too, there are normally no pure types, but for the purpose of instruction subdivisions corresponding to the models developed will be retained, as no specific treatment of salt deposits is intended.

Only in a few cases are indications of primary precipitates preserved in salt deposits (see, for example, gypsum p. 155). Yet the primary bedding, even in spite of intensive metamorphism, remains astonishingly well preserved. In chloride rocks well developed crystalloblastic textures are always observed. The shape of the grains found, as in crystalline silicates, is controlled by the form energy of the respective minerals, as has been explained by STURMFELS (1943). The salt minerals too can be arranged in an approximate idioblastic order (although the order of the KMg and NaMg double salts needs to be checked in further examples).

Idiomorphic anhydrite (additionally carbonate, borate, phosphate, oxides, fluorides),
glauberite, polyhalite,

kieserite,
langbeinite, glaserite,
leonite, bloedite,
kainite,
halite (rock salt),
sylvite, carnallite, tachhydrite,
Xenomorphic bishofite.

The extension of grain boundary planes therefore only proves a replacement, with a mineral lower in the order growing at the expense of one higher up. Even the inclusions are not unique, they are usually of the same age, often older and may even be younger than the host. Because of the narrow stability range of many double salts and the rapid recrystallization (in comparison with silicates), unstable relicts are seldom preserved. Also the chances of the preservation of original grain form is small. Occasionally the existence of pseudomorphs can be demonstrated, but in many cases, as for example after carnallite, they are problematic. DUBININA (1954) reviewed the structures found in evaporites.

The crystalloblastic texture should not be taken as an indication of recrystallization in the solid state. This is highly unlikely in salt deposits. There are always traces of pore and intergranular solutions present, so that the recrystallization proceeds via the soluble phase. Only after plastic deformation can recrystallization in part occur in the solid state, in so far as it, too, is not facilitated by the almost ever-present solution inclusions.

The nomenclature of evaporites, thanks to the practical proposals of RINNE (1908), is based upon mineral composition. This nomenclature has generally been preferred to artificial or local names. STURMFELS (1943) more closely defined RINNE'S nomenclature quantitatively and his proposals are followed here[38]. The major components determine the rock name with the most important mineral at the end. Subsidiary components in the proportions 5–20% are used in an adjectival form. Accessory components of less than 5% are only referred to (e.g. kieserite-bearing) when for some particular reason they are important in the rock. For example in a kieserite-bearing anhydritic-sylvite-halite there is $<5\%$ kieserite, 5–20% anhydrite, $>20\%$ sylvite and more halite than sylvite. Monomineralic rocks should contain $>95\%$ of the major component, but local variations in the accessory minerals up to 20% can occur. Quantitative data can be given as an index.

[38] The suffix "ite" (halitite, etc.) at the end of the principal mineral is not generally accepted (save in the case of sylvinite) and will not be used here. When there is the possibility of error or confusion, the term anhydrite rock, or carnallite rock etc. will be used.

Of the older names, in German the term "Hartsalz" is retained. BOEKE (1910) pleaded for it and for its usage to distinguish kieseritic (= kieserite-sylvite-halite) from anhydritic Hartsalz (= ± kieserite-bearing anhydritic sylvite-halite), a distinction which unfortunately is often neglected in geological literature. It is abundantly clear that, without reliable data on composition, of which the most elementary requirement is a qualitative indication of mineral composition but one which permits a quantitative analysis of mineral composition, it is useless to try to assess the genetic history.

1. Calcium Sulphate Precipitation

In the earliest stages of evaporation the formation of different types of salt sequences has, so far, not been shown. For this reason they must provisionally be treated together. The most important question at this stage is the question of primary gypsum and anhydrite precipitation.

a) Primary Gypsum Precipitation

There are no extensive petrographic data available for Recent gypsum precipitation in saline lakes. In fossil deposits the gypsum shows clear evidence of diagenetic recrystallization, although relict primary deposition structures are recognizable. For example, a rhythmic fine banding can be observed in the Upper Miocene of Sicily in which there is an increase of grain size within the layers (lower part more fine-grained than the upper) (L. OGNIBEN, 1955). This textural variation in the gypsum perhaps reflects original grain size (p. 32). Clearly of diagenetic origin, according to OGNIBEN, are the large gypsum porphyroblasts or swallow-tail twins whose *c* axis is perpendicular to the bedding planes with the swallow-tail pointing upwards. As, however, these may also originate by the slow evaporation of a gypsum-saturated NaCl solution, this textural feature does not have a unique interpretation. The same texture (Mottura's rule, OGNIBEN, 1954) is characteristic of large radial gypsum aggregates in metre-thick banks which overlie the fine rhythmically banded basal layers, each a few tens of centimetres thick, of gypsum or anhydrite (p. 157). These gypsum crystals often contain, on the cleavage planes, anhydrite inclusions which are not crystallographically oriented. They were considered by L. OGNIBEN (1957) to be relicts of primary anhydrite. In the long radial aggregates, sectors with a spherulitic form (the "cauliflower forms" in geological descriptions) not infrequently occur; these OGNIBEN assumed to be due to the volume increase resulting from an early diagenetic alteration of anhydrite to gypsum. Against this, there must certainly be normal polar fibrous growth in which these bulging upper surfaces represent simultaneous radial growth planes. Even the anhydrite inclusions are probably not relicts of primary anhydrite. They are either developed during the fibrous growth of gypsum or are subsequently formed in the gypsum (for similar examples, see NOLL, 1934). Perhaps the secondary origin of fibrous growth aggregates should not be accepted without qualification. It is possible they could develop under special conditions from a fine gypsum mud, or even from a strongly oversaturated solution. They could then be conceived of as a further development stage of the isolated swallowtail twins which are presumably likewise formed directly in the solution. On account of the extraordinary thickness of the gypsum formation of about 400 m and its connection with the swell regions (RICHTER-BERNBURG, Colloq., Hannover 1962), capillary migration of solutions from below cannot play an important part in the growth of fibrous gypsum. Presumably in this (and in other) cases the gypsum precipitation took place over a much smaller surface than the area of evaporating surface, in other words the concentration and over-saturation processes may have to be distinguished from the gypsum precipitation and consideration given to flows of oversaturated solution from the main basin to the place where precipitation occured. No well-grounded conclusions can be drawn about the factors which caused precipitation at the preferred localities. When the temperature coefficients of gypsum solubility in seawater have been re-investigated, it may be possible to give some indication.

b) Anhydrite Rocks

Nowhere in Recent salt lakes is anhydrite precipitated, instead gypsum is always formed (MURSAJEV, 1947). Even in salinas only gypsum is formed[39]. Yet anhydrite is the most important sulphate in fossil evaporites. It occurs as a monomineralic rock in many textural varieties, massive, finely-banded, in flasers, spherulitic (Hauptanhydrit) and in many grain sizes, in rhythmic or irregular alternations with calcite, dolomite, magnesite or clay, etc. and accompanies many other salt minerals. The fineness of texture has been used stratigraphically

[39] The geologically interesting occurrence of Recent anhydrite has been reported from the Trucial Coast, Persian Gulf. This anhydrite is an early diagenetic mineral, developed in supratidal areas of marine lagoonal carbonate sediments, by displacement (KINSMAN, 1966). KINSMAN, D. J. J.: Gypsum and anhydrite of recent age, Trucial Coast, Persian Gulf. Second Symposium on Salt, Vol. 1. Cleveland/Ohio. Northern Ohio Geological Society, 302–326 (1966).

many times (RICHTER-BERNBURG, 1960, and earlier; JUNG, 1960, etc.). This will not be considered here in much detail. Microscopically it is not very characteristic (but see LANGBEIN, 1961).

There is an important morphological peculiarity in the anhydrite pseudomorphs after gypsum. ZIMMERMANN (1909 and earlier) was the first to describe it thoroughly from the pegmatitic anhydrite (A 4), but was uncertain as to its significance.

Fanlike or radially ordered aggregates occur, several centimetres in length, sharply pyramidical or with cone-shaped ends. A single individual has six sides and is often elongated (rhombic) in cross-section. They do not consist of compact anhydrite, but rather of ribs with the internal spaces filled with halite. The ribs have often a fish-bone like arrangement from the centre and run parallel to the swallow-tail ends. The outer parts often show a complete zonal arrangement. The anhydrite itself is very fine-grained, fine-columnar to lamellar, but mostly in idiomorphic crystals in contact with the halite.

For this special form SCHALLER and HENDERSON (1932, p. 18) also had no conclusive diagnosis:

"No conclusive evidence as to the original mineral can be offered, nor is it certain that such an occurrence truly represents a replacement of some earlier mineral. These intergrowths of anhydrite and halite may be original in these minerals".

STEWART (1949–1953) interpreted them as pseudomorphs after gypsum (swallow-tail twins). The correctness of this view was shown by BAIER's angular measurements (BORCHERT and BAIER, 1953). The agreement with the primary or early diagenetic gypsum texture is obvious (L. OGNIBEN, 1954–1957). The secondary growth of anhydrite in rocks such as Hauptanhydrit with its predominantly spherulitic texture is also certain (LANGBEIN, 1961).

On purely petrographic grounds, it is not possible to generalize about the origin of anhydrite, as often anhydrite rocks do not contain pseudomorphs. In particular, the fact that many anhydrites originate from gypsum says nothing about the time when the alteration occurred. The assumption of "post-diagenetic metamorphism of gypsum to anhydrite" (cf. BORCHERT, 1959) cannot be proven petrographically. Since no change in the conditions of formation is necessary for the gypsum-anhydrite alteration, it is perhaps better not to speak of metamorphism (p. 92). Since there are often distinctive anhydrite pseudomorphs after gypsum, the distinction can be made between metastable primary precipitates and the (? early) diagenetic transition to a stable paragenesis.

The rate at which the alteration occurs cannot yet be determined with certainty. The youngest anhydrite beds known to the author are in the Sarmatian of Sicily (Upper Miocene, about 12–15 m. years old, according to KULP, 1960) and there, according to RICHTER-BERNBURG (Colloq., Hannover, 1961) the anhydrite is already closely associated with halite, even though it frequently occurs in bands alternating with massive fibrous gypsum. These fine-grained anhydrite bands are assumed to be due to the alteration of fine-grained gypsum. The coarse fibrous gypsum is preserved, a contrast caused either because the percolating NaCl-rich solutions could not circulate or because of the low rate of alteration of the gypsum. Further investigation must show whether there are considerably younger anhydrites or not.

From the physico-chemical standpoint, primary anhydrite precipitation from evaporating seawater is inhibited by its high nucleation energy.

Gypsum also forms in NaCl-saturated solutions, but here it is certainly altered more rapidly to anhydrite. The water of crystallization released serves only to dilute the pore solutions and dissolve already precipitated salts. As long as the pore solutions and the overlying solution form a communicating system and evaporation is continuous, salt precipitation continues, so that the end result of the dehydration of gypsum is merely to slow down salt precipitation.

As a source of post-diagenetic metamorphosing solutions, gypsum need only rarely be considered. Only gypsum rocks which are in contact with NaCl rocks can persist over geological time. In this case they lie in their stability field.

2. Primary Salt Rocks

a) $MgSO_4$-bearing Salt Sequences

Primary is here considered in an extended sense (p. 92) to include early diagenetic and diagenetic changes (although these are already associated with allochemical and allophase recrystallization). In chlorides and Mg sulphates, however, no pseudomorphs after primary metastable primary precipitates or relicts can be proved. It is also not known to what extent the metastable salts precipitated in the small, shallow salt lakes of the present day can be compared with the much more extensive salt seas of the geological past. It is possible that in the deeper salt lakes in some cases products of stable parageneses precipitated directly. An important, if not entirely satisfactory geological criterion for primary precipitation is the absence of changes in form over a very small space.

The most important indication of primary precipitation from unaltered seawater without reaction at the transition points is a thick kainite layer under a halitic kieserite-carnallite (Fig. 26). To this category belong in all probability the Sicilian salt deposits. Here, according to MAYRHOFER (Colloq., Hannover, 1962) a kieserite-bearing kainite rock, with 2–5 cm thick kainite bands (max. 20 cm) alternating

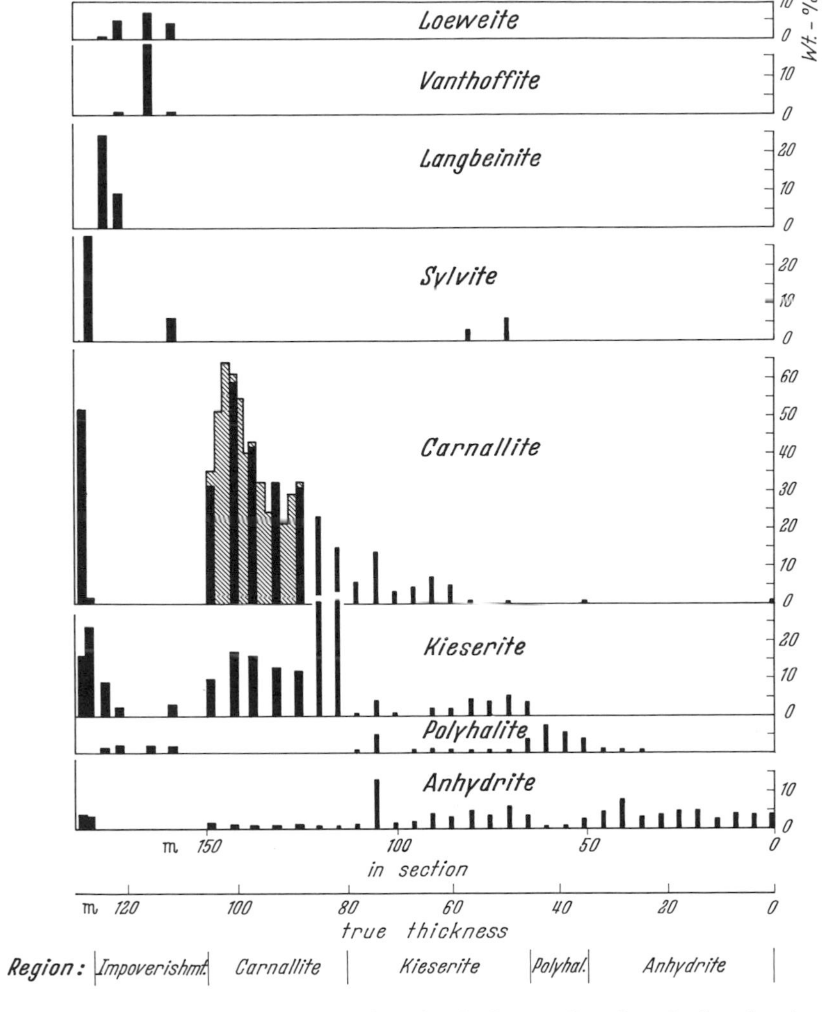

Fig. 38. Composition of the salts of Zechstein 2, Neu Stassfurt, Berlepsch mine, after RIEDEL (1912). Carnallite content after BOEKE (1908) shaded. NaCl = remainder up to 100%, not represented (plus clay and other impurities). At 150 m of the traverse a thrust occurs. The rocks to the right of it are primary, those to the left have been altered by solution metamorphism

with halite, follows above a rock salt series. Above this there follows a kieserite-bearing carnallite above which kieserite-free sylvite rocks should occur. The uppermost horizon is, however, very problematic but in the absence of quantitative data no further discussion is possible.

The most complete primary salt succession, but without the kainite, is the development of the Zechstein 2 in the region of the Stassfurt articline. The structure of the known cross-section is shown in Fig. 38 after RIEDEL (1912). Above the thick carnallite layer at 150 m there occurs a major overthrust (SCHÜNEMANN, 1913; LÜCK, 1913) and the rocks above, strongly altered by metamorphism (p. 178), correspond exactly to the carnallitic horizon. In contrast, the major part of the beds below the overthrust correspond to "primary" precipitates. Above one another follow anhydritic halite, polyhalitic halite, kieserite-halite, kieseritic halite-carnallite.

According to these data, the Stassfurt salt succession corresponds qualitatively to polyhalite precipitation with the complete absence of kainite (p. 102). The complete absence of kainite in this case does not seem probable, and furthermore the amount of polyhalite observed is certainly inadequate. These quantitative discrepancies cannot be explained by the small amount of glauberite precipitated below the polyhalite layer. The glauberite is nevertheless often overlooked.

To explain these discrepancies, it could be asked whether a primary kainite region or a metastable epsomite + sylvite paragenesis might have occured above the kieserite layer. Although these rocks are not present, at least their alteration products should be found. Kainite could have been removed either by thermal metamorphism or by reaction with solution *R* during primary evaporation (p. 93, 94). The metastable epsomite-sylvite paragenesis would also react with solution *R* with the precipitation of kieserite + carnallite (leaving open for the present the question of intermediate steps with metastable $MgSO_4$ hydrates). The structure of the observed salt sequence is thus explicable as primary.

It is not neccessary to assume a subsequent reaction in every case. When concentrated residual solutions of an earlier evaporation cycle are preserved in the same marine basin and are not flushed out because they are overlain by more dilute influxes, then the kainite, too (or metastable epsomite-sylvite) will react with the basal solution. In the Stassfurt area, residual solutions may be considered for the Werra Series. The restricted spatial distribution of the Zechstein 1 potash salts so far known suggests that only a relatively small effect is to be expected. This reaction is certainly the reason for the absence of kainite and other transition layers above the thick rock salt bands within the carnallite seam. These rock salt bands (with anhydrite at the base) are developed from fresh (or partially concentrated) seawater influxes. Gypsum, which was probably altered to anhydrite at a very early stage, and halite are the first to be precipitated from the influx; however, anhydrite and halite are stable in the carnallite-saturated basal solution, while the subsequently precipitated salts react with it, so that above halite carnallite + kieserite + halite repeat.

The processes considered, namely, the syngenetic formation of polyhalite (± glauberite) with partial $MgSO_4$ impoverishment, but more particularly the reaction in concentrated solutions, lead to a decrease in the primary kainite region. Thus, broadly speaking, the Stassfurt salt sequence (below the thrust) is primary in its basic features, only the extent of possible reactions and thermal metamorphism of the original kainite or metastable epsomite-sylvite beds being in question. The assumption of later reaction upon reaching carnallite saturation may perhaps be tested by means of bromine analysis (p. 202).

The petrographic characteristics of polyhalite rocks in most cases point to their being formed at the expense of anhydrite. Noteworthy in this respect are the careful observations of SCHALLER and HENDERSON (1932) on the polyhalites of New Mexico and Texas and of STEWART (1949) on the Yorkshire salt deposits which probably correspond to the German Zechstein I (LOTZE, 1958). As early as 1913 LÜCK had shown the secondary origin of the Stassfurt deposits. The process does not take place in a single step, and proof of this is forthcoming in the descriptions by SCHALLER and HENDERSON (1932) and by STEWART (1949) of the polyhalite alteration of pseudomorphs of anhydrite and halite after large gypsum swallow-tail twins. It follows from this that the gypsum was initially precipitated in large, twinned, loosely packed crystals with perhaps fine-grained gypsum or solution in the intervening spaces. This gypsum was first altered to anhydrite, the large gypsum crystals probably lasting somewhat longer, while the spaces were filled either with halite or, more commonly, with alternating zones of anhydrite and halite. (Skeletal forms, infiltration pseudomorphs according to SCHALLER and HENDERSON, 1932.)

Up to this point the development is the same as that of anhydrite rocks. The conversion of the anhydrite to polyhalite begins after this, with at least partial solution of halite, with polyhalite precipitating in its place.

The direct formation of polyhalite from gypsum has not yet been proved. It is possible that the relatively rare polyhalite spherulites may be so interpreted.

In addition to the pseudomorphs, the other petrographic criteria for the secondary origin of polyhalite according to SCHALLER and HENDERSON (1932) are:

1. convex polyhalite surfaces against anhydrite,
2. irregular penetration contacts,
3. anhydrite relicts in polyhalite,
4. small scattered polyhalite aggregates in anhydrite which do not have immediate contact with neighbouring connected polyhalite masses,
5. formation of large idiomorphic, second-generation anhydrite crystals in polyhalite,
6. continuity of magnesite bands from anhydrite into polyhalite.

These petrographic indications themselves are all too often not clear. However, the initial stages of replacement starting from the former anhydrite crystal boundaries, occur very frequently and are connected by transitions with later stages, which leaves little doubt that the indicators are correctly utilized. For the dense fine-grained polyhalite rocks which often form thick layers, a secondary origin cannot be proven petrographically. The spherulitic and fibrous forms have to be interpreted as secondary because of the nature of their crystal growth.

Several Russian authors (e.g. VALYASHKO and NECHAEVA) have assumed a primary origin for polyhalite in addition to its secondary development. YARZHEMSKY (1954) assumes that the cementing polyhalite in certain thin-bedded sandstone intercalations in the pre-Carpathian salt deposits is of primary origin. (In the majority of these cases, however, he also identifies secondary polyhalite.)

Indisputable examples are certainly rare.

Most examples of secondary (and the more uncertain of the primary) polyhalite formation can be explained by the model of syngenetic polyhalite formation. When high concentrations are reached in the available solution (close to the saturation point of the salts of the five-component system) then, by repeated influxes of fresh seawater, thick polyhalites can form. In a qualitative form this model was used to explain the Yorkshire salt deposits (STEWART, 1949, p. 662) and it could probably also be applied to the Permian of New Mexico.

Secondary polyhalite formed as a result of metamorphism will be treated later (p. 189).

b) $MgSO_4$-free Salt Sequences

There are numerous examples in this category. Their common characteristic was first indicated by LOTZE (1938) and explained by means of a model by VALYASHKO (1958). The important Canadian potash deposits to a significant extent belong to this group, as well as the

Permian salts of the upper Kama (E. Russia, DUBININA, 1954), the upper Devonian salts of the Pripyat salt basin in the Dnieper-Donets region (SHCHERBINA, 1960) and the very carefully investigated Tertiary salts of the upper Rhine. Some examples are also known from the marginal zones of the original Zechstein basin. Unfortunately, in most of these deposits only the potash seams are known in detail. The most important feature of this group is the absence of primary kainite and Mg sulphates both below and within the potash seams. Geologically they seem mainly to be characterized by the small thickness of halite in comparison with the thickness of potash salts. It is probable that within the group at least two sub-groups with transition types can be recognized although at the present time this can only be done very approximately.

α) Without Mg Sulphate, Anhydrite Predominant

This extreme case is found in the Zechstein 2 at Königshall-Hindenburg. The lower section Na 2 α has not been opened up. The middle section Na 2 β contains coarse-grained, indistinctly bedded anhydritic halite with the halite crystals often elongated parallel to the bedding. Clay aggregates occur in occasional flocks, often at the margins of halite crystals, but not as a continuous film, most often in indistinct enrichment zones. The upper section Na 2 γ (~ 20 m) is clearly banded with alternating anhydrite and clay-anhydrite layers; it is also somewhat less coarsely grained.

Since petrological and chemical data confirm that magnesium sulphates are virtually absent from the layer structure of this deposit, representation of the Ca content gives a good indication of its composition (Figs. 39, 40). In the older rock salt only anhydrite and halite are found (conversion factor Ca$\rightarrow$anhydrite $= 3.395$); in the upper part there may be a small amount of sylvite ($<1\%$). The points plotted are mostly the results of block samples (mostly taken over $2^1/_2$ m perpendicular to the bedding). Too little is as yet known about variations in composition within the same horizons of Stassfurt rock salt. Unfortunately no outcrop has yet been found under a seam of carnallite.

Over the greater part of the mining area, the potash layer is much altered as a result of solution metamorphism (p. 183). No extensive occurrence of primary sylvite rocks has been proved with certainty. Possibly the basal layers of the potash deposit furnish an example of this. Yet, they are abnormal in that they have the highest anhydrite content ever found. The potash deposit was formed originally as an anhydrite-bearing halitic carnallite. In the Königshall-Hindenburg mine only

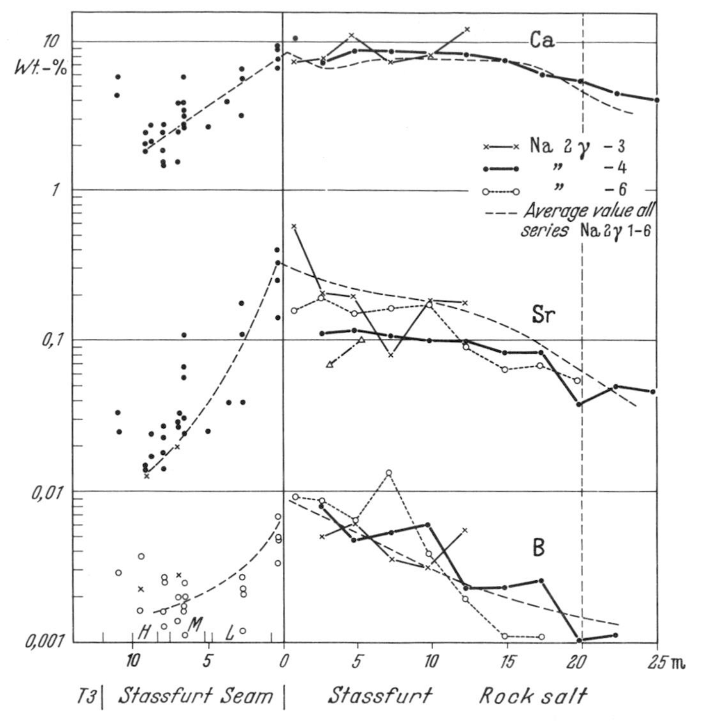

Fig. 39. Variation in B, Sr, Ca in the Stassfurt rock salt and Stassfurt seam in Königshall-Hindenburg. The seam is shown as of anhydritic sylvite-halite. Analyses BRAITSCH (Sr, Ca), HARDER (B)

partial horizons of carnallite form have been explored (see Table 25). But the Eichsfeld bores have revealed similar conditions which, because of their importance, are listed in Table 20. The Felsenfest No. 3 bore reached a thick anhydrite-bearing halitic carnallite layer. The kieserite content throughout is less than $^1/_2$ %. The upper 5 m are developed as a Hartsalz but this gives no clue as to whether it is a primary (in other words a recessive phase, explicitly not represented for potash salts, but see p. 111) or a secondary development. The neighbouring Felsenfest No. 18 bore has also disclosed practically kieserite-free anhydritic halite carnallite.

This type corresponds qualitatively to the model with extreme $MgSO_4$ impoverishment (p. 97). The halite below the carnallite is rather too thin. If a primary sylvinite base below the carnallite is entirely missing, which is not certain, it might be supposed that the composition of the solution had been displaced further towards the $CaCl_2$-containing region (Fig. 28) where carnallite precipitates immediately after halite.

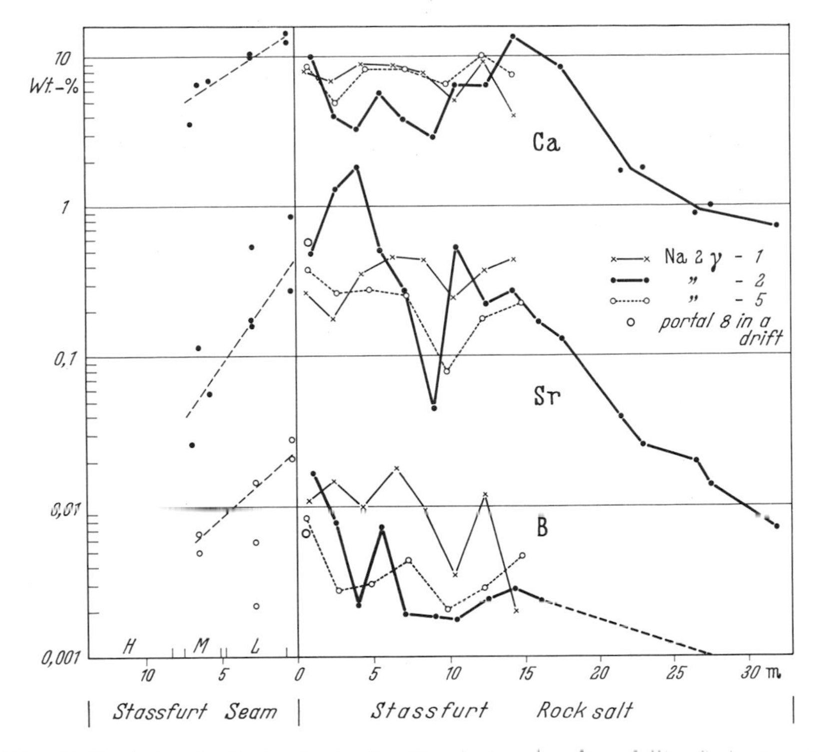

Fig. 40. Variation in B, Sr, Ca in the Stassfurt rock salt and Stassfurt seam in Königshall-Hindenburg. The seam is shown in a K-free (impoverished) form. Analysis BRAITSCH (Sr, Ca), HARDER (B)

But such questions cannot be settled until further geochemical and petrographical data have been obtained.

For comparison the Teistungen No. 1 bore from the Duderstädter Saddle near the leached zone is included in Table 20. It is of special interest because only the lower part of the deposit is kieserite-free. The lowermost part apparently corresponds to a primary anhydritic sylvinitic halite, then follows an anhydritic carnallite-halite. Above that lies a kieseritic halite-carnallite and kieseritic sylvite-halite (probably not primary) (anal. No. 6), overlain by carnallitic halite and kieseritic-carnallitic halite (see note 5 to Table 20).

There are unfortunately no petrographic descriptions of this important deposit. However, based on the available chemical data, it may be regarded as a transition type with partial $MgSO_4$-deficiency corresponding to model III in Table 12. There appears to be a primary sylvinitic basal layer. The middle section, analyses 15–20, might

Table 20. *Analyses of the Stassfurt seam in the near-shore region. Anal. Dr.* P. LÖHR (ALBERTI *and* HEMPEL *Lab., Magdeburg 1906, 1907)*

Felsenfest No. 3 bore (near Hüpstedt, Eichsfeld)

Anal. No.	2–10	12–19	20–23	24–30	31–36	37, 38, 40[b], 42, 43
Depth m.	827.65–832.40	832.85–836.85	836.85–838.85	838.85–842.57	842.57–845.57	845.57–849.00
Rock	Upper anhydritic "Hartsalz"	Upper carn.	carn. halite	carn. halite	halite carn.	halite carn.
KCl	34.87	16.09	7.71	12.03	15.39	15.85
NaCl	47.23	34.23	74.18	53.99	38.36	36.29
$MgCl_2$	0.29	19.24	5.93	13.31	18.07	19.29
$MgSO_4$	0.056	0.18	0.02	0.02	0.07	0.05
$CaSO_4$	9.33	4.63	4.23	2.59	4.38	3.73
Insol.	7.59	1.08	1.00	1.40	2.29[a]	0.88
H_2O + difference	0.61	24.49	7.44	16.58	21.95	23.82
Total	99.98	99.94	100.51	99.92	100.51	99.91

The analyses were carried out on full bore cores, usually on $^1/_2$–1 m sections. The analyses have not previously been published (analyses made for concession purposes at the request of the Gewerkschaft-Hüpstedt-Beberstedt).

Tabulated values were obtained from the individual analytical determinations of ions and averaged.

Table 20 (continued)

Felsenfest No. 18 bore, 1500 m ESE of the Beberstedt shaft			Teistungen I[e], $7^1/_2$ km WSW Bischofferode No. 2 shaft (Duderstädter Saddle)				
Anal. No.	3–9	10–17	6	7–14	15–20	21	22–31
Depth. m.	824.4–849.1	849.1–857.1	556.82–557.32	557.32–561.45	561.45–564.62	564.62–565.14	565.14–569.60
Rock	anhydritic halite-carnallite	anhydritic halite-carnallite	kieserite-halite-carnallite; kieserite-sylvite-halite	carnallite-halite	anhydritic sylvinitic halite	carnallite-halite	sylvinitic halite
KCl	15.36	15.68	17.78	9.23	15.13	6.24	9.33
NaCl	27.46	30.66	25.73	60.98	67.85	74.98	84.15
$MgCl_2$	20.28	20.42	11.68	11.32	0.38	7.72	0.33
$MgSO_4$	0.24[c]	0.01	24.74	0.28	0.00	0.00	0.00
$CaSO_4$	7.44	6.24	2.21	3.37	8.45	1.94	4.21
Insol.	4.20[d]	2.43	0.61	0.95	6.71	0.17	0.46
H_2O + difference	24.98	24.56	17.24	13.93	1.56	8.95	2.01
Total	99.96	100.00	99.99	100.06	100.08	100.00	100.49

[a] Anal. 32: 6.26%, the remainder ~1%.

[b] Anal. 39 and 41 of anhydritic-carnallitic halite not included.

[c] Anal. 4 contained 1.56%, all the others $\leqq$0.05.

[d] Anal. 4: 12.77%, the others <6%.

[e] Anal. 1–3 contained K_2SO_4 instead of $MgSO_4$, $CaSO_4$ without $MgCl_2$; polyhalite probably replaces anhydrite ± kieserite. Anal. 4 carnallitic halite, anal. 5 kieseritic-carnallitic halite.

represent an intercalated recessive phase, in other words indicate a primary anhydritic sylvite-halite, provided that tectonic repetition can be excluded, which cannot be done with certainly from borehole data. A careful bromine examination could exclude possible repetition, since here no great complications introduced by secondary alterations need be feared.

It is to be hoped that these relations may be cleared up in the near future by observations in the nearby Bismarckshall mine (Bischofferode). In the final analysis there is as yet no petrographic evidence for the theoretical model with a primary facies differentiation on carnallites distributed over a wide area. Excessive NaCl and anhydrite contents compared with the models in Table 12 are no counter-argument since these are caused by influxing seawater.

β) Without Mg Sulphate, Dolomite (Ankerite) Predominant

This subgroup is represented in the Tertiary salt deposits of the Upper Rhine valley. Only the potash seams are well known, while for the underlying, relatively thin halite (40 m) quantitative data for the main components are lacking. According to GÖRGEY (1912) it includes intercalations of thinly bedded dolomite marls, namely "fine-grained anhydritic-halitic dolomite", and also "fine-grained dolomitic anhydrite", besides single anhydrite layers. These are therefore related to the pure anhydritic subgroup.

For the potash beds GÖRGEY (1912) gave an excellent petrographic description which was further refined by STURMFELS (1943) and extended by BAAR and KÜHN (1962). The rhythmic alternation of halite and sylvite beds developed in an astonishingly uniform manner over the entire region is characteristic of the Upper Rhine. Usually the underlying halite bed has a nearly planar basal contact but a comb-like interdigitating upper contact with the sylvite layer whose upper contact with the next halite layer is once again smooth. Macroscopically there is a very distinct colour change between the white or gray halite rocks and the orange to intensely red-coloured sylvite layers (see the colour Table 25 in WAGNER, 1953). This formation is constant over many kilometers where individual bands often resemble each other in great detail (STURMFELS, 1943). Several types of sylvite rock can be distinguished according to grain size, anhydrite content, haematite content and distribution:

1. orange-coloured, coarse-grained: 85–95% sylvite, anhydrite in traces,
2. red-flecked, medium-grained: 80–95% sylvite, 1–4% anhydrite,
3. dark red, fine grained: 85–95% sylvite <1% anhydrite,

4. as in 3, but anhydritic: 5–20% anhydrite, in part dolomitic,
5. red-bordered, medium-grained: 20–98% sylvite, 1% anhydrite.

Two further types of halite-sylvite exist, but only in the uppermost layer. They are petrographically distinct and probably not primary (p. 172).

Generally the halite-sylvite (5) forms a transition from the halite layer at the base to the sylvinites; the fine-grained anhydritic sylvinite (4) in contrast often forms at the upper limit of the sylvite layer. The anhydrite content varies in the same sense as the NaCl content.

Only the coarse-grained sylvites (1) are serrated, otherwise the intergrowth increases as the proportion of NaCl increases, with sylvite filling the gap according to its position in the idioblastic sequence. The smaller sylvite grains are often flattened parallel to the bedding. The intensity of the red colour is inversely proportional to the grain size. The thickness relationship between the halite and sylvite layers varies; the fine-grained sylvites, however, are always thin (< 5 cm) and lie above a thicker halite layer, while the coarse-grained sylvites may reach up to 10 cm in thickness. The latter may occasionally contain thin halite layers, and characteristically oriented NaCl growths (up to 1.2% NaCl).

Table 21. *Sylvite rock (hand samples). Wittelsheim, Alsace (after* Görgey, *1912)*

	1	2
Halite, NaCl	8.19	4.09
Sylvite, KCl	90.10	95.62
Anhydrite and residuals, $CaSO_4$	1.63	0.26
H_2O	0.08	0.04

1. Average composition of 7 intensely red, fine grained halitic sylvinites (sample 102 not included, being with 30.96% NaCl, 55.52% KCl, 14.44% $CaSO_4$ considerably different from the average).
2. Average composition of 6 bright red, coarse-grained halite-bearing sylvinites.

Each halite-sylvite cycle begins with a thin clay film, which often also contains anhydrite (and dolomite) and halite. Often these separating beds pass into narrow marl bands. The average composition of the sylvinite hand samples from Alsace is given in Table 21. The chacteristic features are the low NaCl content, the low $CaSO_4$ content and the complete absence of $MgSO_4$.

The average composition of the whole potash deposit in the Buggingen mine according to Sturmfels (1943) is: KCl 34.2%, NaCl 52.8%, $MgCl_2$ 0.3%, $CaCl_2$ 0.3%, $CaSO_4$ 5.0%, carbonates and silicates 7.3% (total 99.9%).

The halite transition layers are thus slightly in excess of the sylvinites. $MgSO_4$ is totally absent.

In Alsace further carnallite rocks occur above the rhythmically banded halite-sylvinite succession. The fine-grained are a pale red, the coarse-grained have an intense red colour, haematite being the pigmenting material. Halite and anhydrite occur in varying amounts even in the thick carnallite beds. The anhydrite sometimes forms as radial aggregates between or in the carnallite grains. Often it occurs together with clay in more or less continuous bands.

The composition of the carnallite hand samples is shown in Table 22.

Table 22. *Carnallite rocks (Wittelsheim, Alsace) according to* GÖRGEY *(1912)*

	1	2	3	4
Carnallite	97.8	84.7	92.1	99.2
Halite	1.8	13.0	4.7	0.6
Anhydrite	0.4	2.3	2.6	0.2
Fe_2O_3	trace	trace	0.1	trace
H_2O	—	—	0.5	—

1. Bright red, medium grained (483 cm above base).
2. Light, fine grained (486 cm above base).
3. Dark red, coarse sparry (493 cm above base).
4. Bright, medium grained (520 cm above base).

The analyses of the carnallites from Alsace and Eichsfeld are not directly comparable, for those from Alsace were carried out on almost pure hand samples, while the Eichsfeld analyses (Table 20) were carried out on averaged samples. In Königshall-Hindenburg, however, almost equally pure carnallite rocks (Table 25) are found.

The absence of Mg sulphate is a primary indicator. Because of that the formation of primary Hartsalz in the Upper Rhine salt sequences can be explained in terms of the model entirely deficient in $MgSO_4$. The high KCl content of the sylvinite bands cannot, however, be explained solely by evaporation. A subsequent mixing of halite and sylvite seems excluded. From the physico-chemical standpoint, the best explanation is of primary precipitation induced by cooling (cf. STURMFELS, 1943). However, the NaCl content of the sylvinite is less than the calculated amount (p. 104, 105). This could be the result of the uncertainty about the temperature coefficient of NaCl solubility or of foreign ions (a low $CaCl_2$ content in the solution, for example). An isothermal evaporation would give much higher NaCl contents in sylvite.

Apart from temperature variations, no other sufficient mechanism is known for the separation of KCl and NaCl in natural salt solutions. The cooling cannot be directly associated with annual temperature fluctuations. In any case, this seems unlikely (cf. LOTZE, 1957, p. 102). A much more plausible assumption is that the solution cooled as a result of flowing into a cooler central part of the basin. Each clay-halite-sylvite cycle would then correspond to the influx of more solution.

Above all, the model of potash precipitation in the center of the basin can explain the quantitative discrepancies between the thinness of the underlying halite layer in comparison with the potash. The NaCl precipitation in total is spread over a wider area, as can be seen from the new palaeogeographic facies map of the Upper Rhine basin (GUNZERT, 1962). BAAR and KÜHN (1962) made their estimate of the reserves from the salt content per 100 cm^2 in the Buggingen mining area. Because of the unequal areas of potash and halite distribution, their calculation, despite its astonishingly close agreement with the observed salt contents, throws little light on the origin of the whole deposit, particularly as even static evaporation with slightly altered evaporation temperatures gives complete agreement with Table 12, No. IV. The highly concentrated influxes, that is, almost at carnallite saturation, assumed by these authors, can, however, be explained by influxes from the marginal areas of a shrinking lagoon along the lines of the explanation given here. There is naturally, agreement concerning the primary nature of the sylvite precipitate, and BAAR and KÜHN (1962, p. 320) themselves speak of a shrinking lagoon, so that really all that is unacceptable here is the interpretation of the influxes as "strongly pre-concentrated" seawater.

VALYASHKO (1958) espoused a similar view, even if his additional assumption of irregular synsedimentary sinking of the basin and drying out of the greater part of the salt lagoon (the so-called "Dry Sea") are not apparently necessary for the Upper Rhine deposits.

Recently the attempt has been made to explain these salt deposits by solution metamorphism (BORCHERT, 1959, 1960). The impoverishment taken as an indication for this, however, is all related to faults or alkali basalt dykes (WIMMENAUER, 1952) and is independent of the origin of the potash deposit. Additionally, in all clear cases of solution metamorphism there is a complete absence of carnallite relicts in the rhythmically banded lower and middle parts of the deposit. Moreover, the origin of the sylvinites by solution metamorphism is unconvincing because of the particularly high KCl and very low NaCl contents. Their interpretation as primary deposits seems preferable. Only the sylvite rocks of the uppermost layers are probably caused by solution metamorphism.

There occur in the upper 30 cm of the seam (according to STURMFELS, 1934) a few bands of dark red and white, coarse-grained halite-sylvite with a considerably higher halite content than in the remaining sylvite of the Buggingen deposit. Also found in these bands are plate-like haematite inclusions which are absent in the other sylvite rocks. STURMFELS (1943) has indicated that these correspond stratigraphically to the carnallite layer of Alsace and regarded them as formed by the alteration of carnallite. Since an $MgSO_4$-free paragenesis is being considered, there is no indication of the possible temperature at which the alteration occurred (see Table 14, Nos. 4, 12, 17). Halite-sylvinite forms at all temperatures. Other independent, clear indications of subsequent infiltrating metamorphosing solutions are absent. Thus in the sense of BAAR and KÜHN (1962) the alteration of the carnallite had to be assumed to take place immediately after precipitation in somewhat diluted solutions (though this must be consistent with a higher KCl content in the altered rock than indicated in Table 14). As this alteration requires a change in the concentration of the solution, in terms of the limitations on p. 92 it can no longer be considered as primary precipitation, especially since there is no criterion for the time at which the alteration occurs. However, this semantic question does not require further discussion here.

Summarizing, it can be shown that for primary salt successions two principal groups can be defined based upon composition, the $MgSO_4$-bearing and the $MgSO_4$-free parageneses between which intermediate types occur. Within the $MgSO_4$-bearing sequences a distinction can be made between the kainite-bearing and polyhalite-bearing sequences, although an unaltered kainite-bearing succession is not yet well known. The subdivision can probably be refined by special petrographic examination of complete salt sequences, analogous to the models given in Table 12. A further group to be considered are the salt sequences without primary sylvite precipitated out of $CaCl_2$-bearing primary solutions.

Within these compositional groups different conditions of formation can be considered, the salt deposits formed by (a) isothermal evaporation or (b) cooling or warming. This distinction is particularly important in sylvite and carnallite rocks. In many cases brine is probably concentrated to saturation by evaporation, but precipitation occurs as the solution cools on sinking to deeper parts of the basin. This will have happened in particular in the deeper salt lagoons of the geological past, especially with chlorides, whereas gypsum appears to be preferentially deposited as a result of warming of the shallower parts of the basin.

Finally a distinction should be made between primary precipitates from stable and metastable equilibria, or alternatively, the transition from metastable to stable primary parageneses. As too little detail is known about chloride salts, both modes of formation have to be considered together in the extended sense of primary precipitation. There may be some characteristic petrographic indication of this. The transition from metastable to stable equilibrium is usually associated with a distinct decrease in volume. Investigations ought above all to be made to test whether in some cases the finely folded clay inclusions in kieserite halite (e.g. the "worms" at the base of the Hesse seam in the Werra potash district) and the finely folded kieserite-rich layers in carnallitic rocks, e.g. of Stassfurt, may represent shrinkage in consequence of the significant volume decrease in the transition from a metastable to stable paragenesis. (In the case of the "worms", a volume decrease in the adjacent rocks might have to be taken into account.) Unfortunately at the present time there are no known specific indicators of contraction structures.

3. Thermally Metamorphosed Salt Rocks

The effects of thermal metamorphism are to be expected only in parageneses which include $MgSO_4$, as the $MgSO_4$-free sequences are stable in the temperature range usually considered, up to about 100° C. Since the temperature mounts steadily both in time and space, the thermal metamorphism of a uniform widely distributed primary precipitate would not be expected to produce small-scale irregularities in paragenesis within the same bed.

Considering the possible primary precipitates, thermal metamorphic effects have to be expected predominantly in kainite-bearing rocks. The Mg sulphate hydrates vanish early at low temperatures; bloedite, therefore, has no real importance as a primary precipitate for quite a slight $MgSO_4$ impoverishment results in its disappearance.

Almost all of the examples of thermal metamorphism quoted in the literature up to the present are controversial. However, in the case of the so-called "Flockensalz" in the Hesse seam it is probably well founded.

The important characteristics of the "Flockensalz" according to ROTH (1953) are: a uniform whole rock development in the Werra and Fulda mining districts without lateral paragenetic changes; the differences which do occur are related only to proximity to the margin of the basin in the Fulda area (a thin sequence, red colouration of the potash salts).

Pure carnallite is not found in the Hesse seam (HOPPE, 1958, p. 49, in contrast to AHLBORN 1953 and older publications), impoverishment is secondary. Transition layers with polyhalite, kieserite, etc. are also absent under the seam. Only in the Neuhof-Ellers mining district have the partial horizons of the Hesse seam been investigated (Table 23).

Table 23. *Average composition of the sylvite kieserite-halite in the Werra Series (weight %)*

Seam	Thuringia			Hesse			Hesse		
Mining district	Winters-hall	Herfa-Neurode	Hattorf	Winters-hall + Herfa-Neurode	Hattorf	Neuhof-Ellers	"Basis"	"Worm"-zone	"Flocken-salz"
							Neuhof Ellers		
Source	WEBER (1961)			WEBER (1961)			ROTH (1953)		
KCl	16	15	15	15	15	21	23	20	23
NaCl	74	64	71	61	59	41	24	45	31
Carnallite	n. o.	n. o.	n. o.	0	n. o.	n. o.	1	1	1
Kieserite	12	21	14	24	26	38	48	33	43
Anhydrite	n. o.	n. o.	n. o.	n. o.	n. o.	n. o.	1	1	1

n. o. = not observed.

The upper, bedded, part of the Hesse seam probably does not belong to this type. It may sometimes be carnallitic (BESSERT, 1933). In hand samples the kieserite may amount to more than 60% of the "Flockensalz", and halite to less than 5%. The kieserite is always $1^1/_2$ times as abundant as the sylvite and this is true also of the analysed samples from the Werra district (BESSERT, 1933, p. 27; WEBER, 1961).

The "Flockensalz" has a loosely packed texture made up of isometric polygonal kieserite grains (up to >1 mm ∅) with rounded margins. In the intervening spaces lie halite and sylvite, with the halite possessing convex or, at most, even grain boundaries against the sylvite. Apparently only one chloride generation occurs. Anhydrite sometimes occurs in the form of large crystals with a polyhalite coating (cf. KÜHN, 1950/51, Fig. 2; 1957, Fig. 3). The form of the kieserite is consistent with its position in the idioblastic sequence.

As the primary kieserite-sylvite paragenesis is only possible above 72° C and then always possesses considerably more sylvite than kieserite, the "Flockensalz" cannot be explained as primary. On the other hand, it cannot derived from an initial pure kainite for this, upon being thermally metamorphosed, would give mainly langbeinite besides kieserite and sylvite (Table 13). Since, however, primary kainite is followed by primary carnallite, which was originally the case in the upper part of the Hesse seam, a combination of thermal and solution metamorphism may be considered (p. 125). This would produce sylvite and kieserite in the appropriate 1 : 1.5 mass ratio.

Since, in addition to kainite, kieserite was present in the initial rock and was preserved during the thermal metamorphism, quantitatively the mass ratio should be greater. This corresponds to the observed ratios (sylvite : kieserite = 1 : 1.6 to 1 : 2.1).

Scarcely likely, and petrographically improbable, is the derivation of "Flockensalz" by solution metamorphism from halitic kieserite-carnallite (e.g. D'ANS and KÜHN, 1960; WEBER, 1961). A calculation shows that the initial rock from which the "Flockensalz" (Table 14, No. 3) must be derived should have 63% carnallite, 26% kieserite, 10% halite, 1% anhydrite + residue. The "basis" (Table 23) gives 63% carnallite, $28^1/_2$% kieserite, 8% halite, $^1/_2$% anhydrite, that is low halite. Such a low halite content is not even to be found with purely static evaporation of seawater, and particularly not with the present kieserite-carnallite ratio (cf. p. 259).

The origin sometimes assumed from transformation of primary sylvite and kieserite deposited in alternating layers (e.g. KÜHN, 1957 and earlier) is excluded. Evidence of erosion is unknown and the corresponding initial rock with sylvite and kieserite in alternating separate bands does not exist. For example, such a kieseritic layer would have the composition 22.73% KCl, 23.82% NaCl, 45.73% $MgSO_4$ (D'ANS and KÜHN, 1940, p. 61). As a primary rock this is impossible. Moreover, if transformed in concentrated solutions, kieserite and sylvite would be altered to kainite, whereas in unsaturated solutions they would be redissolved. The kieserite is not flushed out (KÜHN, 1957, etc.) as solution forms of kieserite are sharp-edged, as may be seen in the examination of residues. If the kieserite were to show solution forms, the sylvite ought to be completely dissolved because of its higher solubility and solubility velocity.

The Zechstein 1 sequence in the Werra district corresponds at least in its upper part, the Hesse seam, to normal precipitation out of almost unaltered seawater. It contains a "primary" kainite region which was perhaps developed from the (early diagenetic) alteration of a metastable sylvite-halite-epsomite precipitate. But this is not an undisturbed evaporation sequence since the intervening rock salt horizon between the two seams, and also the uppermost part of the halite under the Thuringian seam, imply precipitates from less highly concentrated solutions. It is therefore not clear why, on the one hand, in the Hesse seam kainite has not reacted to form kieserite and carnallite once carnallite saturation had been attained, while in the Stassfurt seam this is supposed to have occurred (p. 160). It is, of course, possible that the diagenetic history of the two seams was different. It is also conceivable that the kainite reaction in the Hesse seam was only partially completed when evaporation was interrupted by an influx of fresh solution. By this means as kainite kieserite carnallite initial rock could, without additional assumptions, give rise to sylvite-kieserite by thermal metamorphic processes at 72° C.

According to the average composition of the Hesse seam in the Werra and Fulda districts given by WEBER (1961), the Fulda area is characterized by a higher kieserite and sylvite content, but an unaltered kieserite/sylvite ratio and a correspondingly smaller halite content. WEBER interpreted this difference as a result of a strong thermal gradient (dynamic polytherms of BORCHERT 1940, 1959) with the higher temperatures existing during primary precipitation in the Werra district. The lower NaCl content suggests a somewhat higher temperature in the Fulda district, which is in keeping with its being nearer the coast. The regional facies differences still require more extensive investigation, and in particular the different zones of the seam in the Werra district.

The kieseritic-sylvinitic halite of the Thuringia seam cannot be derived by thermal metamorphism of kainite + carnallite, for the kieserite content relative the sylvite is too low according to Table 23. In this case, WEBER'S interpretation (1961) in terms of solution metamorphism may well be the correct one, and this is consistent with the relatively small lateral facies changes. The assumption of a metamorphic phase I, that is, the alteration of hexahydrite-halite-carnallite to kieserite-halite-carnallite, is hypothetical and should in any case be interpreted as an early diagenetic alteration of a metastable to a stable primary precipitate.

The higher temperatures, above 72° C, required for thermal metamorphism should have been reached during the period of Tertiary volcanism (Vogelsberg, Rhön) at the latest. At that time there should have been an appreciable increase in the geothermal gradient in rocks with 1000 m overburden, of the same order as that found today in the heart of the Urache volcanic area (A. SCHMIDT, 1921) where the gradient is about 1°/11 m. The not uncommon secondary kainitization of basalt dykes must be assigned to the low-thermal volcanic phase which followed.

It is still not certain whether the thin kieserite-sylvite beds below the carnallite (at ~70 m in Fig. 38) in the Stassfurt salt succession (p. 159, 160) originated from the thermal metamorphism of kainite. In the Hanover potash districts the widespread "langbeinite" footwall below the Stassfurt seam could be so interpreted. In this example there is an extensively kieseritized layer with langbeinite porphyroblasts several centimetres in length (KOKORSCH, 1960, and many older publications). The process could be regarded as the normal thermal metamorphism of kainite, which at 83° C gives rise to langbeinite as main product (p. 111), the langbeinite being later altered once more in solutions from the breakdown of carnallite.

It is not sufficient to assume primary hexahydrite was the mineral from which langbeinite was formed (KOKORSCH, 1960, p. 65), as it is impossible to derive KMg sulphates from Mg sulphates alone. Langbeinite can be formed from kieserite + sylvite by solution metamorphism in infiltrating NaCl solutions (p. 123, 124) and it is not necessary to introduce SO_4-rich solutions. The kieseritization of langbeinite, however, is only possible in $MgCl_2$-rich solutions (carnallite alteration solutions, cf. Table 16, No. 5).

4. *Salt Rocks Altered by Solution Metamorphism*

a) Isothermal Solution Metamorphism

Solution metamorphism is the most important type of alteration found in potash seams, for a great number of sylvite rocks were formed in this way. The following geological criteria may be listed:

Frequent small-scale facies changes within one mining area;

Regular lateral and vertical paragenetic changes;

Frequent occurrence of NaCl-filled joint planes or clefts, that is evidence of penetration by solutions along joint planes in the centre of a region of impoverishment (STORCK, 1954);

Unaltered halite key horizons despite facies difference in the seam;

Regular thickness variations between different facies types.

There are no real difficulties in calculating solution metamorphism changes using stable equilibria. Yet little is known about the actual process, and there appears to be no systematic experimental investigation bearing on this question. In all probability, therefore, the calculated equations for the reactions are only a crude and summary approximation. Nothing is known about the distribution mechanism or the distribution velocity of solutions in carnallite, etc. Nothing is known about the role of diffusion nor anything about the effects of differences in solution velocities or the growth rates of the reacting minerals. Clearly it is impossible to assume a sudden penetration by solutions. At least in the case of kainitization (p. 182) it is known that the moisture is present within the rocks is not greater than the normal interstitial amount. It is therefore a question of the slow progression of solution and alteration fronts.

As the reaction kinetics of solution metamorphism are still uncertain, the calculated and observed salt sequences will not always correspond. Models have only been calculated for a few temperatures. The comparison with models, however, provides a basis for a more detailed investigation of the deviations. The investigation of particular examples however must be left for a future work.

The previously established division into $MgSO_4$-free and $MgSO_4$-bearing salt sequences will again form the basis for discussion.

α) $MgSO_4$-bearing Salt Sequences

One undoubted example is the kieseritic Hartsalz of the Stassfurt anticline, that is the upper part of the overthrust zone shown in Fig. 38. The tectonic repetition of the Stassfurt seam was recognized by SCHÜNEMANN (1913). This contradicted EVERDING'S (1907) "Descendenz" hypothesis (p. 256) although it provided no satisfactory explanation. A plane with well developed slickensides was found, and with the help of index horizons it was shown that it represented an altered and tectonically deformed carnallitic kieserite-halite such as is found in the same cross-section between 114–125 m. Further examination (see R. HERRMANN, 1939) revealed a discordant domed or dished distribution of the mineral zones with respect to the bedding (the halite key horizon). Above a halite nucleus (sometimes rich in anhydrite) follow halite with vanthoffite, then loeweite, then langbeinite and above them a kieserite sylvite-halite ("Hartsalz islands") passing laterally to a marginal rim of carnallite. The same zonal sequence in concentric bands can be found within a single stratigraphic horizon. The succession of zones is always the same (the "Hartsalzvorläufer-facies" of R. HERRMANN, 1939). This regularity indicates their secondary nature, although with formation prior to the disharmonic tectonic overthrusting.

The petrographic indications (LÜCK, 1913[40]) also favour a secondary origin of loeweite and vanthoffite from kieserite. The vanthoffite is found in the form of nests or lens within the kieserite layers which may themselves be in part or completely altered to vanthoffite with perhaps some loeweite (partially in porphyroblasts with poikilitic NaCl intergrowths.)

According to LÜCK (1913) the "Hartsalz islands" of the Stassfurt anticline are composed of alternating layers of a reddish halite, fine-grained kieserite and dark anhydrite. The sylvite which fills cavities is milky white in the interior but has an intense red-coloured rim. Large sylvite crystals are associated with tectonically disturbed zones

Table 24. *Changes in composition in the same section of the Stassfurt seam (after* RÓZSA, *1915)*

Locality	Berlepsch shaft				Ludwig II		Berlepsch		Berlepsch	
Form.	kieseritic halite-carnallite				kieseritic sylvite-halite				halite-kainite	
Stratigr. section	I–O	I–O	K–M	I–O	I–O	I–O	K–M		I–O	K M
Carnallite	52.1	46.0	58.6	47.8	—	—	—		—	—
Kieserite	16.3	17.5	12.7	15.9	18.9	17.2	16.4		—	—
Kainite	—	—	—	—	—	—	—		57.2	59.8
Sylvite	—	—	—	—	20.3	25.6	24.7		6.4	4.7
Halite	28.8	34.1	25.8	33.7	57.7	54.8	56.4		33.8	32.4
Clay + anhydrite	2.8	2.4	2.9	2.6	3.1	2.4	2.5		2.6	3.1

Table 24 (after RÓZSA, 1915) contains a quantitative example of lateral composition change in the same stratigraphical horizon taken from a part of the Stassfurt seam on the ENE flank of the Stassfurt anticline. The kainite cap is also included. In considering impoverishment with loeweite, vanthoffite, etc. (see above), no average samples are available in the same stratigraphic section (cf. Fig. 38).

The observed sequence corresponds very well qualitatively with the solution metamorphism of a halite-kieserite-carnallite at about 83° C.

[40] WEBER, 1931 incorporated LÜCK's data and added further doubtful indications such as the occurrence of thenardite and in the potash layer the occurrence of carnallite relicts in vanthoffite (p. 70), and of loeweite in sylvite (p. 69). Further, being unaware of the idioblastic sequence (p. 153, 154), he assumed incorrect age relations between the minerals.

The first step in the alteration is a kieserite-sylvite-halite. Although the quantitative composition (Table 24) does not agree very well with that calculated (p. 118, 119), this may be due to the samples being non-representative. The next stage of alteration is to a kieseritic langbeinite-halite (p. 123). Now as a carnallitic kieserite-halite (Fig. 38) underlies the primary halite-kieserite-carnallite in Stassfurt (as indicated by the stratigraphic succession, see above), an excess of kieserite accompanies the breakdown of carnallite in the deeper horizons, so that as an additional alteration product a loeweite-halite is formed from langbeinite-kieserite-halite (Table 16, No. 3) and from this finally a vanthoffite-halite. From the vanthoffite itself further alteration products (e.g. d'ansite) could be formed, but none have so far been recorded in the Stassfurt deposits. The upper part of Fig. 38 thus clearly represents a secondary carnallite-hartsalz impoverished transition zone.

The impoverished salts in the South Harz can be distinguished from the form of impoverishment described above, which is found in many localities in the Stassfurt-Bernburg anticline, by the amount of anhydrite. BAAR (1952), using solution equilibria, first suggested that it was derived from an anhydritic-kieseritic carnallite by solution metamorphism.

It is also probable that the loeweitic-halitic vanthoffite layer of Wilhelmshall (FULDA, 1935) was derived by solution metamorphism from a halite-kieserite (Table 16, Nos. 9 and 10). Unfortunately, this occurrence is no longer accessible for more detailed study.

The origin of the grey sylvite-kieserite-halite in Zechstein 2 of the Hannover district has not yet been conclusively explained. From average samples KOKORSCH (1960) gave the following as the average composition of the normal grey kieseritic Hartsalz of Hildesia-Mathildenhall (range of individual components in parentheses):

45.4 (43.5–50.4)% halite,
29.0 (26.5–32.3)% kieserite,
19.8 (17.2–22.3)% sylvite,
1.5 (0.4 – 2.7)% anhydrite,
0.5 (0.0 – 2.0)% polyhalite,
0.2% langbeinite,
0.7% carnallite,
0.2% boracite,
0.3% $(Fe, Al)_2O_3$,
0.5% HCl-insolubles.

From Table 14, No. 3 it appears that they could be derived by the effects of a saturated NaCl solution on a kieseritic halite-carnallite with 20.9% NaCl; 18.7% kieserite; 58.5% carnallite; 1.8% anhydrite, polyhalite, boracite and insolubles. In fact, such carnallite rocks do occur in the Hildesheim region (according to ENGEL in LOTZE, 1938, p. 457), although their lateral transitions are still little known. However, the distribution of the sylvite kieserite halites appear to be very uniform which is not in agreement with solution metamorphism without some modification. Because of considerable tectonic activity, only a small part of the seam is exposed, and knowledge of the regional relationships is consequently restricted.

The "altered sylvinites" („Umwandlungssylvinite") of the Werra district which occur both in the upper part of the Thuringia seam and locally in the Hesse companion seam are equally products of solution metamorphism (D'ANS and KÜHN, 1940, p. 79). They are often marked by strong enrichment in sylvite close to carnallite, in large nests or more or less discordant layers (as in anhydritic sylvite-halites, p. 186) and veins, but often parallel to the bedding of clay inclusions. In the proximity of such clay inclusions against the otherwise grey carnallite rocks a distinct red coloration of sylvite by very fine haematite may be observed. In the Werra district (and according to current information also in the South Harz), although langbeinite has been found, the loeweite and vanthoffite types have not been observed. The thin langbeinite layers at the basalt contacts in the Hesse seam belong here. They form in comparison with the thermal metamorphism of the Hesse seam (p. 173) a younger solution metamorphism phase (according to Table 16, No. 2).

Despite extensive recent geological and petrological work (SYDOW, 1959; SIEMEISTER, 1961), the origin of the Ronnenberg seam K3Ro is still uncertain. SIEMEISTER sought to derive it from the solution metamorphism of a halite-carnallite, but nowhere in the closely studied Salzdetfurth mining region has the assumed initial rock been found. The occurrence of kieserite in the sylvinite, even though in small amounts, supports the idea of a metamorphic origin. Yet according to SIEMEISTER several formation processes are superimposed and an attempt has to be made to find a suitable location to decipher these. There also arise important tectonic complications and secondary precipitation in joints and fracture zones so that, in the present state of knowledge, a genetic interpretation of the composition is premature.

The cap zones must also be regarded as a solution metamorphism product. The vast majority of kainite rocks are secondarily formed in this way. They occur in cap formation on salt levels and salt slopes, that is in the transition zone between the water-bearing roof rock and the salt deposits.

The cap zones show a regular arrangement with an outermost gypsum rock as a residual product of older anhydrite; it is usually more than 100 m thick, much jointed and faulted or brecciated and filled with solutions (and because of this much feared in shaft sinking). This gypsum cap ends against a nearly horizontal plane bounding the rest of the salt deposit (hence the term salt level). The kainite cap above potash layers and especially carnallite, usually begins with a thin schoenite crust which may be one to a few metres in thickness. Then follows the kainite in which nests of schoenite may still occur. The kainite cap, usually 20–50 m thick, is white and saccharoidal in texture. It comprises a mixture of fine-grained kainite and halite in varying amounts. The original bedding is destroyed or obliterated. Only thick rock salt index horizons remain. As regards other minerals in the kainite cap, good crystals of leonite or bloedite are often found. Relicts of kieserite-sylvite-halite are not uncommonly observed in the kainite cap, and often very coarsely crystalline sylvite nests (with blue halite) may occur. The Stassfurtite nodules are altered to pinnoite (± kaliborite), and exceptionally hydroboracite is found.

The major part of the Stassfurt cap kainites were derived from kieserite carnallites. Applying the examples from Table 14, Nos. 9, 11, 14, 16 to the Stassfurt kainite cap (Table 24) gives a calculated kainite content which is low but a halite amount which is high, while the calculated and observed sylvite content is in relatively good agreement. Before quantitative deviations can be discussed in detail, further analytical data are required as too little is known about the variations in the composition of the seam.

Qualitatively there is agreement between observation and calculations from the 25° C isotherm and from the $MgCl_2$ polytherms of the five-component system (Figs. 19, 20). Kainite and schoenite do not occur together in paragenesis but schoenite forms at the expense of kainite. Glaserite also does not occur in true paragenesis with kainite.

In the salt edge of the Werra district the kainite has often resulted from the alteration of kieserite + sylvite (KÜHN, 1957; ROTH, 1957). KÜHN's confirmation that the kainite rock contains no more than the usual interstitial water is important for all similar alteration processes. The alteration therefore must proceed by means of very small amounts of solution.

Kainitization begins at the boundaries of kieserite grains. Finally all the kieserite and sylvite is replaced by a fine-grained felted mat of kainite crystals. Of interest is KÜHN's further confirmation that the grain size of the kainite increases from an initial 20 μ to about 80 μ with increasing distance from the kieserite-sylvite contact (accretive crystallization).

Kainitization is also possible in solutions which penetrate from below as, for example, from many basalt dykes and sections in the Werra district. They were named thermal-water kainite (Thermalwasserkainit) by D'ANS. Petrographically they are indistinguishable from the cap kainites. Geologically there may be an important age difference.

According to VAN'T HOFF'S data (1905, p. 84) and the older polytherms, the kainite-bloedite paragenesis is not stable (p. 73, 74). Recently both were unquestionably observed in the Ischl salt deposit as a thick crust around a kieserite-bearing langbeinite (MAYRHOFER, 1955). There is an outermost thin bloedite skin while between the langbeinite and kainite a thin $MgSO_4$ hydrate layer is intercalated (hexahydrite, according to MAYRHOFER, 1955, although the determination is uncertain for the refractive index is too high). True, there are no quantitative data about either the chemical or mineralogical composition. Yet it is certain that the kieserite-bearing langbeinite grains form a metamorphic paragenesis. It is only stable above about 50° C (Fig. 20) and can equally well form from kieseritic kainite as from sylvite + kieserite. The rims of $MgSO_4$ hydrate, of kainite + bloedite and of bloedite are in contrast cap formations, that is derived from kieseritic langbeinite by solution metamorphism below about 37° C according to Fig. 20, but probably still above 28° C.

The reporting of the kainite-bloedite paragenesis by MAYRHOFER is an important support for the correctness of the new polytherms. Its occurrence is, furthermore, an indication that there was originally a potash seam with the normal $MgSO_4$-bearing precipitation sequence in the Alpine foreland. This had been suggested also upon the basis of the long-familiar polyhalite inclusions. Of these $MgSO_4$-bearing salt sequences, only the most slowly reacting metamorphic products remain.

β) $MgSO_4$-free Salt Sequences

In terms of mineral composition these are much simpler and only three types of formation occur, always in the same order, namely

anhydrite-bearing halite-carnallite
anhydritic sylvite-halite (anhydritic Hartsalz)
anhydritic halite (impoverished)

where the carnallite thickness diminishes in the impoverished salt. A good example of this is found in Königshall-Hindenburg (Stassfurt seam).

In the impoverished material the secondary formation of halite at the expense of sylvite is indirectly indicated by nebulitic relicts of very fine haematite inclusions which are typical of sylvite (see below) but are absent in halite or sylvite-halite. This results in a distinct brownish coloration of the secondary halite. In the proximity of the Hartsalz rounded sylvite relicts also occur in the brownish halite. The intensity of the brown colour diminishes with increasing distance from the Hartsalz.

The anhydritic sylvite-halite is usually well banded due to a layered enrichment of anhydrite, halite and sylvite. It does not occur, however, in any regular rhythm. The individual bands vary in composition and thickness. Furthermore brecciated or massive Hartsalz (in limited regions) may frequently occur, and often the sylvite is enriched in lenses a few inches in extent.

In the majority of sylvite rocks, the sylvite is coarse-grained and red-rimmed. It possesses strongly denticulate margins caused by the formation of small anhydrite crystals on the edges and smooth, concave grain boundaries where it is in contact with halite. There is generally a clear grain elongation parallel to the banding. There is also a characteristic enrichment in fine haematite inclusions, which under the microscope are not recognizable as individual platelets in the outermost parts of the crystal. In clay-rich layers the sylvite has often an intense

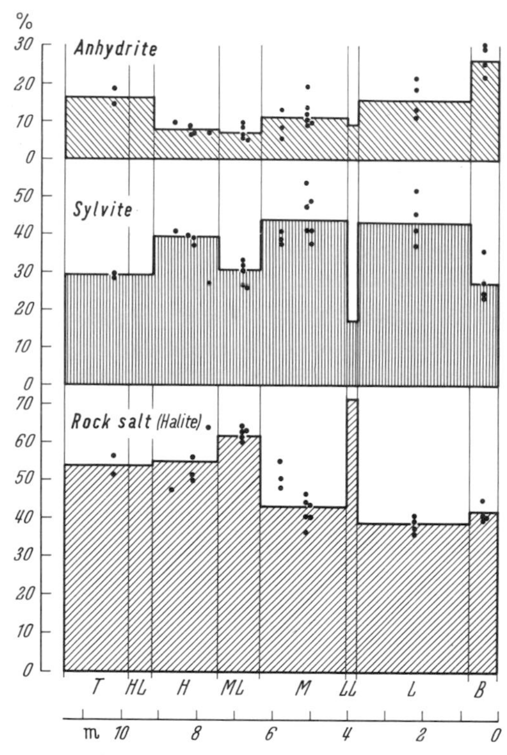

Fig. 41. Average composition of the anhydritic sylvite-halite from the Königshall-Hindenburg mine (points represent individual analyses of average samples. Analysis Braitsch). *H.M.L.* upper, middle and lower zones of the Hartsalz; *HL*, *ML*, *LL* = upper middle and lower parts of the key horizon; *B* = basal beds, *T* = transition to Salzton *T 3*

red colour, but in thin section the haematite has always a streaky-cloudy distribution. The haematite may also occur in nests or zonally in the form of large plates, mostly without texture, but many times with the same skeletal framework observed in carnallite. Close to the carnallite boundary, carnallite relicts may still be observed in the sylvite so that in the transition region there is petrographic evidence for the secondary formation of Hartsalz from carnallite.

Closely connected with the well-banded sylvite halites, particularly near the carnallite, there often occur discordant red-flecked halitic sylvinite veins a few centimetres thick. They are made up of 80–90% sylvite with halite and 1% anhydrite in fine crystallites. The sylvite is often in the form of long columnar crystals (~1 cm long ~1 mm diameter) oriented at right angles to the vein walls. The columns are isometric-polygonal in cross-section but often grow oblique to the cleavage planes. Halite is found either as thin columns with rounded ends or as elongated grains confined between sylvite grains. The sylvite is coloured red by finely divided haematite dust. Characteristically, however, the concentration of large haematite flakes on the margins of sylvite grains gives a distinctive honeycomb texture in cross-section. The large haematite flakes originate from the dissolution of carnallite and were crowded on the grain boundaries by the sylvite crystallizing out. This structure also is indicative of the secondary origin of Hartsalz at the expense of carnallite.

Fig. 41 indicates the average Hartsalz composition of some parts of the Stassfurt seam in the Königshall-Hindenburg mine. Because there are well-defined rock salt key horizons above and below the upper part of bed M in Fig. 41, it was possible to examine samples of all three structural forms of the seam (Table 25).

Table 25. *Composition of the Stassfurt seam (middle part) in different forms; Königshall-Hindenburg mine (from* Braitsch, *1960, Table 4)*

Sample	2982	2990	3387	3133	3389
Thickness	1.6 m	0.65 m	0.9 m	0.25 m	0.7 m
Type	halitic carnallite	anhydr. sylvite-halite		(tectonically thinned)	anhydr. halite
Halite	11.7	50.6	48	55.0	86
Sylvite	1.2	40.7	37	38.5	0.4
Carnallite	81.5	—	—	—	—
Kieserite	0.9	< 0.05	—	—	—
Anhydrite	2.6	8.2	13	5.1	12
Sr	0.020	0.029	0.033	0.027	0.026
Fe	0.017	0.028	0.037	0.015	0.053
Residues	0.23	0.44	0.91	0.46	0.96

All the recent workers (STORCK, 1954; KÜHN, 1955; HENTSCHEL, 1958; BRAITSCH, 1960) agreed that the present structural differentiation is the result of solution metamorphism of a halitic carnallite.

According to Table 14, No. 3, the initial rock 2982 of Table 25 will give the following after incongruent carnallite decomposition in saturated NaCl solution at 83° C: 41.4% sylvite; 51.7% halite (23% newly precipitated, 28.7% from the original carnallite rock); 6.4% anhydrite; 0.5% residues.

The kieserite of the original rock would be fully taken up in the resulting solution, as it is still undersaturated with respect to $MgSO_4$. The agreement with the observed Hartsalz composition is satisfactory. The further breakdown of the (calculated) Hartsalz formed in later infiltrating NaCl solutions at 83° C according to Table 15 gives an impoverished salt:

90.5% halite; 8.8% anhydrite; 0.7% residues.

Here also there is satisfactory agreement between calculation and observation. Unfortunately there is only a single block sample available for both the carnallite and the impoverished salt so that nothing is known about variation.

At a higher horizon (the lower part of bed H of Fig. 41) the carnallite was completely kieserite-free. The bed, 1.3 m thick, was composed of

16% halite; 1.7% sylvite; 76.4% carnallite; 0% kieserite;
3.8% anhydrite; 1.1% residues (clay).

According to Table 14, No. 4, from the breakdown of the carnallite in an NaCl solution the following anhydritic sylvite-halite can be calculated:

53.8% halite; 35.8% sylvite; 8.1% anhydrite; 2.3% clay.

In the same stratigraphic section, about 10 m from the carnallite margin, an 0.85 m anhydritic sylvite-halite was observed with the composition

46.8% halite; 41.0% sylvite; 9.1% anhydrite; 2.3% clay.

There is thus in the halite and sylvite a distinct difference between the calculated and observed compositions. If the carnallite alteration is considered in two stages (p. 120) then the calculation gives for the enriched sylvinite so formed the composition (BRAITSCH, 1960):

45.1% halite; 45.2% sylvite; 7.5% anhydrite; 2.2% clay.

In so far as the differences between the calculated compositions based upon Table 14 and the observed compositions are real, it can be assumed that the natural process corresponds to a partial enrichment and thus lies between the two calculated alteration processes. At the present time, however, only a single block sample of the carnallite rock and the Hartsalz has been investigated, and as the horizon is no longer accessible the variation in the composition remains unknown.

During the last few years the occurrence of anhydritic sylvite-halite has frequently been considered to have been derived from the alteration of a primary kieseritic halite-carnallite by $CaCl_2$-rich NaCl solutions (D'ANS, 1947; BAAR, 1952; STORCK, 1954; KÜHN, 1955; HERRMANN, 1961 a), where anhydrite is formed by the reaction

$$CaCl_2 + MgSO_4 \cdot H_2O \rightarrow CaSO_4 + MgCl_2 + H_2O\,.$$

In Königshall-Hindenburg this assumption is contradicted by the petrography. In the present mining area, which lies between the older, almost kieserite-free district (carnallite rock of Table 25) and the Holzerode bore in which kieseritic halite-carnallite is found (STORCK, 1954), there occurs a poor kieserite-bearing anhydritic sylvite-halite close to a poor kieserite-bearing carnallite. The kieserite occurs as untwinned crystals up to $^1/_2$ mm in size equally in the fine-grained anhydrite layer and in the enclosing coarse-grained anhydritic halite layer which rests directly upon a coarse-grained 2 cm thick intensely dark red sylvite layer. The kieserite crystals show smooth long faces but strongly indented end faces. Those in the anhydrite layers contain numerous anhydrite inclusions, those in the halite layers few, with occasional large halite inclusions. There is no noticeable replacement of kieserite by anhydrite, and the inclusions are of the same age or older than the kieserite. This invalidates KÜHN's conjecture that the kieserite occurred in small round agglomerates. Attention was called earlier (BRAITSCH, 1960, p. 6) to the occurrence of younger kieserite in an ascharite nodule in the Older rock salt. The anhydrite in this case preceded the kieserite formation.

Thus in the present mining area a primary transition can be observed from the pure anhydrite-bearing halite-carnallite to the kieserite- and anhydrite-bearing halite-carnallite types and, with a closer approach to the Holzerode bore, notable amounts of kieserite in the Hartsalz are to be expected.

The formation of additional anhydrite can be observed but not at the expense of kieserite, in the zone of impoverished salts where anhydrite spherulites sometimes occur. In the vicinity of the joints along which the metamorphosing solutions were introduced (reported by STORCK, 1954) anhydrite coats the joint faces and occurs in solution cavities (see KÜHN, 1955, Fig. 3).

There is still a further clear argument against the effects of $CaCl_2$ solutions on carnallite. If we take the average composition of the carnallite rock from the Holzerode bore 4 km east of the Königshall-Hindenburg mining district (Storck, 1954; Kühn, 1955): 69% carnallite, 19.2% halite, 6.3% kieserite, 3.5% anhydrite, 1.5% insoluble residues in a thickness of 26.7 m, then upon secondary saturation of this carnallite with a $CaCl_2$-rich solution in the sense of Storck 1954 and Kühn 1955, the kieserite in the carnallite rock ought to be replaced by an equivalent amount of anhydrite. In this case the carnallite rocks of Königshall-Hindenburg should contain about 10% anhydrite, which is more than double the amount observed. Although with very rich $CaCl_2$ solutions (see p. 126) all the kieserite of the Holzerode carnallite rock could be replaced by anhydrite, this particular model is not applicable to the conditions which pertained in Königshall-Hindenburg, as is shown by the low anhydrite content of the carnallite and the almost complete absence of kieserite. The upper 20 m of the Older Rock Salt immediately underlying the Stassfurt seam, which has 15–25% anhydrite without kieserite or polyhalite, must be included in considerations of genesis. These large amounts of anhydrite have been secondarily derived from kieserite without the entire potash seam being dissolved (p. 127). The kieserite zone was therefore already replaced by anhydrite at the primary stage. No kieserite-carnallite could have occurred above the missing kieserite-halite zone, at least not without the intercalation of an anhydrite-bearing carnallite (p. 98, 99) unless a very sudden change of solution is assumed. Thc claim that "all primary carnallites contain kieserite with anhydrite" (d'Ans and Kühn, 1960) is thus not generally valid.

These calculations, of course, can exclude only the hypothesis of the impregnation of primary carnallite rocks with carnallite-saturated but still $CaCl_2$-containing solutions. Whether the anhydritic sylvite-halite originated from the alteration of an $MgSO_4$-free or $MgSO_4$-poor carnallite by $CaCl_2$-bearing or $CaCl_2$-free solutions can only be decided by the use of additional criteria (see above and p. 212, 213). The occurrence in carnallite rocks and in the transition beds of rocks totally free of $MgSO_4$ is an important point in considering the question of the primary salt rock.

The salts of the Königshall-Hindenburg mine and of Eichsfeld (see Table 20) are particular examples which can only be adequately explained in palaeogeographic terms. The salts belong to an area in the vicinity of the ancient coast (for map see Lotze, 1938, p. 464) and overlap the area of distribution of the main dolomite (Hauptdolomit). Whether dolomitization or continental influxes were responsible for the changes in solution composition is not yet certain (p. 250). The high clay content in comparison with central area of deposition is an argument in favour of the relative proximity of the shoreline. There is, however, no simple relationship between the alteration of the saline solution and the clay content. For Königshall-Hindenburg, it was shown that the anhydrite and clay contents vary inversely with respect to each other (BRAITSCH, 1960). Probably the flocculation of the clay material is independent of $CaSO_4$ precipitation.

The South Harz district itself, where kieserite and anhydrite always occur in carnallite rocks, forms the primary transition region from the pure kieseritic to the pure anhydritic facies (cf. Teistungen No. I bore, p. 165). This primary facies differentation concerns only the carnallite rock and the underlying Stassfurt rock salt. In contrast, the anhydritic and kieseritic sylvite-halites almost always originate from the solution metamorphism of carnallite. On account of the presence of primary kieserite in the carnallite rock varied alteration products can result. BAAR (1944/45) described as normal the following lateral sequences.

1. anhydrite-bearing kieserite halite-carnallite,
2. anhydritic kieserite-sylvite-halite,
3. kieserite-bearing sylvite-polyhalite-halite, or
3a. kieserite- and polyhalite-bearing sylvinitic langbeinite-halite,
4. anhydritic sylvite-halite,
5. anhydrite-halite,
6. halite-anhydrite.

These he derived first qualitatively but later quantitatively (BAAR, 1952) by solution metamorphism. The most important feature in this paragenetic sequence is the proof that polyhalite can be a metamorphic product. According to the still somewhat uncertain data on the polyhalite stability field (p. 78), at 83° C the stability limit lies between points Q and R (Fig. 23) so in solution metamorphism polyhalite should occur between the kieserite-sylvite and the kieserite-langbeinite parageneses. In deducing this sequence, it is unnecessary to assume $CaCl_2$-bearing metamorphic solutions for possibly the $CaCl_2$ would be completely used up on the formation of zones 4–6 (zone 6 corresponds to the region of solution influx).

There are however, in other areas in the South Harz other examples which indicate infiltrating $CaCl_2$ solutions, as in the Glückauf mining district, Sondershausen, where within the carnallite facies transitions can occur from a predominantly anhydrite-bearing to a predominantly kieserite-bearing carnallite (BAAR, pers. comm.; HERRMANN, 1961a) and from anhydrite to kieserite bands. It is unfortunate that there are no quantitative chemical or petrographical data for these important exposures. In spite of this, on geochemical grounds (HERRMANN, 1961a, also p. 212, 213) it seems that the assumption of $CaCl_2$-bearing solutions is justified, for primary transitions are unlikely in a 10 m zone.

Thus, although secondary anhydritization by means of $CaCl_2$ solutions should not be doubted, it should not be regarded as a definitive indication of facies differentiation. It has nothing to do with primary precipitation. It can however, emphasize the local primary differences which are still preserved to some extent in the Stassfurt seam, but only in the carnallite rocks.

The spatial facies differences in the carnallitic Stassfurt seam are also expressed in terms of the total anhydrite-kieserite content. In the central parts of the basin (Stassfurt, Hildesheim) with an average of 18% it is substantially higher than in the South Harz with about 10% which in turn is greater than the 5% or so found in the marginal areas of the basin (Eichsfeld and Königshall-Hindenburg). More important perhaps than these quantitative changes, which may only reflect the higher temperatures near the basin margins, is the deficiency in primary $MgSO_4$ which this represents.

The possibility of the secondary formation of tachhydrite by solution metamorphism will only be briefly discussed. Where primary bischofite is present, which is still uncertain in the case of the German Zechstein salts in the light of present information, tachhydrite can precipitate from percolating, unsaturated $CaCl_2$ solutions with the resolution of the bischofite (D'ANS, 1961). In contrast, the formation of tachhydrite from the incongruent alteration of carnallite occurs only under particularly favourable circumstances, in particular at high temperatures and at high $CaCl_2$ concentrations ($\gtrsim$ 40 mol $CaCl_2$, 1000 mol H_2O) with the precipitation of appreciable amounts of sylvite. KLING (1915) observed in the Stassfurt seam of the Mansfeld basin the paragenesis carnallite-sylvite-tachhydrite which seems to indicate this possibility, even with cooling of the solution formed. Nevertheless, many of the details of the formation of tachhydrite are obscure and certainly not all tachhydrite occurrences can be so explained. In the German salt deposits, it seems that the formation of primary tachhydrite cannot be assumed, for although tachhydrite occurs in the primary kieserite-carnallite paragenesis all experimental data indicate it cannot be in equilibrium with them. The formation of primary tachhydrite is not excluded under different circumstances, as for example in the salt deposits of the former French Congo (p. 42) which to the writer's knowledge have not been described in detail.

b) Polythermal Solution Metamorphism

Secondary precipitation caused by temperature fluctuations is of subordinate importance in nature. It concerns only those salts with large temperature coefficients of solubility. Of the commoner salt minerals, these are carnallite and sylvite. As metamorphic solutions, the voluminous Q or E solutions (Table 14) resulting from the incongruent alteration of carnallite, and the NaCl–KCl solutions (Table 15) resulting from impoverishments have to be considered. In this context the solutions generated by thermal metamorphism and squeezed out of adjacent rocks are of lesser importance.

Examples are:

a) Carnallite joint fillings in anhydrite (± kieserite)-sylvite-halite in the neighbourhood of the transition to unaltered carnallite, as, for example, in Königshall-Hindenburg. Here it occurs as a colourless filling a few millimetres in thickness. On a still smaller scale are the precipitated idiomorphic carnallite crystallites in secondary halite from the same area. The halite is precipitated from the infiltrating solution (Table 14), the carnallite in contrast forms as a result of slight cooling of the Q solution which develops. To this latter category belongs the secondary, fibrous carnallite precipitated in fissures of the Hartsalz, especially in tectonic shatter zones. Locally brecciated carnallite completes the textural pattern (LÖFFLER, 1960). These are not a replacement of sylvite but rather a precipitate in cavities. Large occurrences of this kind, some even commercially exploited, are found in the Werra district according to BAAR (1958) on bedding slip planes (see also p. 194, 195). However, the origin of the largest outcrop of this kind, the so-called carnallite companion seam to the Hesse seam, is still in doubt (HOPPE, 1958). The origin of the fibrous carnallite found in the gray salt clay (Salzton) may perhaps be explained in another way (p. 240).

b) The occurrence of nests of coarsely crystalline sylvite, usually with blue halite. The sylvite is commonly a milky white with a somewhat blue opalescence. Occurrences of this kind have been reported by STORCK (1954) near the transition from anhydritic sylvite-halite to anhydritic halite in Königshall-Hindenburg. They are explained as the result of the cooling of the KCl–NaCl equilibrium solutions produced during the impoverishment processes. No quantitative data are available.

c) Bischofite inclusions in the salt diapirs of North Germany, although there is no certain proof of examples of this kind. As bischofite occurs in large amounts only in the Vienenburg mine, where over 30,000 tons were mined (FULDA, 1931), and in lesser amounts in the Wilhelmina mine (lower Aller), both being in the form of concordant inclusions in the carnallitic Stassfurt seam, a primary origin is improbable. If this were the case, bischofite should be found more commonly in the sub-Hercynian basin. Unfortunately, these outcrops are no longer accessible so that more detailed investigations, geochemical and petrographic in particular, are not possible.

Secondary formation of bischofite during the cooling of solutions squeezed out after thermal breakdown of carnallite (Table 13, solution H′) seems possible, although the solubility data are so unreliable (p. 111, 112) that a quantitative calculation provides no proof for this assumption. Carnallite would be the first to precipitate from such a solution as it cooled. At about 130° C, $MgCl_2 \cdot 4\,H_2O$ could also form, changing to bischofite when the solution reaches 117°. Upon further cooling bischofite precipitates out with very little carnallite. Very much smaller quantities of bischofite are derived from the solution resulting from the alteration of carnallite. In this case the carnallite must be broken down at above 140° C (p. 129) and would produce a heavy carnallite precipitate upon cooling. Characteristically, large amounts of secondary carnallite are found in the sub-Hercynian basin under the Ronnenberg seam (SIEMEISTER, 1961), namely in Vienenburg and Salzdetfurth.

It cannot for the time being be decided which of the two explanations best fits the Vienenburg bischofite. Occasional natural salt solutions are known, as for example, from Neustassfurt (HERRMANN, 1961 b, solution XIII) which could perhaps be derived from such a process. Without any doubt the high temperatures required are unusual and can only be assumed for deeply buried salt deposits (3000–5000 m). However, within the sub-Hercynian basin this is locally quite possible.

Since the stability fields of many salt minerals and parageneses are strongly temperature-dependent, reactions of the salts which occur in incompletely removed solutions must also be considered. In this category belongs, for example, the secondary carnallitization of sylvite-bearing rocks (p. 105). It was shown, however, that the process from a Q or E solution is not possible isothermally. The geological extent of these processes is often much exaggerated. Definite examples are not known and according to the data on p. 129 they are only to be expected in thin reaction rims.

5. *Dynamic Metamorphism of Salt Rocks*

Dynamic metamorphism has purposely been excluded from consideration up to the present, because it has no significant effect upon solution equilibria (p. 129). A few textural studies of salt tectonics already exist (BALK, 1949; TSCHOEPKE and KARL, 1957; FRIEDRICH, 1959), and also of grain structure (LAMCKE, 1937; LÖFFLER, 1960; SCHWERDTNER, 1961; CLABAUGH, 1962), which indicate the highly metamorphic character of salt rocks.

Many evenly bedded carnallite rocks show a well developed mosaic texture. The carnallite crystals are certainly irregularly allotriomorphic but in cross-section are clearly longer than broad with the long axis in the plane of the bedding. They show an impressive axially symmetrical grain structure with $[001] = n_\alpha$ perpendicular to the layer texture, without any preferred direction in the plane, that is to say the pseudo-hexagonal (0001) lies in the control plane (LÖFFLER, 1960). The orientation of the lamellar twinning has not yet been statistically examined (other textural types see PEI-KENG-LENG, 1945) so that a genetic interpretation is still premature. The currently known indications can be explained in terms of crystallization foliation according to principle of pressure solution of anisotropic crystals exposed to an imhomogeneous pressure distribution (p. 107). However, because of the high plasticity of carnallite the origin of the structure by flow processes is more likely ("Plättungsgefüge", etc.).

In strongly deformed salt bodies halite occurs in a variety of forms depending upon its tectonic position. In the cores of large isoclinal folds the so-called "Stausalz" (SEIDL, 1913)[41] is observed. It is coarse-grained and so poorly consolidated that on handling it readily crumbles into small pieces or even single crystals. In some cases stemlike fragments also occur. Clay aggregates can be found both in the halite grains and on the grain boundaries, usually without any recognizable bedding. The anhydrite etc. content of true Stausalz is much less than in the same beds away from the anticlinal core.

On the flanks of folded salts "Zerrsalz"[41] is found. It shows a well-developed banding with alternating NaCl and anhydrite-rich bands with small grains and good cleavage. It contains rather more anhydrite than in its original condition, with the anhydrite displaying a good linear texture with [010] within the plane of the banding. It is not yet known whether the [010] direction is parallel to or at right angles to the gliding direction (parallel a or b structures). Qualitative thin section observations of the halite suggest that (110) is oriented parallel to the glide planes, as the cleavage planes (or fluid inclusions on these planes) are predominantly inclined at angles of less than 45° to this plane. In the Grand Saline salt diapir it was possible to prove the ordering of halite by translation and preferentially by gliding parallel to {001}, less commonly parallel to {110} (CLABAUGH, 1962).

Such rocks are extreme cases and between them there are all transition stages. In the "Stausalz" the NaCl is enriched ideally by pure plastic flow without the participation of solutions; in "Zerrsalz" on the other hand there is a tectonic unmixing with the anhydrite passively segregated in the gliding plane direction. These processes strongly modify the original salt thickness and the quantitative primary composition.

[41] No good translations of these words can be found and it seems better to retain the German expressions.

Unfortunately, no systematic quantitative investigations of compositional questions have yet proved possible. Perhaps the different anhydrite contents of a series of samples from the "Older halite" (Älteres Steinsalz) of Königshall-Hindenburg (Figs. 39, 40; the NaCl content corresponds with a good approximation to the residues analysed after the calculation of all Ca as anhydrite) can be so explained, as the especially anhydrite-rich samples come from the steep anticlinal flanks.

Qualitative examples are known from many salt diapirs as well as from the major Königshall-Hindenburg anticline.

In carnallite rocks also a separation of the other components results from preferential plastic flow of the carnallite. The extreme case of separation by dry plastic deformation, as indicated earlier, is scarcely likely. It is much safer to assume that solutions always participate in mass transport, with particular reference to pressure solution of inhomogeneously stressed crystals with reprecipitation at the points of least stress. Another possible cause of carnallite precipitation which ought to be considered is the squeezing of metamorphic solutions in the higher crust levels, e.g. diapirs, and the associated cooling.

Through these processes large secondary carnallite enrichments, of especially pure carnallite can be formed as, for example, in the so-called "Hemeling carnallite" in Salzdetfurth with more than 99% carnallite in large regions (SIEMEISTER, 1961) or in the "Staukuppen" in the Thuringian part of the Werra district which has more than 96% carnallite (GOTTESMANN, 1962). In both cases only a qualitative interpretation is possible. In the first case solution transport must have predominated and in the second mechanical transport. Moreover, it is not yet entirely clear what conditions are necessary for the plastic deformation of carnallite and what for the formation of brecciated carnallite (Trümmercarnallite). Two parameters to be considered must be the ratio of directed to hydrostatic pressure and the rate of deformation. According to BAAR (1958), the formation of brecciated carnallite may perhaps be explained in some if not in all cases by tectonic adjustments succeeding a phase of solution metamorphism.

In salt deposits, because of pronounced material, and the corresponding mechanical, inhomogeneities under gravitational stresses, dislocation planes and cracks often form parallel to the bedding, which fill with solutions (and gases) and naturally become the preferred site of secondary salt precipitation. Whether the principle of pressure solution (p. 106) or cooling effects predominate is not yet clear. In the first case only a small amount of solution is necessary and cavities can be closed almost simultaneously with their opening. With precipitation during cooling, however, the cavities, with an initial volume several times greater than their present filling must be assumed to have remained open, which seems unlikely in the large occurrences. The significance of secondary potash salt lenses parallel to the bedding was first pointed out by BAAR (1958).

In the chloride salt rocks deformation is always obliterated by contemporaneous or subsequent recrystallization (precrystalline deformation), so that it is impossible to deduce from the texture any information regarding the history of its formation, and in particular regarding tectonic deformation at different times. Relict internal structures occasionally remain visible (e.g. the haematite from carnallite in sylvite, p. 183), but this tells us nothing about the tectonic history. Texturally, the chloride salt rocks correspond to highly metamorphosed crystalline schists. Only in the most favourable circumstances can different joint generations be recognized (SIEMEISTER, 1961).

As anhydrite does not recrystallize under conditions of salt metamorphism, traces of deformation such as translation lamellae, bent cleavage planes and undulose extinction can be recognized. In such cases, generally found in monomineralic anhydritic rocks, deformation is post-crystalline with regard to the anhydrite. Such anhydrite tectonites (= tectonically deformed anhydrite rocks) are known from salt domes. They show a regular grain structure with a statistical parallel orientation of the [010] direction of the anhydrite crystals within the bedding (SCHWERDTNER, 1961). In part this regularity is a reflection of the grain form, in part a result of translation (the translation direction in anhydrite is [010], see Table 3).

The few examples discussed are insufficient for a comprehensive review of the behaviour of salt rocks during formation. Attention can only be drawn to the importance of further investigations in this area.

II. Minor Components

1. Bromine

The behaviour of bromine can to a great extent be explained by simple models. The application of these models to the genetic problems of salt deposits is particularly justified because the relative initial bromine content (that is the amount with respect to the total salt content) of present-day seawater is virtually constant. Seventy new analyses from the China Sea, Indian and Antarctic Oceans gave for the weight-ratio 100 Br/Cl the following (frequency in brackets): 0.342 (1); 0.343 (36); 0.344 (30) and 0.345 (3) (analyst; UENO, pers. comm. K. SUGAWARA). The same uniformity is also seen in a review by E. GOLDBERG (pers. comm.). Thus in terms of standard chlorinity the bromine content of seawater lies between 0.0065–0.0065_5 %. From this a complete mixing of bromine in seawater can be assumed for the geological past. It follows that with the help of models important geological conclusions can be drawn from the observed bromine content as, for example, the relative bromine content of the initial solution, in certain cases the temperature of primary precipitation, and additionally the course of salt precipitation (progressive and regressive compositional changes etc.). The practical applications, which include its possible use for stratigraphic correlation and tracing horizons in salt sequences, and prospecting for potash salts, can be seen without further explanation from the models (Table 18, and Figs. 33, 34). The necessary preconditions for the application of the models are exact analyses and petrographic criteria to recognize secondary alterations. In primary salt sequences the bromine content of the chloride is a unique measure of the concentration reached in the solution and also of the total salt content of the solution.

a) Progressive Evaporation

The first test of the models (Table 18) is in the comparison of the bromine content of Recent salt lakes and salinas. The observed bromine contents are given in Table 26.

Table 26. *Bromine content of Recent salt lakes and salinas*

Concentration stage	Wt. %-Br/Soln.	Reference
Halite saturation (Siwasch Salt Lake)	0.048	BURKSER et al., 1958
Halite saturation Perekop Salt Lake)	0.019–0.055	BURKSER et al., 1958
Halite saturation (various salt lakes and salinas)	0.033–0.062	VALYASHKO, 1956
Beginning of epsomite-sylvite precipitation	0.222–0.342	VALYASHKO, 1956
Beginning of carnallite-sylvite precipitation	0.259–0.342	VALYASHKO, 1956
Beginning of bischofite-sylvite precipitation	0.431–0.608	VALYASHKO, 1956

According to Table 18 the bromine content at the beginning of NaCl saturation is 0.054% Br/solution and is largely independent of temperature. The remaining values are not directly comparable with the models, for in Recent salt lakes metastable equilibria prevail. In addition, in the smaller salt lakes an appreciable shift of solution composition results from the influx of weathering solutions.

For salt deposits considerable quantities of data have been collected. Unfortunately in many cases it is impossible to guarantee the reliability of the data either from the sampling or analytical viewpoint (because of

Br-rich liquid inclusions, or sylvite impurities, etc.). The most useful are the series of Br analyses of halite rocks. The content (wt %/NaCl) in Na 2 observed from the margin to the interior of basin is given in Table 27.

Table 27. *Lower and upper bromine content in fully investigated halite rocks*

Stratigr. horizon	Locality[b]	Transition		Reference
		Anhydrite/Halite Wt. % Br/NaCl	Halite/Potash Wt. % Br/NaCl	
Na 1	Saxony-Weimar (Werra)	0.007	~0.023–0.028	D'ANS and KÜHN, 1940
Na 1	Aschersleben-Sangerhausen, 10 bores	(0.002–0.004)	—	SCHULZE[a], 1960
Na 2	Königshall-Hindenburg	not determined	0.025	HERRMANN (pers. comm.)
Na 2	Bischofferode	0.007	~0.025	BAAR, 1954
Na 2	Sondershausen	0.007	0.019	BAAR, 1954
Na 2	Udersleben	0.0065–0.007	not determined	SCHULZE[a], 1960
Na 2	Rossleben	0.007	not determined	BAAR, 1954
Na 2	Golbitz-Könnern	0.005	not determined	SCHULZE[a], 1960
Na 2	Bernburg-Gröna and Plömnitz	0.006	~0.023–0.024	SCHULZE[a], 1960
Na 2	Aschersleben V Schierstedt 1/55	0.006	~0.028	SCHULZE[a]. 1960
Na 2	Aschersleben IV Schierstedt 3/56	(0.003) 0.006	~0.03	SCHULZE[a], 1960
Na 2	Stassfurt	(0.003) 0.0065	0.025–0.028	SCHULZE[a], 1960
Na 2	Wilhelmshall	not determined	0.03	SCHULZE[a], 1960
Na 2	Upper Allertal	(0.003) 0.006	0.033–0.035	SCHULZE[a], 1960
Lower	Buggingen	0.007	0.03	GUNZERT, 1961
Oligocene	(halite under the lowest layer)		0.026	BAAR and KÜHN, 1962

[a] Analyses from the Geological Survey Jena and the Sonderhausen Potash Research Station.

[b] See also: RAUP, O. B.: Bromine distribution in some halite rocks of the Paradox member, Hermosa Formation, in Utah. Second Symposium on Salt, Northern Ohio Geological Society, Cleveland, Ohio, 1966.

WARDLAW, N. C., WATSON, D. W.: Middle Devonian salt formations and their bromide content, Elk Point Area, Alberta. Canadian J., Earth Sciences **3**, 263–275 (1966). – WARDLAW, N. C., SCHWERDTNER, W. M.: Halite-anhydrite seasonal layers in the Middle Devonian Prairie evaporite formation, Saskatchewan, Canada. Geol. Soc. Am. Bull. **77**, 331–342 (1966).

These figures confirm the older data of D'ANS and KÜHN (1940) and KÜHN (1955), from which the following boundary values were obtained:

anhydrite region	0.003–0.017% Br/NaCl,
polyhalite region	0.017–0.023% Br/NaCl,
kieserite region	0.023–0.028% Br/NaCl,
carnallite region	0.028–0.048% Br/NaCl.

According to SCHULZE (1960) the range in the kieserite region is perhaps larger (0.015–0.028% Br/NaCl), but in this case the lower value is certainly not primary.

In all the series from Zechstein 2 there first occurs a slow but steady increase in the bromine content, but below the potash layer (somewhere within the kieserite region) the bromine content increases rapidly. The bracketed bromine values (Tables 27) are certainly not of primary origin. They are found in the first 5 m above the basal anhydrite where the bromine curve shows a rapid rise and then a kink. Extrapolation of the steadily rising part of the bromine curve gives the unbracketed figures in Table 27. The low values, below 0.006% Br/NaCl, also found in the Werra rock salt in the southern and eastern Harz foreland, can presumably be attributed to partial diagenetic recrystallization of NaCl in Br-poor solutions, perhaps from the early diagenetic gypsum-anhydrite dehydration. Thus SCHULZE'S assumption of regional differences in the initial bromine content above the basal anhydrite becomes unnecessary. However, the samples do not appear to have been petrographically examined for signs of recrystallization. The primary lower limit should be 0.006–0.007 Br/NaCl. The theoretical value at the point where the first halite crystallizes from present day seawater is ~0.0075% Br/NaCl (Table 18). VALYASHKO (1956) also found this value in the first precipitated halite. He also calculated this value from new determinations of bromine partition equilibira.

Both in Zechstein 2 and in the halite series under the Lower Potash layer of the Upper Rhine, there is evidence of progressive evaporation without the intercalation of dilution phases. However, the increasing bromine content in Zechstein 2, at least in series with a thickness greater than 300 m, certainly does not show a logarithmic increase. But the bromine partition factor b_{NaCl} decreases with increasing $MgCl_2$ content (Table 17), so that at the beginning of NaCl precipitation a slower increase than the logarithmic one must occur. Moreover, the assumption of uniform evaporation matched by inflows of the same order, is untenable (p. 143). Furthermore, during the greater part of NaCl precipitation unknown amounts of more concentrated brine refluxed back into the ocean so that the increase in concentration occurred at a slower rate than during purely static evaporation. In Zechstein 1 under the Thuringian seam the intercalation of a short dilution phase can be deduced from the bromine curve according to D'ANS and KÜHN (1940). The values of % Br/NaCl at first increases steadily from 0.007% to 0.03% at which point carnallite could occur. Thereafter it decreases noticeably and only begins to increase again immediately below the Thuringian seam.

The bromine content of halite both at the beginning of NaCl saturation and at the beginning of the deposition of potash salts corresponds approximately to the theoretical values. At the beginning of the polyhalite region the figure of 0.01% Br/NaCl (not given in Table 18) was calculated from somewhat inexact older data. The reason for the major discrepancy between this and the observed figure may be found mainly in the insufficiently known solution composition at the beginning of polyhalite precipitation and partly in the lack of precision in the model itself.

The bromine content of primary sylvite and carnallite precipitates is far less accurately known, as usually the total bromine is determined and then, using the paragenetic bromine ratio, the bromine content of the halogenides occurring is calculated. As the paragenetic ratio so far assumed for carnallite was incorrect, all bromine calculations based upon it are also incorrect, apart from which, a paragenetic bromine partition should not be assumed in advance when this is what one is trying to prove. The exact determination of the partition ratio cannot be explained here; a graphical, extrapolation method has been described by D'ANS and KÜHN (1940, p. 59) and by KÜHN (1957). When two or more salts containing bromine occur, it is more convenient to use the determinant scheme to which reference has so often been made in this book. From the individual values in the literature and our own determinations, we find that relative errors of up to 50% in the Br/NaCl value can easily occur, particularly in the NaCl-carnallite paragenesis, according to the assumed compensating straight lines. Hence it is pointless to give more than one significant figure from the preceding analyses and more than two significant figures for the Br/KCl or Br/carnallite value. Thus the partition ratio is not exact to the first decimal place. With these restrictions, there are only a few values for the bromine content of primary potash salts which are applicable (Table 28).

These few values are insufficient to judge the validity of the models. The agreement in the sylvinites is good, for the observed values lie within the region of the calculated figures (cf. Fig. 35). However, in the Upper Rhine salts the bromine ratio between halite and sylvite is not the paragenetic one, but because both chlorides occur in separate layers within the bed they did not crystallize together (BAAR and KÜHN, 1962). In future investigations the NaCl part of the

Table 28. *Bromine content of chlorides in potash seams (figures in wt. %)*

Rock	Br/NaCl	Br/KCl	Br/carnallite	Reference
Carnallite from the Thuringian Seam, Kaiseroda	0.03	—	0.22	D'ANS and KÜHN, 1940, p. 62
Carnallite companion seam, Neuhof Ellers	0.02_7	?	0.20	KÜHN, 1957
Carnallite comparison seam, Neuhof Ellers lower part (sample 13)	0.02_2	(0.35)	0.15	KÜHN, 1957
Carnallite companion seam, Neuhof Ellers upper part (sample 15)	0.02_4	—	0.19	KÜHN, 1957
Carnallite from the Thuringian Seam, Herfa-Neurode (samples C 12 + C 13)	0.01_3	(0.17)	0.15	KÜHN in WEBER, 1961
Carnallite from the Thuringian Seam, Wintershall (samples C 12 + C 13)	0.01_8	(0.14)	0.14	KÜHN in WEBER, 1961
Carnallite from the Hesse Seam, Wintershall (sample "Lagercarnallit")	0.03_4	—	0.20	KÜHN in WEBER, 1961
Carnallite from the Stassfurt Seam, Königshall-Hindenburg (two samples)	0.03_6	—	0.26–0.30	KÜHN in STORCK[a], 1954
Carnallite from the Stassfurt Seam, Glückauf Sondershausen, S. Harz				HOFFMANN (pers. comm.)
Carnallite between Unstrut bank 9–10	0.03_9	—	0.26_4	
8–9 top	0.03_9	—	0.29_5	
8–9 middle (see note p. 212)	0.04_5	—	0.33_8	
8–9 base	0.04_2	—	0.30_3	
7–8 top	0.03_7	—	0.26_3	
7–8 middle	0.03_8	—	0.25_7	
7–8 base	0.04_1	—	0.27_8	
Carnallite from the Stassfurt Seam				HOFFMANN (pers comm.)
Berlepsch-Maybach shaft (Stassfurt anticline)				
III. Main level section at 2100 m				
top of bank P	0.04_3	—	0.34_2	
base of bank N	0.04_2	—	0.34_3	
top of bank L	0.04_1	—	0.32_9	
top of bank K	0.04_2	—	0.33_8	
middle between bank J–K	0.04_5	—	0.34_6	
top of bank J	0.04_1	—	0.33_1	
bank H	0.03_9	—	0.31_9	
base of bank H	0.04_0	—	0.32_3	
Carnallite from the Ronnenberg Seam, Salzdetfurth, 634 m level	0.03	—	0.24	KÜHN in SIEMEISTER, 1961, Fig. 46
Carnallite, Amelie mine, Alsace	0.03_8	?(0.23)	0.20	BAAR and KÜHN, 1962[b]
Sylvinite layer B 1, Buggingen	0.03	0.37	—	KÜHN, 1955, p. 65[b]

Table 28 (continued)

Rock	Br/NaCl	Br/KCl	Br/carnallite	Reference
8 Sylvinites, Buggingen	0.03	0.32–0.39	—	BAAR and KÜHN, 1962[b]
3 Sylvinites, Marie Luise mine, Alsace	0.02_6	0.28–0.35	—	BAAR and KÜHN, 1962[b] (excl. no. 7)
Sylvite and Carnallite rocks from Saskatchewan[c]				
Kainite rock (71.6% kainite), Miniera Sambuco Sicily	0.02	?	0.08 Br/Kainite	KÜHN, 1955, p. 65
Bischofite, Vienenburg			0.44_0 Br/Bischofite	HERRMANN (unpublished)

[a] Value of Br/carnallite calculated from the given paragenetic ratio and the Br/NaCl figure.

[b] Further data for carnallite and sylvinites from the potash mines of Alsace and Buggingen see BRAITSCH and HERRMANN (1964); footnote p. 142.

[c] SCHWERDTNER, W. M.: Genesis of potash rocks in middle Devonian Prairie evaporite formation of Saskatchewan. Bull. American Ass. Petrol. Geol. **48**, 1108–1114 (1964). – MCINTOSH, R. A., WARDLAW, N. C.: Barren halite bodies in the sylvinite mining zone at Esterhazy, Saskatchewan. Canadian Journal of Earth Sciences **5**, 1221–1238 (1968). – WARDLAW, N. C.: Carnallite-Sylvite relationships in the Middle Devonian Prairie evaporite formation, Saskatchewan. Geol. Soc. Am. Bull. **79**, 1273–1294 (1968).

sylvite layer must be separated. Whether the high values of up to 0.39% Br/KCl indicate an almost carnallite-saturated solution and a high evaporation temperature (40° C) cannot yet be determined[42].

In the few Zechstein 1 carnallite rocks so far investigated the Br/carnallite ratio is somewhat smaller than is anticipated from the model. This is also true of four new determinations from the Wintershall mine (analyses of KÜHN in WEBER, 1961). All these results pertain to the Thuringia seam or to the companion seam to the Hesse seam. Below the Thuringia seam there is a clear dilution phase indicated by the fall in the Br/NaCl curve below the seam (see above) and it is therefore possible to relate that to the low bromine content of the carnallite, perhaps by the assumption that a potash layer was dissolved (occurring in the region above the recorded maximum of the Br/NaCl curve) and its subsequent re-precipitation. By solution and re-precipitation the absolute bromine content of the salt is reduced without altering the paragenetic bromine ratio Br/NaCl : Br/carnallite $\approx$ 1 : 7. The observed bromine ratios are in accord with the preceding statement.

[42] Regarding these questions see the following paper: BRAITSCH, O., HERRMANN, A. G.: Zur Geochemie des Broms in salinaren Sedimenten, Teil II: Die Bildungstemperaturen primärer Sylvin- und Carnallit-Gesteine. Geochim. et Cosmochim. Acta **28**, 1081–1109 (1964).

Several carnallite samples from the Stassfurt seam have been investigated. In the interior of the basin (Stassfurt anticline) BOEKE (1908) found 0.4% Br/carnallite. He did not investigate the mineral concentrates, and calculation from his total bromine content using the paragenetic bromine ratio shows this value to be high. Recent, unpublished data were made available to the author by R. O. HOFFMANN (Berlin). Here mineral separates were examined and the bromine content extrapolated to the pure mineral phase. The series of samples from the Stassfurt anticline do not show a regular curve from below [bank H in RÓSZA's (1915 and others) description; first marked occurrence of carnallite] to above. This, according to HOFFMANN, is not the case in other samples. Most values lie between 0.32 and 0.37% Br/carnallite, a few are higher up to a maximum of 0.47% Br/carnallite, rather more are lower down to a minimum of 0.22% Br/carnallite. It is still too early to form a final interpretation of the Stassfurt salt sequence on the basis of available data. It is certain, however, that the too-high bromine content of carnallite cannot be of secondary origin. The observed relations cannot be simulated with sufficient accuracy with any of the calculated models. Nevertheless, they are best explained qualitatively in terms of a model with reaction of an original kainite (or a metastable primary precipitate of epsomite + sylvite), both with respect to absolute values and to the irregularity. The halite must to a large extent be recrystallized for it shows very nearly the paragenetic bromine ratio with respect to carnallite. The lack of a trend in the bromine content within the seam probably reflects the fact that reaction did not run from below to above, or the reverse, but was more or less simultaneous in the whole kainite layer so that several carnallite recrystallizations must be considered. Much more detailed data on this and the distribution or other elements must be awaited, so that the reaction model described can only be considered as a working hypothesis.

In contrast to the Stassfurt anticline, the values for the primary carnallite rocks of the South Harz district (Glückauf, Sondershausen) are lower so that here the bromine does not correspond to the reaction model. In the Königshall-Hindenburg mine on the edge of the Zechstein basin the author found 0.3% bromine/carnallite in primary bedded carnallites. This figure corresponds to those given by KÜHN (Table 28). It lies in the region of the theoretical values in a completely $MgSO_4$-deficient system (Table 18c).

Table 28 also contains a sample from the base of the Ronnenberg seam although it represents carnallite precipitation from a metamorphic solution. For the partition of bromine the nature of the solution is naturally unimportant and the paragenetic bromine ratio is here observed.

All the carnallite samples in Table 28 approximate to the expected bromine ratio

$$Br_{halite} : Br_{carnallite} = 1 : 7 \pm 1 \, .$$

Only samples C 12 + C 13 show a noticeably higher value. Since, however, experience shows that there can be considerable errors in the Br/NaCl value, this discrepancy is not particularly important. The paragenetic ratio given by D'ANS and KÜHN (1940) is not found in any sample. KÜHN (1955a) refers to a tachhydrite-bearing sample from Krügershall-Teutschental in which the ratio 1 : 13 is found between the halite and carnallite. In this sample the minerals certainly do not belong to one generation, as the presence of tachhydrite (p. 190) indicates. Consequently a paragenetic ratio is not to be expected. On account of the errors associated with the determination of Br/NaCl values, it is advisable to use the absolute bromine content of the pure mineral instead of the ratio, especially since for the minerals halite, sylvite, carnallite, for example, cannot form simultaneously from the same solution, that is they are not "paragenetically" formed.

There is no doubt about the reliability of the bromine models for primary precipitates. They may therefore be used to resolve doubtful genetic problems. The present values, however, can only serve as a starting point. They indicate that the naturally occurring relationships are generally much more complex than the theoretical models. Nevertheless the usefulness of the models in genetic matters is already apparent.

In questions regarding the primary bromine content in the total salt, and for the determination of evaporation temperature in the case of a $MgSO_4$-free potash layer, only the bromine contents of the primary chlorides are considered, and this only where there is a change of paragenesis, for only they correspond to particular points on the precipitation curve. The initial bromine contents in the NaCl indicate that the Oligocene sea and the Zechstein sea both had the same relative bromine content as the present-day ocean. As the greater part of the bromine in the residual solutions is refluxed back into the ocean, a small increase in the oceanic bromine might be anticipated after each major period of salt formation, but the salt content of the world seas is so high compared with that of the salt deposits that such an increase is scarcely noticeable.

For temperature determinations see p. 261.

b) Varying and Recessive Concentration Changes

The younger Zechstein series (Z 3 and Z 4) of North Germany are characteristic examples of deposits formed under very variable concentration conditions. They all have in common a marked decrease in the bromine concentration immediately above the Hauptanhydrit from 0.02% Br/NaCl to less than 0.01% Br/NaCl within a few metres. Then follow frequent increases and decreases in the bromine concentration, which vary considerably from outcrop to outcrop. A large part of these fluctuations may be attributed to secondary recrystallization in relatively bromine-rich metamorphic solutions squeezed out of the compacting salts (BAAR, pers. comm.). In all the descriptions of the younger salts indications of secondary crystallization are found as, for example, the occurrence of crystal nests, augen salts (p. 257) etc.

From what is known at the present time, it seems that there was a fairly uniform, retrograde concentration change in Zechstein 4. It was caused by strong seawater influxes with respect to the total evaporation (p. 144). The excess of influx over evaporation was compensated by a reflux of more concentrated solution.

The upper halite sequence above the upper potash layer of the Upper Rhine also corresponds to a recessive concentration change with individual marked dilution phases. In the uppermost part of the section in general the concentration decreases steadily, and in the last cycle returns almost to the initial value at the beginning of halite precipitation (GUNZERT, 1961).

c) Bromine Content and Fine Bedding

KÜHN (1953a, b) showed that at the beginning of each halite layer a noticeably lower bromine content is found than at the end of the same layer. There is therefore a rise in concentration within each layer. These relationships can easily be explained by influx of fresh seawater, as was shown on p. 144. KÜHN (1953b), too, explained the decrease in the bromine content above the clay-anhydrite layers by a solution dilution, however, he suggested rainwater which was assumed to have dissolved the already precipitated halite. The amount of rain necessary for this process is excessive. BORCHERT (1959, p. 76) sought to explain the bromine minimum in each halite layer by secondary recrystallization in the water resulting from the dehydration of gypsum to anhydrite. If this were true, then the halite under the anhydrite layer should be similarly recrystallized, and in addition the bromine content should decrease more than is actually observed.

d) Bromine Content and Water Depth

The bromine content of rock salt is a reliable measure of the stage of concentration reached by the solution. It is in general quite independent of water depth. True, there are models in which the water depth can be given for each stage of bromine concentration. This category includes the purely static evaporation of seawater and the evaporation of fresh or partially concentrated seawater influxes which restores the original sea level without refluxing of concentrated solutions. These conditions are not realized in nature. As a result attempts to use this model lead to widely scattered (by a factor of 10 within the same series) values and in part to implausible depth figures. It was also found that the bromine concentration in the influxing brine used by KÜHN was too high (p. 145). Even the assumption itself, the influx of a pre-concentrated solution, is most unlikely, for pre-concentration up to the stage of halite saturation requires a 1/10 reduction and to the beginning of gypsum saturation a 2/7 reduction in the brine volume (Table 8). Even with continual solution influx, limited variations in solution concentration occur. During the precipitation of rhythmically banded halite series, especially in the case of the Zechstein salts, a partial reflux of more concentrated solution is particularly likely. The amount of the reflux is unknown and in consequence the water depth cannot be determined.

e) Distribution of Bromine during Metamorphism

The application of bromine partition to genetic problems is still in its infancy, and has frequently led to contradictions. Moreover, there are very few cases where the distribution ratios have been correctly determined. Then again, the bromine contents have often been determined only for the products of alteration, being assumed for the initial rocks. The bromine content of the infiltrating solution has always to be assumed. Even when using solutions whose composition has been analytically determined (HERRMANN, 1961 a), there is usually a certain arbitrariness in the selection of the solution. Calculations based upon models cannot therefore claim to be sufficiently exact and certainly cannot be accepted as the sole proof of an assumed genesis.

The simplest relations are those associated with thermal metamorphism where no foreign solutions are involved. Even in the case of the alteration of kainite at 83° C and the breakdown of carnallite associated with it, authigenic melting solutions are produced. This combined process has already been discussed in some detail for "Flockensalz" (p. 173).

From the present-day sylvite content of "Flockensalz" (23%), the following initial rock can be calculated (from p. 125) in grams:

$$60.5\,\mathrm{k} + 22.2\,\mathrm{c} + 0.4\,\mathrm{n} \xrightarrow{83^\circ} 23\,\mathrm{sy} + 33.8\,\mathrm{ks} + 26.1\,\mathrm{Q}_{83}.$$

Since the "Flockensalz" contains 43% kieserite, the initial rock must have contained 9.2% kieserite, 31.4% halite (with the kainite under, or in alternating bands with, the carnallite).

According to p. 146 the metamorphic solution will contain (with $p_k = 0.1$; $p_c = 0.26$; $p_n = 0.03$; see Table 18):

$$p_{Q\,83} = \frac{60.5 \cdot 0.1 + 22.2 \cdot 0.26 + 0.4 \cdot 0.03}{26.1 + 23 \cdot 0.8} = 0.26\%\ \mathrm{Br}$$

thus the sylvite formed will contain:

$$0.8 \cdot 0.26 \approx 0.21\%\ \mathrm{Br/KCl}$$

while in the halite the initial value of 0.03 (in the range 0.023–0.037) is retained because no second-generation halite is precipitated. However, the textures show that the halite has been recrystallized to a large extent so that the bromine content will have been determined by the bromine content of the solution.

According to KÜHN (1957), the observed bromine contents of samples from the Neuhof-Ellers mine in the Fulda district are (bracketed values from the Wintershall mine, Fulda district from KÜHN in WEBER, 1961):

	Base	"Worm" part	"Flocken-salz"	Overlying zone	NaCl intermediate material
% Br/NaCl	0.015 (0.027)	0.020 (0.022)	(0.021)	0.023 (0.023; 0.025)	0.022
% Br/KCl	0.206 (0.233)	0.211 (0.205)	0.208 (0.196)	0.215 (0.205; 0.198)	0.195

These bromine figures are thus in agreement with those of the basic model. The bromine content of the NaCl is somewhat less than the theoretical values, either because of partial recrystallization in the metamorphic solution or because of smaller bromine contents in the NaCl of the initial rock (cf. p. 201, 202).

The derivation of the "Flockensalz" by solution metamorphism from a halitic kieserite-carnallite (p. 175) is in less good agreement with the model. Calculation under the same conditions (with a bromine-poor infiltrating NaCl solution, $p_c = 0.25$) gives

$$\sim 0.15\%\ \mathrm{Br/KCl}$$

and only with a value of 0.35 for p_c, which is certainly too high, is

$$\sim 0.21\%\ \mathrm{Br/KCl}$$

obtained. The high bromine content of the sylvite can only be explained by the introduction of bromine from infiltrating solutions and by the carnallite having a higher bromine content than observed. These additional hypotheses do not make the solution metamorphism assumption more probable.

2. *Strontium*

a) Strontium in Sulphate Rocks

Although the behaviour of bromine in the formation and alteration of salt deposits can be represented with sufficient accuracy by means of simple models, this is not true of strontium. Even in present-day seawater the strontium content varies between 6.8–9.8 mg Sr/litre in water of normal salinity, and certain regional irregularities in the distribution can be recognized (SUGAWARA and KAWASAKI, 1958). In isolated lagoons greater relative fluctuations are to be expected (related to the same salt content), although at the present time no quantitative data are available.

In Recent gypsum precipitates in salinas only diadochically included strontium is known, as for example in the gypsum from the saltpans of Trapani, Sicily, with 0.28% strontium (CORRENS in discussion, D.M.G. meeting, Bonn 1960). Just as large is the strontium content of the Recent gypsum precipitates from Lake Inder (HERRMANN, 1961 a). In the Zechstein anhydrites strontium occurs in the same order of magnitude (Table 29). There are, however, some notable differences. There

Table 29. *The strontium content of the German Zechstein anhydrites*

Layer	Locality	No. of samples[a]	% Sr Average	Spread	Analyst
A 3	Siegfried Giesen nr Hildesheim	10 H	0.081	0.07–0.10	author
A 3 α–ζ	S.E. Harz foreland 3 bores	17 BA	0.11	0.07–0.17	HERRMANN in JUNG and KNITZSCHKE, 1961
A 2	S. Harz foreland	18 H	0.15	0.11–0.42	HERRMANN, 1961 a
A 2 γ	Wiehe	2 BH	0.16	0.12–0.19	author
A 2 β	S.E. Harz foreland	2 BH	0.34	0.30–0.39	author
A 2 α		4 BH	0.21	0.07–0.45	author
A 2 γ	S.E. Harz foreland 3 bores	3 BA	0.43	0.35–0.53	HERRMANN in JUNG and KNITZSCHKE, 1961
A 2 β		3 BA	0.38	0.27–0.45	HERRMANN in JUNG and KNITZSCHKE, 1961
A 2 α		3 BA	0.26	0.24–0.27	HERRMANN in JUNG and KNITZSCHKE, 1961
A 1 α–λ	S.E. Harz foreland 3 bores	37 BA	0.21	0.13–0.29	HERRMANN in JUNG and KNITZSCHKE, 1961

[a] H = hand sample, B = bore core, A = averaged sample from a full core with exactly defined boundaries.

is an average about 0.1% strontium in the Hauptanhydrit A 3, the Werra anhydrite A 1 has about 0.2% with remarkably small sample scatter. The basal anhydrite (Basalanhydrit A 2) shows a considerably greater scatter in the strontium values with an increase in the Sr content from bottom to top. Nothing certain can be concluded at the present stage of investigation regarding the dependence of the strontium content upon distance from the former coastline but there appears to be less in the interior of the basin.

Free celestite was found only in the samples richest in strontium (with 0.3–0.4% Sr) after dissolving away the anhydrite in 10% NaCl solution. Local (or secondary) celestite enrichments were also found in the upper part of the Emsland Werra anhydrite (FÜCHTBAUER and GOLDSCHMIDT, 1956) and in the Permian anhydrites of Oklahoma (HAM et al., 1960). Otherwise it is always a case of diadochic inclusion of strontium in anhydrite[43].

These results demonstrate that during the course of evaporation of normal seawater the conditions for the formation of large celestite enrichments do not arise. The occurrence of such enrichments in the Malm of northwest Germany (Hemmelte-West, ~75 km WSW of Bremen, MÜLLER, 1960, 1961) and the historically known occurrences from Obergembeck (KIPPER, 1908 and others) and similar occurrences are evidence of abnormal Sr concentrations under special and as yet unknown conditions. [Negative temperature coefficient of $SrSO_4$ solubility? Diadochically included strontium in a primary metastable aragonite with the precipitation of celestite upon altering to stable calcite (on proto-dolomite)? The effect of organisms? see, for example, SIEGEL, 1961.] Special palaeogeographic preconditions are important in this respect, also according to MÜLLER (1961). Yet it is not evaporation but the strong temperature increase which promotes circulation in shallow water regions. MÜLLER similarly assumed without specific grounds, an increased strontium concentration in the brine to explain the relatively high celestite content of the Malm rock salt, and then attributes this to influxes derived from continental areas (see discussion comments to MÜLLER, 1960). A rapid calculation shows that in the whole Malm rock salt of north Germany there is at least $1000 \cdot 10^6$ tons Sr (probably more that $3000 \cdot 10^6$ tons in $2 \cdot 10^{12}$ tons halite, cf. RICHTER-BERNBURG, 1953, p. 602) so that the deposit of Hemmelte-West of itself is without real significance for the geochemical balance of salt basins. That is to say that by far the greater part of the total Sr, despite local

[43] Secondarily formed celestite was found in the main dolomite (Hauptdolomit) of the North-German Zechstein by THEILIG and PENSOLD (Über das Vorkommen von Cölestin im Hauptdolomit des norddeutschen Zechsteins. Chemie der Erde **23**, 215–218, 1964).

enrichment, remains in the solution. The question of celestite will not be further discussed here. Our knowledge of celestite enrichment should be furthered when the results of the investigations of some Australian lagoons become available. H. SKINNER (1960) reported there was an occurrence of up to 3% celestite in recent proto-dolomite precipitates, that is, in the carbonate precipitation stage. These, however, are not comparable with conditions in the large seas where salt deposits were formed at different stages of the Earth's history.

The average strontium content of anhydrite in the absence of celestite is about 0.2%; this shows that in comparison with the Sr content of the solution (p. 148) an enrichment occurs in the solid phase. For this reason, the partition coefficient b was assumed to be greater than 1 (p. 148, 149); see also footnote on this page. The strontium concentration scarcely changes during the alteration of gypsum to anhydrite as the strontium content of primary gypsums is equally large. The different result of NOLL (1934) was essentially based upon secondary gypsums.

Although the partition factor is greater than 1, according to HERRMANN'S (1961 a) calculation[44], some 90% of the strontium originally present remains in solution during the precipitation of gypsum. During the precipitation of gypsum the Sr content of the solution increases by a factor of three (p. 148, 149). There should therefore also be a rise in the Sr content of the precipitate from below to above. However, 1. the primary gypsum precipitates no longer exist in the rocks investigated and 2. in none of these examples was static or even purely progressive evaporation realized. The latter is most closely approximated in the Stassfurt series and in this series an upward increase in strontium indeed seems to occur.

b) In Halite and Potash Salts

Soon after saturation in halite, saturation in celestite is attained and accordingly celestite is observed in the halite beds of most Zechstein salts. In addition, the chlorides and calcium-free sulphates themselves carry traces of strontium, as the following table (Table 30) shows.

From this it is apparent that the strontium content of the halite is higher than in the other salt minerals (MÜLLER, 1962). This may be an expression of the fact that in solutions of nearly the same Sr concentration considerably less Mg is diadochically replaced by Sr than is Na.

These Sr traces in the Ca-free salt minerals form a very small part of the total Sr content of salt rocks and may be excluded from further

[44] According to the old data of JÄNECKE. However newer determinations introduce only small corrections and the result is hardly altered.

Table 30. *Strontium content of halogenides and Ca-free sulphates of salt deposits (in wt. % Sr)*

Mineral	Number of samples	Average	Spread	Author
Halite	25	0.0003	0.00007–0.0007	MÜLLER, 1961
Sylvite	2		0.0001; 0.00002	MÜLLER, 1961; NOLL, 1934
Carnallite	4	0.0001	0.0002–0.00002	MÜLLER, 1961; NOLL, 1934
Bischofite	3	0.00006	0.00008–0.00003	MÜLLER, 1961
Tachhydrite	1		0.178	HERRMANN, 1961d
Kieserite, hexahydrite, epsomite	5	0.0001	0.0003–0.00003	MÜLLER, 1961
Bloedite, loewite	2		0.0003; 0.0001	MÜLLER, 1961
Kainite, leonite, langbeinite	4	0.0002	0.00009–0.0005	MÜLLER, 1961

consideration. As tachhydrite (p. 190) is formed secondarily in Sr-rich $CaCl_2$ solutions, it may also be dismissed from further examination. Its high Sr content depends upon the ready replaceability of Ca.

The total Sr content of the Stassfurt salts shows a very characteristic and real variation: 1. in its dependency upon stratigraphic horizon, 2. in its dependency upon palaeogeographic position. Figs. 39 and 40 illustrate the total Sr content in a series of samples from the Königshall Hindenburg mine investigated by the author. Despite the wide scatter in the results, a uniform tendency is apparent: the strontium content increases within the Stassfurt rock salt series. The maximum values are reached just below the potash layer. In the potash layer the strontium content again decreases. Calcium and strontium both vary in the same sense but the rise and fall of the strontium content is much more rapid. The average values from six investigated sample series from the Stassfurt rock salt are given in Table 31.

There is little point in calculating average values for the whole Stassfurt seam for this would conceal its characteristic feature, an important increase within the salt sequence.

The marked differences in Ca and Sr contents between individual series are not easily explained. They are, in part at least, of secondary origin, some being due to diagenetic alterations in solution and some to later plastic flow of NaCl (p. 193), more rarely to solution metamorphism.

In Königshall-Hindenburg the Sr/Ca ratio within the potash layer falls from 31 to 6. In all samples isolated celestite is found and in ratios >15 always in appreciable amounts. HERRMANN (1961a) observed celestite as soon as the ratio was $\gtrsim 10$. There is no discernable difference in the Sr/Ca ratio in the various mineralogical associations in the potash

Table 31. *Average strontium and calcium content in Stassfurt rock salt from the Königshall-Hindenburg mine*

Metres under the Stassfurt seam	Wt. % Sr	Wt. % Ca	$1000 \cdot \frac{\% Sr}{\% Ca}$
1	0.3	8.5	35
3	0.24	6.4	37
5	0.22	7.2	31
7	0.20	7.5	27
10	0.18	7.5	24
15	0.13	7.5	17
17	0.095	6.6	14
20	0.06	4.6	13
23	0.04	3.5	11

layer in Königshall-Hindenburg, yet there are differences in the absolute contents which are due to solution metamorphism (p. 186).

The researches of HERRMANN (1961 a) in the South Harz show that the strontium content in anhydrite during the precipitation of halite was about as high as in pure anhydrite rock, namely (number of samples in parentheses)

in the basal anhydrite 0.12–0.21; average 0.17% Sr/anhydrite (4),
in the Stassfurt rock salt 0.09–0.41; average 0.19% Sr/anhydrite (13),
in the Unstrut banks 0.08–0.32; average 0.23% Sr/anhydrite (38).

On account of the wide scatter of the measurements, the apparent increase in the mean values still lies within the limits of statistical error. The higher values recorded by NOLL (1933) probably included free celestite, as in stratigraphically equivalent samples after the separation of celestite only 0.25% Sr/anhydrite was found (BRAITSCH, 1960)[45]. Polyhalite, too, includes similar amounts of Sr in its lattice (HERRMANN, 1961 a; MÜLLER, 1962).

A similar conclusion, namely that the Sr–Ca ratio increase from below to above, was reached for the Stassfurt rock salt in the Magdeburg-Halberstadt region (Gröna bore) and the South Harz by HERRMANN (1961 a). In addition, a dependence upon palaeogeographic position was also demonstrated. HERRMANN (1961 a, Tables 12 and 13) found the following average values (Table 32).

It would be desirable to demonstrate the regional distribution within similar stratigraphic horizons more precisely. Nevertheless the differences between the deeper part of the basin (Gröna) and marginal region are certainly real.

[45] In 100 anhydrite samples from different horizons in the Stassfurt Series of the South Harz area an average content of 0.21% Sr/anhydrite was found (HERRMANN, 1961 a).

Table 32. *Average strontium content at different horizons in the Stassfurt salts (after* HERRMANN, *1961 a)*

Strat. horizon	Locality	Rock	Number of samples	Wt. % Sr	$1000 \cdot \frac{\% Sr}{\% Ca}$
Stassfurt seam	S. Harz	Kieserite-carnallite	7	0.03	101
		Anhydrite-carnallite	4	0.045	26
		Anhydrite-sylvite-halite	10	0.14	27
		Anhydrite-halite	11	0.45	76
		Unstrut banks[46]	44	0.45	77
Stassfurt rock salt	S. Harz		12	0.01	17
	Gröna		30	0.0044	7.5

Although in the South Harz area, both in the Stassfurt rock salt and in the transition zone to potash beds (Unstrutbänke[46]), the same regularity exists as is found in Königshall-Hindenburg, relationships within the Stassfurt seam are less clear. HERRMANN (1961 a) could not carry out a complete sampling programme of stratigraphic horizons above the 10th Unstrutbank because of the absence of index horizons, so that no direct comparison is possible. Furthermore in the Glückauf mine, Sondershausen, strong secondary tectonic effects have to be taken into account. Here, too, the large strontium variations suggest secondary redeposition and to some extent secondary introduction of strontium.

As regards the secondary introduction of strontium in metamorphic solutions, there are considerable differences of opinion, particularly with respect to its extent, but because of the lack of experimental data, the question cannot be settled at present.

According to HERRMANN (1961 a), the greater part of the $CaSO_4$ in the carnallite and sylvite rocks of the southern Harz originated from the interaction of $CaCl_2$-bearing metamorphic solutions and kieserite carnallite (p. 187). The more recent investigations of HERRMANN (1961 b) indicate that these $CaCl_2$-bearing solutions contain appreciable amounts of strontium (up to 0.14% Sr) and other trace elements, and indeed, the calculation of the incongruent decomposition of carnallite with $CaCl_2$-containing solutions does give approximately the composition of the anhydrite-sylvite-halite observed.

[46] Also described in the literature as the "Liegendgruppe" of the Stassfurt Seam. It contains 10 characteristic anhydrite-halite index horizons, the so-called "Unstrutbänke", which make possible fine correlation over the whole South Harz and Unstrut region (TINNES, 1928). In Königshall-Hindenburg an exact matching is not possible. According to HENTSCHEL (1958), Na 2γ corresponds to the "Liegendgruppe". In the South Harz the potash begins to occur locally in the beds between the index horizons at quite different places, usually just above bank No. 7.

In contrast with this it was shown (p. 187) that in the Königshall-Hindenburg mine, in which is a typical, i.e. almost wholly $MgSO_4$-free equivalent of the South Harz facies, the anhydrite of the anhydrite-sylvite-halite is predominantly of primary origin – more strictly, early diagenetic, see p. 157. Furthermore, the incongruent decomposition of carnallite with NaCl solutions also produces the present anhydrite-sylvite-halite. In special cases as, for example, in the Glückauf mine, Sondershausen, the secondary formation of anhydrite and celestite as described by HERRMANN may occur to a limited extent. This is certainly not generally true. However, it serves to suggest that, with strontium, secondary redeposition and perhaps also later introduction should not always be neglected.

This view is supported by the heavy-metal content of sylvite and rock salt, which cannot be derived from the carnallite rocks (HERRMANN, 1961 d), although the absolute quantities are too small to allow quantitative proof.

Unlike HERRMANN (1961 a), we consider the unidirectional increase in the strontium content and strontium ratios in the Stassfurt rock salt from below to above, and from the centre of the basin towards the basin margin, to be primary, like the replacement of kieserite by anhydrite in the shallower marginal regions of the basin. The regularity of the strontium increase in the Stassfurt rock salt would not be explicable if the celestite were subsequently precipitated from metamorphic solutions. Three possible causes may be considered for the increase in the strontium content from the basin centre towards the margin:

1. an increase in the strontium content of the solution in the marginal areas due to continental influxes;

2. an increase in the strontium content of the solution in the marginal areas due to a stronger $MgSO_4$ deficiency, which may also be due to continental influxes (p. 250). In SO_4^{--} poor Cl^--rich solutions the $SrSO_4$ solubility is greater, so that when such a solution evaporates more $SrSO_4$ can be precipitated. The vertical increase in the strontium content, on the other hand, is presumably caused by an decrease in $SrSO_4$ solubility with an increase in the total salt content of the solution.

3. a decrease in $SrSO_4$ solubility with increasing temperature and therefore preferential precipitation in the shallow nearshore regions. In this case continental influxes might not be necessary. Whether one or other of these factors is dominant, can only be determined by further examination and in individual cases. They should, however, suffice to explain the primary facies differences between the basin interior and rim.

A consequence of the ready mobility of strontium in chloride saline solutions is that the primary distribution is obscured and additionally disrupted by the local introduction of strontium in metamorphic solutions.

In undisturbed progressive evaporation, as in the Stassfurt cycle, the strontium promises to be an important additional indicator of conditions of formation; however, before significant geological conclusions can be drawn, the important basic physico-chemical facts must first be established. In salt deposits with irregular progressive and recessive sections the interpretation is naturally more difficult, and such relationships cannot be considered here. There is also little point in investigating randomly collected samples without regard to their stratigraphic and palaeogeographic background, for this permits no conclusions about their origin.

E. Other Components of Salt Deposits

The behaviour of other trace elements in the formation of salt deposits is still not clarified on a physico-chemical basis. With each component added, the system becomes more complex and in consequence very little experimental work has been carried out and that only with respect to the solid phases found in nature. It is therefore impossible to begin to compare theoretical models and observed distributions.This, of course, diminishes the reliability of the conclusions for the interpretation must now rest upon general genetic considerations and analogy. Despite this, other important minor components must be considered here, within the limitations indicated above, because they give further information about conditions of formation. However, a review of the geochemistry of salt deposits is not contemplated here (cf. the review by HERRMANN, 1958).

The selection of minor components was made with particular reference to their origin. As the composition of seawater is sufficiently well known, abnormally high values of impurities yield reliable indications of primary fluctuations in the initial solution, of influxes, or of the secondary introduction or mobilization of salts by solutions which penetrated the deposits at a later stage. However, in many cases the distinction between primary and secondary deposition is uncertain or controversial, so that in practice it is often impossible to distinguish between syngenetic and epigenetic "semisaline" impurities (LEONHARDT and BERDESINSKI, 1949/50).

In the context of the stated problem, boron and iron will be discussed in some detail. In addition, the clay minerals and salt clays (Salztone) deserve general discussion but without going into detail about their mineralogy.

1. Boron

a) The Behaviour of Boron during Salt Precipitation

The application of the simplified Boeke equation (1a) p. 149 indicates 0.0012% B in seawater at the onset of gypsum precipitation, 0.0037% B at the beginning of halite precipitation, 0.037% B at the beginning of carnallite precipitation and 0.05% B in solution at the beginning of bischofite precipitation.

There are still many gaps in our knowledge of the boron-bearing system in oceanic salt solutions. VAN'T HOFF (1909) determined the approximate stability limits of some borates in the five-component system at 25° and 83° (cf. also JÄNECKE, 1912). The establishment of equilibrium in the borates is particularly slow and in primary precipitates no stable solid phases occur. D'ANS and BEHRENDT (1957) determined the solubility of boracite at 46° C in an almost carnallite-saturated solution as 0.025% boracite (=0.005% B). For other solutions, other temperatures, and other natural borates, no recent solubility data are available on the five-component system. The experiments of E. SCHMIDT (1959) show that the solubility of boracite clearly increases with rising temperature, but in contrast decreases as the $MgCl_2$ content of the solution increases.

According to VAN'T HOFF'S data, boracite occurs in $MgCl_2$-rich solutions, ascharite has the largest stability field in the five-component system, kaliborite occurs only in the K_2-corner and pinnoite only in the region of point *Q* (Fig. 16), while the NaCa-borates occur only near the SO_4–K_2 side. This indicates a characteristic dependence of the borates upon adjacent rocks, which is also reflected in natural occurrences (and to some extent first inferred from them). The borates therefore suffer substantial transformations in secondary reactions, especially in kainitization (BOEKE, 1910a) and gypsification and soon disappear completely with the increasing effects of water. A famous example of this kind, which will not be discussed further here, is the Inder deposit of S. E. Russia (GODLEVSKY, 1937; NIKOLAEV, 1946; who gives a paragenetic scheme essentially according to VAN'T HOFF, 1909; and YARZHEMSKY, 1945, 1953).

The main trends established by VAN'T HOFF'S work are still valid today. Only Ca and CaSr borates have been added. In former times these were thought to be characteristic of terrestrial salt lakes while in contrast the Mg borates were considered characteristic of oceanic salt deposits. This rule, however, must now be changed since CaSr borates also occur in $MgSO_4$-free salt parageneses, that is, in salt deposits derived from altered seawater. At the present time they are much too rare to be of use as index minerals.

The strongly oversaturated forms make it understandable that borates are not among the first salts to precipitate out. In the artificial evaporation of natural brines, NIKOLAEV and CHELISHCHEVA (1940) found that even with solutions enriched to 1.4% B still no borate precipitated. Saturation with respect to boracite is reached during the course of NaCl precipitation. Yet boracite is almost always subsequently precipitated from pore solutions through a series of intermediate steps. The stassfurtite (and ascharite) concretions are certainly secondary but not post-diagenetic in origin. Otherwise they could not be preserved in carnallite rocks which are unstable in the subsequent infiltrating solutions.

Present knowledge does not permit the correlation of borates with a definite sequence of formation stages of the adjacent rocks except for the borates in the cap rock. The formation and alteration of the borates occurs much more slowly than that of chloride salts, so that often relicts of older parageneses are observed. Probably the formation of the borates extends over a long time interval so that it is often impossible to distinguish early diagenetic from post-diagenetic, including metamorphic, changes. This is also true of the concretionary danburite deposits and of the occasional formation of tourmaline which is doubtless also authigenic (p. 24).

To produce well-crystallized pseudo-cubic boracites, it used to be thought that a temperature of formation above the transition point to the cubic high temperature form (>265° C), still higher in Fe-containing boracite, HEIDE et al., 1961) was necessary and in consequence a secondary boron influx in hydrothermal or even pneumatolytic solutions. These temperatures never existed in boracite-bearing salt deposits. The pseudo-cubic forms can best be explained as mimetic twins which is understandable from their crystal structure (ITO et al., 1951). Syntheses well below the transition temperature have also yielded at times pseudo-cubic crystals (HEIDE, pers. comm.). Occasionally, too, well-formed pseudo-cubic boracite is found in carnallite rocks as, for example, at Königshall-Hindenburg, which certainly originated at temperatures well below the transition temperature.

In addition to the boron in borates and borosilicates, other boron-containing minerals have been found in the last few years in salt deposits. In sulphate rocks, particularly anhydrite, a relatively high boron content has been reported (HARDER, 1959; HAM et al., 1961). In the German Zechstein anhydrites (see below) it was not always possible to distinguish in which form the boron occurred, especially since boracite (in the Basalanhydrit and Hauptanhydrit) and danburite (in Basalanhydrite, p. 24) have repeatedly been recorded. In contrast, in the Permian anhydrites of Oklahoma the borates and borosilicates are absent (HAM et al., 1961). They contain 18 to 200 g B/ton with an average 75 g B/ton of which up to one-third is in a water-soluble form, presumably as fluid inclusions. According to HAM et al. (1961) the boron seems to be incorporated in the anhydrite lattice, perhaps as a BO_4 complex, and these authors succeeded in preparing a B-containing anhydrite, although not under conditions of natural salt formation. At least, the existence of the complex $[B(SO_4)_4H_4]^-$ (GREENWOOD and THOMPSON, 1959) and the compounds $Me^I [B(SO_4)_2]$, and $Me^{II} [B(SO_4)]_2$ are known (SCHOTT and KIBBEL, 1962). However, the crystal chemistry is not yet clear. In gypsum, on the other hand, boron is not included in recognizable amounts, and in secondary gypsification of anhydrite the boron separates out as individual mineral phases (probertite concretions, more rarely ulexite[47] and priceite). The individual Ca borates found in the residual deposits of the Inder salts should also be examined to determine whether their boron content did not originate from an older anhydrite which contained boron.

The apparently homogeneous distribution of boron in anhydrite is no proof of the primary precipitation of anhydrite, as the gypsum-anhydrite alteration, in the presence of an NaCl saturated solution, probably occurs relatively early (p. 158). As gypsum does not take up boron, the anhydrite can only have incorporated boron from the solutions present. This speaks against a post-diagenetic gypsum-anhydrite dehydration.

In addition to anhydrite the clay minerals, particularly the micas, are important boron hosts in salt deposits. This illustrates the importance of the impurities in the total material balance in salt deposits, as subsequently will be more fully discussed. According to HARDER (1959), the micas of salt deposits contain from twice to twenty times as much boron as normal marine clays. The boron content of the micas also appears to increase as the concentration of the salt solution rises (FREDERICKSON and REYNOLDS, 1960) since the higher boron contents are found in the salt clay (Salzton) above primary carnallite rock (e.g. in T 3). Also, according to HARDER (1961), the boron is fixed in the mica lattice and not adsorbed, possibly occurring in fourfold coordination in Al sites.

No quantitative relationships for boron are known either for the micas or anhydrite, so that no conclusions can be drawn from the boron in the solid phases as regards the boron content of the solution. Furthermore, there is no simple proportionality between the boron content of the solution and the micas because of interference due to the metabolism of lower organisms (FREDERICKSON, pers. comm.). It is clear, however, that a considerable part of the boron refluxed in the non-evaporated residual solutions, just as was the case with bromine. With the precipitation of the chlorides, the B/Cl ratio in the solution increases. As the equilibrium partition is not known, no conclusions can be drawn about the boron content of the solution from the boron content and the B/Cl ratio of the precipitates, nor can a clear distinction be made between "primary" and "secondary" (introduced by metamorphic solutions) boron. In this respect the inferences that can be drawn are more restricted than in the case of strontium where, because of the close

[47] See: KORITNIG, S.: Ulexit-Konkretionen ("cotton balls") im Zechsteingips der Werra-Serie. N. Jb. Miner. Abh. **103**, 31–34 (1965).

geochemical relationship, the Sr/Ca ratio was relatively little influenced by salt precipitation.

b) Boron Content and its Regional and Vertical Distribution

For the German Zechstein anhydrites the analyses of HARDER (1959) give the following data (Table 33).

According to this, the boron content of the Basalanhydrit (A 2) is generally somewhat greater than that of all the other anhydrites. It is also clear that there is a greater boron content in the salt rocks from Zechstein 2 than in the other Zechstein series. It is not yet possible to distinguish regional variations in distribution within the same bed. In the Permian of Oklahoma, where only the sulphate phase was fully developed, the higher boron content is linked to regions of strong evaporation close to the ancient coastline (HAM et al., 1961).

Table 33. *Boron content of the German Zechstein anhydrites (after* HARDER, *1959)*[a]

Stratigraphic position	Number of samples	Average B content g B/ton	Scatter g B/ton
A 1	26	20–25	1–70[b]
A 2	11	200	100–500
A 3	16	50–60	24–90
A 4	4	40	3 80

[a] For new data see: FABIAN, H.-J., KLENERT, G.: Der Bor-Gehalt der Zechstein-Anhydrite. Erdöl und Kohle, Erdgas, Petrochemie **15**, 603–606 (1962).

[b] HARDER's (1959, p. 168) samples (739) and (740) were incorrectly labelled and are not included in the determination of the mean value. In the Königshall-Hindenburg mine the Basalanhydrit was not penetrated so that A 1 and Na 1 are not known from there.

There has been relatively little systematic investigation of the boron content of salt rocks (BILTZ and MARCUS, 1911). From Königshall-Hindenburg there are now further, unpublished analyses of HARDER (Figs. 39, 40). There is, in the Stassfurt rock salt, on average a relatively uniform increase in the boron content from about 10 g B/ton in the middle of the Stassfurt rock salt to about 70 g B/ton immediately under the Stassfurt seam. In the series in which ascharite is found (Na 2 $\gamma - 1$) the boron values are somewhat higher than elsewhere. In the Stassfurt seam itself the boron content decreases to about 20 g B/ton (with the formation of sylvite; about 60 g B/ton with the formation of halite). In total the boron content varies in the same sense as the sulphate content (Figs. 39, 40, Ca and Sr occur as sulphates).

Through these results, the microscopic investigations of insoluble residues (Braitsch, 1960) which established a uniform distribution of boron within the Stassfurt Series, are quantitatively supplemented and the assumption of a high boron content in the impoverished Stassfurt seam confirmed. However, only a few impoverished samples were examined so that, apart from a relative boron enrichment due to the solution of sylvite (p. 122) an absolute increase in boron is not finally assured. The calculation of boron in borates usually gives values somewhat lower than the analytically determined total boron. The deficit can probably be explained in part by the boron in the clay residues and in part perhaps by the boron in the anhydrite.

Besides the uniformly distributed macroscopic, but not usually visible, boracite within the beds there occurs in the basal zone of the potash layer, originating from an NaCl fissure filling, an enrichment in macroscopically visible boracites (pseudo-cubic 1–3 mm edge length). Contrary to older ideas (Storck, 1954; Kühn, 1955; Hentschel, 1958), these are quantitatively unimportant in the study of boron distribution and represent for the most part merely recrystallization of finer boracite crystals already present. Only a small part of the boron was introduced by infiltrating solutions.

In the interior of the basin (Stassfurt, Vienenburg) the boron content, according to the analyses of the Stassfurt seam by Biltz and Marcus (1911), appears to be noticeably higher (up to >100 g B/ton). Macroscopic boracite (Stassfurtite) and ascharite concretions, which are absent in Königshall-Hindenburg always occur here. From more than a hundred analyses of the Stassfurt seam in the Hildesia mine an average of about 400 g B/ton is found, with a spread in individual analyses from <100 g B/ton to about 2000 g B/ton (Kokorsch, 1960, and pers. comm.). The grey sylvite-kieserite-halites of the Stassfurt seam in the Hannover potash district according to Kühn (1955a) contain an average of 600 g B/ton, both ascharite and boracite concretions being found. In the South Harz area, in the transition zone between the pure anhydrite marginal facies and the kieserite basin facies, boracite and ascharite concretions occur less frequently or are absent, but macroscopic boracite crystals are both larger and more frequent than at Königshall-Hindenburg. The macroscopic boracites appear to be particularly enriched in the "Liegendgruppe" of the Stassfurt seam (see p. 212) which ought to correspond to the zone Na 2 γ in Königshall-Hindenburg (Hentschel, 1958). No quantitative data have yet been published for the South Harz district.

As with strontium, so, too, in the case of boron, regional variations in concentration are found, although in the reverse direction. The higher concentrations occur in the centre of the basin.

c) Origin of the Boron

The dependence of the boron content on palaeogeographic position, the increase of boron within an evaporation sequence and, in particular, the uniform distribution of the boron within the beds are undoubted indications of primary origin. Thus, the boron comes overwhelmingly from the evaporating brine. The borates, however, are not in the form in which they were originally precipitated. This is apparent from the occurrence of inclusions of anhydrite and clay in all Ca–Sr borates. In addition the formation of all concretions took place in the sediment. This can have occurred, however, in an early diagenetic stage when the precipitated salts were still impregnated with the mother solution. In most well-developed boracite crystals there are usually no unique age criteria, and they appear to belong to several generations. In carnallite and perhaps in the less thoroughly impregnated Stassfurt rock salt, they were probably formed, at the latest, in the diagenetic phase, when solution circulation was still possible. No concrete data are available to indicate how long the diagenetic phase may have persisted. There is also secondary redeposition of boron as a consequence of metamorphic processes. In individual instances some boron may have been introduced by infiltrating solutions, for an appreciable boron content in some salt solutions has recently been demonstrated (HERRMANN, 1961b; HERRMANN and HOFFMANN, 1961). This is, for instance the explanation of the macroscopic boracite in the halite-impoverished salts of Königshall-Hindenburg. Even in such cases, however, the predominant process is a recrystallization of the boracite already present. In contrast, the occurrence of lueneburgite in paragenesis with fluorite in a secondary anhydritized thrust zone in the Stassfurt rock salt of Königshall-Hindenburg presumably represents the introduction of boron and fluorine by foreign solutions (BRAITSCH, 1961a). In other circumstances the ascharite concretions in carnallite rock may be dissolved incongruently and the boron removed as, for example, in the Aschersleben area (LÖFFLER, 1960, p. 31). In this case the penetration of boron-poor or boron-free solutions is assumed.

The boron content in Zechstein 2 is one or two orders of magnitude greater than in the other Zechstein series. As the salt solutions of Zechstein 2 certainly contain an large part of the boron-enriched residual solutions of Zechstein 1, a somewhat higher boron content in Zechstein 2 is not unexpected, but, because of the relatively small amount of salt precipitated in Zechstein 1, this effect is not a sufficient explanation of the high boron content in Zechstein 2.

It is therefore necessary to assume special conditions for Zechstein 2, whether with respect to the initial concentration of the solution or with respect to conditions of precipitation. To throw some light on the special conditions of Zechstein 2, a complete geochemical and mineralogical investigation of other minor components is necessary. The basic assumption that by far the greater part of the boron in the German Zechstein salts resulted from evaporation and early-diagenetic circulating solutions should probably be retained.

In contrast to this view is the assumption of "thermal" or post-volcanic introduction of boron supported by D'ANS in particular (in D'ANS and KÜHN, 1960, and older works). There are now, however, direct counters to such an argument. For example, in the Upper Rhine Buggingen potash mine datolite[48] has recently been discovered in alkali basalt dykes (ankaratrite, WIMMENAUER, 1951). What forms is thus not borate but borosilicate in small nodes and only at the contacts with salt rocks. Large volumes of post-volcanic solutions penetrated as a consequence of the basalt intrusion and their effects are observable in the impoverished zones at the basalt contacts (WIMMENAUER, 1951), yet away from the immediate contact zone little or no borate precipitation occurred. While the boron content of the unaltered sylvinite and halite in average samples from banks *A*, *B*, *C* (fine stratigraphic divisions of WAGNER, 1953), is 1 g/ton, and in bank *D* 5 g/ton, in the intervening "marl horizons" it averages about 200 g/ton, and in the impoverished zone bounding the dykes with macroscopic datolite at the contact, the boron content again clearly increases, although with a wide scatter of individual values (bank *A* about 10–80 g B/ton). In one sample from the impoverished zone individual datolite crystals were found in the insoluble residues[48]. Post-volcanic solutions are usually too dilute to effect anything but solution metamorphism in salt rocks. They cannot possibly explain concretions of ascharite etc. in carnallite. In the Werra area no boron minerals have yet been found at the basalt contacts.

2. *Iron*

a) The Behaviour of Iron during Salt Precipitation

Present-day seawater with only 0.002 g Fe/ton contains the least iron of any naturally occurring salt solution. At the beginning of gypsum precipitation there is in solution about 0.005 g Fe/ton, at the beginning of NaCl precipitation 0.017 g Fe/ton, and at the beginning of carnallite precipitation about 0.17 g Fe/ton. If colloidal iron oxide sols are taken

[48] See also: BRAITSCH, O., GUNZERT, G., WIMMENAUER, W., THIEL, R.: Über ein Datolithvorkommen am Basaltkontakt im Kaliwerk Buggingen (Südbaden). Beiträge zur Mineralogie und Petrographie **10**, 111–124 (1964).

into account there is then about a tenfold increase. From this it is abundantly clear that salt deposits originating from pure seawater must be practically Fe-free. As Fe chlorides are readily soluble, they cannot originate from the evaporation of normal seawater. From the spread of marine sedimentary iron ores at various geological epochs, however, it is apparent that under special conditions in bordering seas significantly higher iron concentrations occurred (originating from continental weathering solution products). It is therefore not possible in general to compare the conditions in present-day seawater with those existing in far smaller basins.

A further complicating factor is introduced by the oxidation stage of iron. Larger amounts of dissolved iron can only be present in the ferrous form, that is under reducing conditions. D'ANS and FREUND (1954) have shown qualitatively that, in the presence of sufficient reducing material (organic matter) in concentrated $MgCl_2$ solutions, $Fe(OH)_3$ is reduced. According to the determinations of WHITMAN et al. (1925), the solubility of Fe^{II} in $MgCl_2$ solution is noticeably increased. In primary salt solutions reduction by bacteria (FLINT and GALE, 1958) and algae (MORRIS and DICKEY, 1957) also plays a part.

Since the upper limit for the partition coefficient of iron between precipitated carnallite and the solution is known to be $^1/_4$ (p. 82), an estimate can be made of the minimum iron concentration of the solution from the diadochically included iron in the carnallite. Yet in the carnallite of fossil salt deposits the iron occurs as haematite in the form of oriented intergrowths and there are no quantitative data about the $FeCl_2$ content of the rock-forming carnallite.

Now PRECHT (1905 and earlier, cited in BOEKE, 1911), JOHNSEN (1909) and later BOEKE (1911) and in more recent times MARR (1957) assumed that the oriented haematite inclusions were caused by the subsequent demixing and oxidation of primary diadochially included Fe^{II}. JOHNSEN (1909) formulated the oxidation process as follows:

$$6(KCl \cdot FeCl_2 \cdot 6\,H_2O) \rightarrow 4\,FeCl_3 + Fe_2O_3 + 6\,KCl + 3\,H_2 + 33\,H_2O\,.$$

The $FeCl_3$ would hydrolyze in the water formed, producing more haematite and free hydrochloric acid. Johnson permits $Mg(OH)_2$ to participate in the reaction:

$$4\,FeCl_3 + 6\,Mg(OH)_2 \rightarrow 2\,Fe_2O_3 + 6\,MgCl_2 + 6\,H_2O$$

with the $MgCl_2$ produced reacting with the KCl of the first stage to produce carnallite. The total reaction then is:

$$6(KCl \cdot FeCl_2 \cdot 6\,H_2O) + 6\,Mg(OH)_2 \rightarrow 3\,Fe_2O_3 + 6(KCl \cdot MgCl_2 \cdot 6\,H_2O) + 3\,H_2O + 3\,H_2\,.$$

No reason is known for the supposed demixing and oxidation. Further, the production of hydrogen according to D'ANS and FREUND (1954) would drive the equilibrium back to the left. Finally, despite PRECHT'S assumption, the necessary $Mg(OH)_2$ does not occur (p. 234) in salt deposits.

Setting aside these ideas for the time being, then the iron content of the haematite present averages 0.03% Fe (p. 228) and, using the D'ANS-FREUND partition coefficient, we obtain a solution iron content of over 0.12% Fe during the precipitation of carnallite, that is an enrichment of 10^3 or 10^4 times as against a carnallite-saturated solution from pure seawater.

The oriented haematite intergrowths of haematite in carnallite may also be explained by contemporaneous growth, in other words, epitaxy, since in solution the nucleation energy of oriented growths is lower (p. 36). This possibility, however, does not yet appear to have been experimentally verified. The oxidation of an Fe^{II} primary precipitate would seem to be simpler. This might be accompanied by biological metabolic processes (e.g. FLINT and GALE, 1958) but it need not be. As in normal seawater all Fe^{II} in equilibrium with atmospheric oxygen is immediately oxidized to Fe^{III} and precipitated as the "hydroxide"[49] (COOPER, 1937), similar equilibria must be assumed for $MgCl_2$-rich solutions, even though they may contain higher Fe^{II} concentrations. In neutral to slightly acid media in the presence of Mg^{++} and Ca^{++} ions direct precipitation of haematite may be possible (SCHELLMANN, 1959).

The direct precipitation of haematite from the solution requires in any case a lower Fe content in the solution since the partition coefficient is no longer the decisive factor.

In addition to carnallite, sylvite also often has an appreciable haematite content, so that the crystals are a strong dark red colour. In sylvite, the haematite is often concentrated in the form of very fine flakes at the crystal margins, so that the interior has a bluish to milky opalescence and the rim has a dark red colour. Only close to the boundaries of sylvite-carnallite rocks are relicts observed of the typical haematite scales found in carnallite (p. 184). It may be concluded from this that in the process of the secondary formation of sylvite from carnallite (p. 186) the haematite is also recrystallized. This would help to explain the occasional skeletal forms of haematite found in sylvite (RICHTER, pers. comm.) although the different character of haematite in sylvite and carnallite has not yet been particularly well explained. BOEKE (1911) assumed that haematite in sylvite could also originate

[49] This primary precipitate changes with the loss of water into products containing fewer OH molecules, the so-called condensed hydroxides (GLEMSER, 1961 and earlier) and finally into α Fe_2O_3 (haematite).

subsequently from a soluble ill-defined iron compound. This should, however, produce a sylvite with uniformly distributed haematite.

In neither carnallite nor sylvite is the primary incorporation of Fe^{II} or an Fe^{II} compound in the lattice of natural salts during precipitation sufficiently certain. However the diadochic inclusion of Fe^{II} does occur in bloedite, tachhydrite and certain boracites, among other minerals. In boracite the iron appears to be enriched in comparison to the solution, and is often zonally different in abundance. None of the named minerals is, however, primary. This is especially true of all rinneite occurrences which originate from subsequent infiltrating solutions, generally associated with the incongruent alteration of carnallite. In a few salt solutions from potash mining districts notable $FeCl_2$ contents have been determined, e.g. about 5.8 mol $FeCl_2$/1000 mol H_2O ($\approx$35.1 g $FeCl_2$/litre) from Mathildenhall nr. Hildesheim (BAUMERT, 1928), that is, about one-quarter of that necessary for the formation of rinneite (Table 7). In the salt solutions investigated by HERRMANN (1961b) the Fe content lay between 0.02–2 g Fe/litre ($\approx$0.04–4.4 g $FeCl_2$/litre). The $FeCl_2$ content of liquid inclusions has still to be systematically investigated. Its occurrence is clearly indicated by the frequent observations of Fe "hydroxide" precipitates produced on fresh fracture planes of many salt rocks after only a few days' exposure to the atmosphere.

When considering the total iron in salt deposits, the Fe chlorides and the diadochically included iron in bloedite, boracite etc. can be neglected. Under "primary" authigenic formations, in addition to the dominant haematite, some magnetite usually has to be considered. It generally does not occur as good crystallites but rather as irregularly formed aggregates. The form of the iron mineral depends upon the redox potential (p. 263) and often pyrite and occasionally pyrrhotite (p. 20, 21) are found in salt deposits. It was observed in Königshall-Hindenburg that haematite (and a little magnetite) and pyrite often occur together. The haematite is found in sylvite and carnallite, and the clay residues of the same samples contain the pyrite and often smaller amounts of magnetite. In (impoverished) anhydrite-halites, haematite is generally absent save in the form of fine infilling of hair-like fissures in the rock salt (perhaps altered to goethite?) giving the latter a brownish colour. The total iron content is usually somewhat higher than in potash salts (BRAITSCH, 1960) particularly in the Stassfurt anticline (RICHTER, 1961)[50]. It is still not clear whether relative enrichment due to the leaching out of sylvite is a sufficient explanation.

---

[50] See also the following publications: RICHTER, A.: Die Rotfärbung in den Salzen der deutschen Zechsteinlagerstätten. 1. Teil: Chemie der Erde **22**, 508–546 (1962); 2. Teil: Chemie der Erde **23**, 179–203 (1964).

This difference in the distribution of iron minerals between potash salts, rock salt and clay residues requires detailed investigation with particular respect to redox potential. It appears that the precipitation and alteration of the relatively insoluble Fe oxides and sulphides covers a long time period in the formation of salt deposits. In the Cambrian salt stocks of the Persian Gulf haematite and pyrite crystals up to 2 cm in length have been observed (in addition to equally large crystals of authigenic quartz, dolomite, and apatite). Here a long period of growth with the circulation of small amounts of pore solution (or higher temperatures?) may be assumed.

In addition to the named iron minerals, the iron content of the clay residues should not be neglected since in most cases it contains an important part of the total iron (p. 227).

b) Distribution, Frequency, and Origin of the Iron in Salt Deposits

In anhydrite rocks the iron is predominantly bound to the clay minerals, as can be seen from the comparison of the Fe, Al_2O_3 and SiO_2 contents of a series of samples from the Zechstein anhydrites A 1–A 3, analysed by JUNG and KNITZSCHKE (1960, 1961). This dependence is particularly well illustrated in the Werra anhydrite where the SiO_2, Al_2O_3 and Fe contents vary in the same direction as the carbonate content and enable the progressive evaporation phase A 1 α to ϑ, to be distinguished from the recessive phase A 1 ϑ_2 to A 1 $\lambda - \nu$. The absolute Fe values are (number of samples in brackets):

in the Werraanhydrit A 1 0.03–0.44 % Fe (34),
in the Basalanhydrit A 2 0.16–0.34 % Fe (9),
in the Hauptanhydrit A 3 0.07–0.51 % Fe (17).

In all cases these represent average samples from stratigraphically well defined sections (see also Table 29). Considering only the samples with $\leqq 0.8\,\%$ SiO_2, then the following are the absolute iron contents (number of samples in brackets):

in the Werraanhydrit A 1 0.03–0.11 % Fe with an average of 0.07 % Fe (27),
in the Basalanhydrit 0.17 % Fe (1),
in the Hauptanhydrit 0.07–0.29 % Fe with an average of 0.17 % Fe (13).

From this it appears that the samples from the Werra anhydrite contain less iron than the remaining anhydrites. The significance of this difference (in the event that it is real) does not yet have an adequate explanation. The proportion of clay minerals is not known. Yet the iron content in comparison with the Al_2O_3 and SiO_2 is 2–3 times higher than in the salt clay (Salzton), from which an appreciable pyrite content can be deduced. However, special investigation of this is still lacking.

In the chloride salt deposits the iron is mainly distributed in the clay minerals (chlorite in the widest sense), in pyrite and haematite with subordinate amounts in magnetite and dolomite (ankerite p. 234).

The distribution of iron in the different minerals is quantitatively known in very few cases. The sulphur in the sulphide was determined by RICKE (1960) in four clay residues from the Stassfurt rock salt and the Stassfurt seam of the Königshall-Hindenburg mine; from these figures can be obtained the maximum pyrite content, i.e. the amount of iron bound to sulphur (Table 34).

Table 34 *The pyrite content in clay residues from Königshall-Hindenburg (calculated from the S determinations of* RICKE, *1960)*

Anal. No. (RICKE, 1960)	Sample	max. pyrite %	Fe in pyrite %
186	Clay residue from anhydrite-bearing halite-carnallite	0.004	0.002
185	Clay residue from anhydritic sylvite-halite	1.6	0.7
183[a]	Clay residue from the Stassfurt rock salt some 2 m below the Stassfurt Seam	2.4	1.0
184[a]	Clay residue from the Stassfurt rock salt some 40 m below the Stassfurt Seam	3.4	1.6

[a] The numbering of samples 183 and 184 in RICKE (1960, p. 61) is misleading as dark gray clay desalted residues from the Stassfurt rock salt were being considered here.

The total iron content in the Stassfurt rock salt lies between 0.1–0.2% Fe which, normalized to the clay residues, is slightly over 3% Fe (Fig. 42). Approximately half the iron is contained in pyrite and half in the clay minerals, the other iron-bearing minerals in the Stassfurt rock salt being without significance. Whether there is an upward decrease in the amount of pyrite in the salt sequence, as there appears to be, still requires closer investigation. On the basis of the well-known colour change from gray to a slightly reddish rock salt in the region of the transition to the Stassfurt seam, a decrease in pyrite might be anticipated.

In the potash salts of the Stassfurt seam the total iron content is somewhat better known. The total iron content does not contribute much to the understanding of the genesis of the deposit since it varies with the insoluble residues (Fig. 42). Yet it is possible to estimate, with

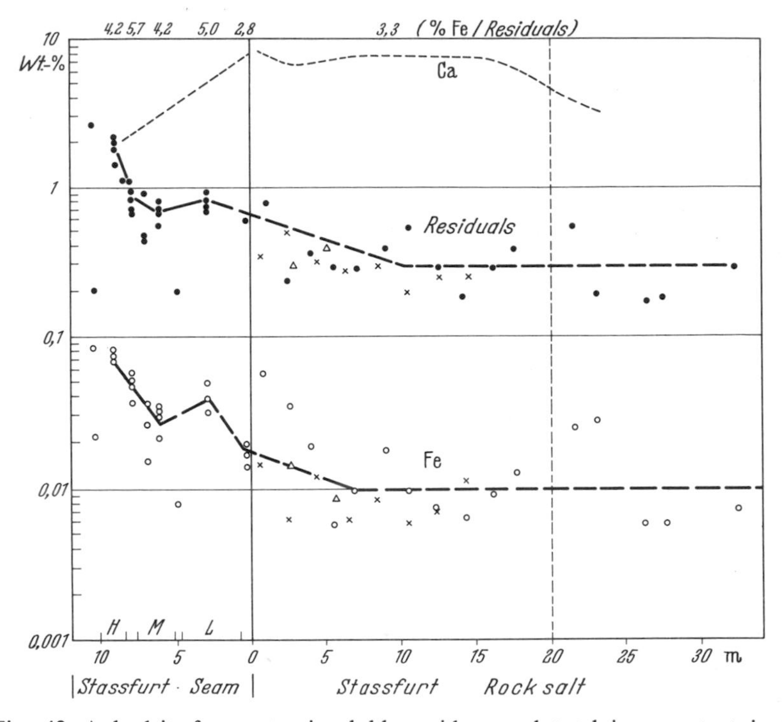

Fig. 42. Anhydrite-free water insoluble residues and total iron content in the Stassfurt seam and the Stassfurt rock salt from the Königshall-Hindenburg mine (Braitsch, 1960 with additions)

the help of the total iron content of the haematite-free clay residues of the Stassfurt rock salt (Fig. 42) and the gray salt clay (Salzton) (3.5–4% Fe, Niemann, 1960), the amount of iron brought in with the suspended clay fraction. This gives at least an indication of the maximum amount of iron in solution. If this correction is applied to the widely scattered values for the Stassfurt seam in Königshall-Hindenburg, the values for Fe which could have been precipitated from solution range from 0 to a maximum of 0.03%. It is certainly true that more haematite is found in the Stassfurt seam than corresponds to this figure, namely:

in anhydrite-bearing halite-carnallite % haematite-Fe	in anhydritic sylvite-halite % haematite-Fe
upper part of the bed 0.05	upper part of the bed 0.02 (max. 0.06)
central part of the bed 0.015	central part 0.01–0.02

These values were determined from carnallite and sylvite concentrates and recalculated (by a method corresponding to a bromine partition analysis) in terms of carnallite and sylvite to 100%. It is still uncertain whether there is a real trend within the seam. In all samples the amount of iron occurring in the clay residues or as pyrite is similar or greater.

In the Stassfurt anticline area an increase in the iron content within the Stassfurt seam is reported. The investigations of PARCHOW (1910) on the same section as shown in Fig. 38 show an increase in the iron content of the carnallite from about 0.002 to about 0.035% Fe (after applying an approximate correction for the iron content of the clay residues which can be estimated using PARCHOW'S MgO determinations). In the rock salt intermediate layers in the carnallite seam the Fe content also increases upwards (from 0.00015 to 0.0003% Fe in the lowermost part to 0.0003–0.0007 in the upper part of the seam, RICHTER (1961)[51]. BOEKE (1911) gave as the average for the red Stassfurt carnallite 0.027% Fe with 0.09% Fe as the maximum. The values given by MARR (1957) for the Unstrut and Halberstadt districts are mostly a half to a full tenth power higher and should be re-examined with particular reference to their dependence upon the increasing content of insoluble residues.

The overall iron content in the German Zechstein salts is small. If the Stassfurt seam were to be completely dissolved only a few mm haematite would remain distributed over a somewhat thicken clayey layer. About other deposits insufficient is currently known. A somewhat greater iron content may be anticipated in the Cambrian salts of the South Iranian salt stocks; in the insoluble residues of the rock salt large amounts of coarse platy haematite occur together with pyrite. In secondary deposits at the margins of the salt stocks, enrichments of several 10^5 tons of the finest submicroscopic platy haematite are found, which must have originated in part at least from the solution and reprecipitation of the iron in the salt stocks (cf. also H. W. WALTHER, 1960).

The agreement between the variation of the iron content and the quantity of the insoluble residues, and the higher iron content in near-shore areas of salt precipitation (e.g. in the Fulda potash district in contrast to the Werra potash district, ROTH, 1953), show convincingly that the iron was introduced into the basin along with the insoluble residues. The iron, however, is certainly not fixed in the form in which it was originally introduced. Much more likely is that a certain breakdown of the detritus occurred in the concentrated salt solutions with the result that the brine has a higher concentration of dissolved Fe than normal seawater. In the presence of a small amount of H_2S, which can be assumed with reasonable assurance in the Werra and Stassfurt rock

[51] See also footnote p. 225.

salt, the iron must have immediately precipitated as the sulphide. From petrographic observations it can be concluded that the iron continues to be leached out of the detritus after deposition, since the sylvite in clay inclusions often possesses a high haematite content (see SCHALLER and HENDERSON, 1932), and the salt clay (Salzton) and argillaceous anhydrite-bordering salt beds have a deeper red colour (RICHTER, 1961)[52]. This is also true of many fissure fillings and other secondary salts. The iron of many salt solutions (p. 225) probably originated only through reaction with adjacent rocks.

In comparison with the terrigenous iron, which was often mobilized only during diagenesis and metamorphic processes before being finally fixed, the iron precipitated directly out of solution probably plays only a subordinate role. Still, the apparently real upward increase of iron in the Stassfurt carnallite rocks may probably be attributed to an increasing amount of iron in the solution. It is unknown, however, whether this iron was dissolved from terrigenous detritus before or during carnallite precipitation, or whether it was introduced by continental weathering solutions. The iron of salt deposits is probably derived almost completely from terrestrial sources, most of it mobilized from the detritus during formation and alteration of deposits.

3. Salt Clays and Insoluble Residues

In the case of iron the importance of continental impurities in the composition of salt deposits was indicated. Although these impurities are not readily soluble when compared with salt minerals and on a laboratory time scale and for convenience are referred to as insoluble, this is not true on a geological time scale. Commonly an intensive material exchange occurs between the salt rocks and the impurities through the medium of the solution. This is clear from the formation of authigenic minerals such as quartz, potash feldspar, talc, various clay minerals of the chlorite group and many others (p. 20).

Definite indications of authigenic origin are:

in quartz and others, the idiomorphic form and the occurrence of irregularly distributed anhydrite inclusions;

for talc, its occurrence on halite cleavage and possibly also translation planes, in individual cases idiomorphic form and growth zones;

in magnesite in addition to idiomorphic form, the common occurrence of haematite inclusions, etc.

For the majority of clay minerals only the probability of authigenic origin can be demonstrated with the help of distribution and geochemical arguments.

---

[52] See also footnote p. 225.

a) Distribution of the Clay Minerals

The distribution of the clay minerals has been examined in the Zechstein anhydrites and the gypsums resulting from them (FÜCHTBAUER and GOLDSCHMIDT, 1959; DREIZLER, 1962), in both potash seams in the Werra Series (BRAITSCH in WEBER, 1961), in the Stassfurt salts of Königshall-Hindenburg and variously from the South Harz district (BRAITSCH, 1960) and in the salt clays (FÜCHTBAUER and GOLDSCHMIDT, 1959; NIEMANN, 1960). There has unfortunately been no systematic investigation of the Stassfurt salts in the central part of the basin so that no generally valid statements can be made. The following is known from investigations made so far:

Talc is the predominant, and in the central part of the basin often the only clay mineral in anhydrite rocks and the gypsums formed from them. In the chloride salt rocks it occurs in the anhydrite-rich horizons but is irregularly distributed or represented by chlorite minerals. It is rare in the potash seams, but is known to occur in many impoverished salt samples, as in the Thuringia seam and the Stassfurt seam in the South Harz. It is perhaps more abundant in the central part of the basin, as in three correlated samples from Hansa (No. 23 in Fig. 1), in the solution residues from the Stassfurt rock salt and from kieseritic sylvite-halite from the Stassfurt seam, mainly talc with some chlorite occurs with predominant ascharite. The talc is in the form of idiomorphic flakes up to $> 1/2$ mm diameter. The regional distribution of talc thus requires special investigation. The talc from saline sediments is less well crystallized than that in slates and shows small differences in its refraction and its behaviour upon heating. In some cases it is replaced by "talc serpentine" (p. 25) or accompanied by poorly crystallized serpentine.

Chlorite is the predominant clay mineral of many salt clays (but only in the non-carbonate parts) and many chloridic salt rocks, at least in the vicinity of the ancient strand line. It is also an important companion mineral to talc, also in anhydrite rocks. It is found in two varieties.

1. Normal chlorite in the penninite-clinochlorite group. It predominates in the Stassfurt rock salt of Königshall-Hindenburg (where it is often extensively replaced by corrensite, see below), and in the anhydritic sylvite-halite and anhydritic halite (impoverished) varieties of the Stassfurt seam. Particularly striking is its dependence upon the kind of accompanying rock in the Thuringian seam where penninite-like chlorite occurs in the sylvite-kieserite-halites, when the main seam is predominantly kieseritic Hartsalz.

2. Chlorite of the amesite-berthierine group. It predominates in the anhydrite-bearing halite-carnallites of Königshall-Hindenburg and in

the sylvite-kieserite-halite of the Thuringia seam where the upper part of the bed in the seam is formed of fragmental carnallite. In the South Harz mines the predominance of the amesite-type chlorite rather than penninite in the carnallite rocks is less pronounced but is still recognizable. In the older literature this variety of chlorite was often confused with kaolinite.

Unfortunately the separation of the two chlorite varieties is never quite complete. Both occur together in the gray salt clay, (Grauer Salzton) too. In the lower part the amesite-type chlorite seems to predominate; however, a reinvestigation of this question is necessary. As both varieties vary somewhat in their chemical composition, both possess somewhat variable optical properties.

Corrensite 2, in the form described in detail by BRADLEY and WEAVER (1956) with a regular alternate chlorite-vermiculite layering, is the predominant clay mineral in various series of samples from the Stassfurt rock salt in Königshall-Hindenburg and from the South Harz, from many impoverished samples of the Stassfurt seam, and in particular from the Werra potash seam. It is always associated with varying amounts of a penninite-like chlorite. In the salt clays and anhydrites corrensite is rare. Instead of corrensite, clay minerals with an irregular layering often occur. Recently corrensite was reported from the evaporites of the Salado formation in New Mexico (FOURNIER, 1961).

Muscovite (and *illite*) are normally ubiquitous. Exceptionally, authigenic muscovite has been reported (HOFFMANN, 1961) but, as no tests were made for talc[53], the diagnosis requires confirmation. The muscovite (illite) strongly predominates in the carbonate part of the salt clays, but in the noncarbonate salt clays its abundance relative to the other clay minerals diminishes.

In muscovite primarily the 2-M structural type is being considered.

The relative abundances of talc and muscovite are inversely proportional, when a lot of talc occurs, muscovite diminishes or is absent. To a somewhat lesser extent the same applies to muscovite and the chlorite group (including corrensite).

Kaolinite is absent in true saline sediments. Only in the carbonate rocks (Zechstein limestone on the western margin of the Harz) was it reported in isolated cases (FÜCHTBAUER and GOLDSCHMIDT, 1958). By analogy, it could occur in the strongly carbonate horizons of the salt clay, but it has not yet been reported.

As an example, the distribution of the clay minerals in the Königshall-Hindenburg mine is given in Fig. 43 with each given as a percentage of

[53] The author has generously made available the information that talc aggregates were later identified (using an axial angle of 0° C). The idiomorphic flakes described could, however, be muscovite. No X-ray data are available.

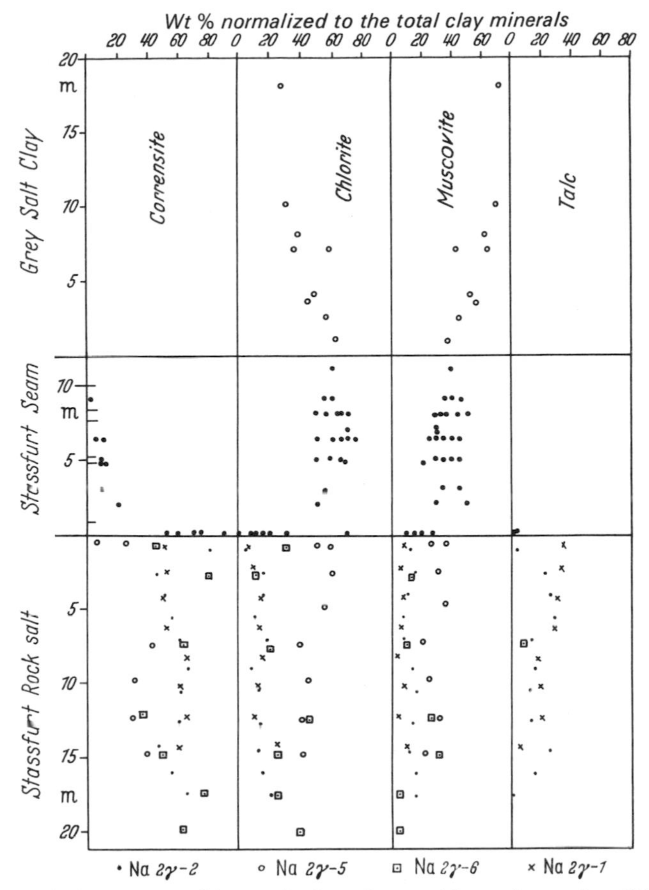

Fig. 43. Relative composition of the clay residues from the Königshall-Hindenburg mine. Stassfurt rock salt and Stassfurt seam, author's determinations, all on average samples. In the Stassfurt rock salt, series of samples from the same section are indicated by the same symbol. Gray Salzton from NIEMANN (1960)

the total clay mineral content. The clay minerals make up about two-thirds of the insoluble residues (Fig. 42) with considerable fluctuations. The rest consists of very variable quantities of quartz, potash felspar, magnesite, celestite, borates, pyrite, etc. which will not be discussed here in detail. The relationships found show a characteristic dependence of the clay minerals on the accompanying minerals and their independence of age of the rock. This dependence is seen over a very restricted region within similar beds yet occurs with different forms of the bed. It is therefore not possible to make fluctuations in the influx of primary detrital material responsible for the composition of the clay residues.

b) Chemical Characteristics of the Salt Clays

The predominance of Mg silicates is a characteristic of salt clays. They also contain authigenic potash felspar, and muscovite. The chemical composition of the salt clays can be displayed by means of a ternary diagram with K_2O, Al_2O_3 and MgO (in weight percent) (Fig. 44). SiO_2 is not used because the very variable detrital and authigenic quartz content would result in a wide scatter of the points represented. The exceptional position of the salt clays would be just as clear from the ternary diagram SiO_2, Al_2O_3, and MgO.

In an attempt to draw unambiguous conclusions, only the dolomite-poor or dolomite-free samples were considered. The dolomite [in GÖRGEY'S (1912) analyses in part as ankerite] was excluded from the calculations. In addition analyses with a predominance of anhydrite or salts were excluded, as small analytical errors in the remainder strongly influence the representative point.

The field of salt clays and salt residues falls clearly within the triangle clinochlorite (representing the salt clay chlorites)-muscovite-potash felspar, and is concentrated in the clinochlorite half of the diagram. The Keuper clays which were subject to strong saline influences also occur here (ECHLE, 1961). Pure talc residues were not included. For comparative purposes the following were used: 1. the clays of the Russian Platform (VINOGRADOV and RONOV, 1956), 2. the mean composition of four clay types according to RONOV and KHLEBNIKOVA (1957). In these the Al_2O_3:MgO ratio is much higher than in salt clays. Nevertheless, the points representing some of the Russian Platform clays fall in the field of the salt clays. These are, however, without exception samples from formations subject to a strong saline influence (upper Devonian and Permian). The dolomite fraction was not excluded so that the MgO content is a maximum compared to the clay fraction. Unfortunately RONOV et al. (1957) combined the clay samples from salt marshes with marine clays. However, in comparison with marine clays only, there is a clear displacement of the average value towards the salt clay field. The geochemical peculiarity of the salt clays lies in their significantly higher relative MgO content. This occurs only in the silicates (except in carbonates) and not as the hydroxide[54]. The occurrence of rare impurities such as boracite, wagnerite, sellaite, etc. can be excluded from considerations of the MgO balance.

The absolute value of the Al_2O_3 content in normal marine and continental clays of the temperate zone does not show important changes.

[54] PARCHOW (1910) sought to prove this old assumption of PRECHT'S (cited in BOEKE, 1911) by determining the MgO and haematite contents in the Stassfurt anticline, but as he included the kieserite-free solution residues, he included the MgO of the clay minerals.

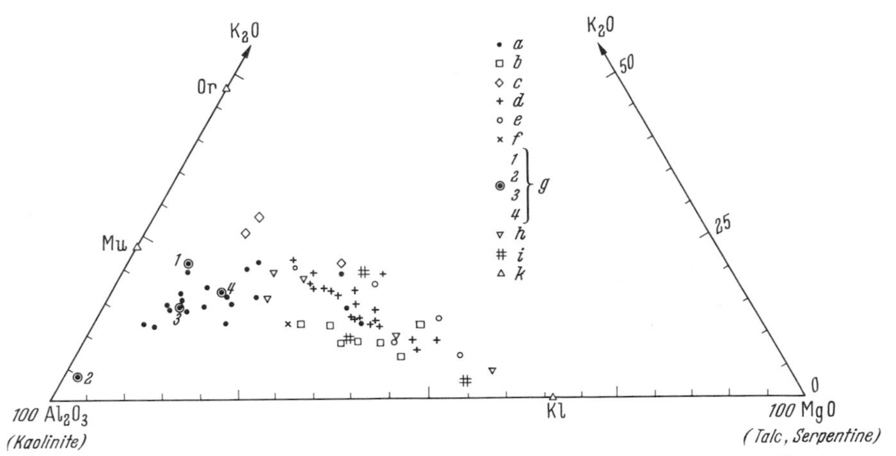

Fig. 44. Normal clays and salt clays (weight % $K_2O + Al_2O_3 + MgO$, normalized to 100%). *a* clay from the Russian platform (VINOGRADOV and RONOV, 1956); *b* salt clay from Stassfurt (BLITZ and MARCUS, 1910, analyses 22, 23, and 1912 analyses 2–5); *c* slat clay from Alsace (GÖRGEY, 1912, analyses 2, 3, 9, p. 444); *d* salt clay T 3 from Königshall-Hindenburg (NIEMANN, 1960); *e* clay residues from the Stassfurt seam and Stassfurt rock salt, Königshall-Hindenburg, analysis R. BLANK; *f* Berchtesgaden salt clay (GÜMBEL cited by SCHAUBERGER and RUESS, 1951); *g* average composition of clays (RONOV and KHLEBNIKOVA, 1957), *1* continental clays from the cold and temperate zones, *2* continental clays from the tropical zone, *3* marine clays, *4* marine clays and clays from salt marshes and lakes in arid regions (averaged); *h* Keuper clay (ECHLE, 1961, Table 11); *i* clay residues from salts of the English Zechstein (ARMSTRONG et al., 1951); *k* theoretical composition of pure minerals: Or = potash feldspar, Mu = muscovite, Kl = clinochlorite $Mg_5Al(OH)_8AlSi_3O_{10}$

In contrast to the continental clays of the tropical zone, there is a marked Al_2O_3 deficiency in the salt clays. As not infrequently talc is the predominant or the only Mg silicate (e.g. in anhydrite residues; locally in the Leine rock salt, etc.) the Al, at least in part, must have been dissolved out of continental detritus. Al-free continental detritus does not, in practice, need to be considered. It is thus not legitimate to assume that the Al remains constant during the alteration of continental detritus into salt clay.

Only the clay residues which have not been greatly affected were chosen for consideration. The index mineral for salt deposits can probably be regarded as talc, so that in the final stage the clay composition is displaced still further in the MgO direction. The formation of koenenite would follow as will be briefly discussed on p. 241. Like talc, the amesite-type chlorite, on account of its preferred association with carnallite and rocks derived from it, may be an index mineral for these rocks although perhaps only in certain specific palaeogeographic situations.

c) Origin and Material Balance of Salt Clays

The distribution as well as the chemical criteria of the salt clays and clay residues of salt rocks is evidence of the formation of new minerals, in particular of magnesium silicates. In a few cases it can be demonstrated that the newly formed minerals first occur during metamorphism of salt rocks. In this category belongs, for example, the formation of normal chlorite at the expense of the amesite-type during the alteration of carnallite. HOFFMANN (1961) was able to show that the amount of the newly formed minerals in the sylvite rock which results from the alteration of carnallite is appreciably greater than in the original carnallite rock. This demonstrates the importance of the presence of circulating solutions for recrystallization. In comparison, the time of the alteration is less important. FÜCHTBAUER and GOLDSCHMIDT (1959) suggested that (syn- or epigenetic) circulating solutions were responsible for the formation of chlorite on the basis of the differences in the mineral assemblages in carbonate and non-carbonate clay rocks. They assumed that the rapid compaction and closure of pore spaces in carbonate rocks very early inhibit recrystallization, so that the residues from the calcareous salt clays very probably represent a conservation of the original clay minerals of the Zechstein Sea. From this it follows that, of the clay minerals, muscovite in particular (and the kaolinite reported very exceptionally in the Zechstein limestone) is to be regarded as relict. Some of the chlorites may also be relict, but the greater part is probably authigenic. It is certain that alteration proceeds very slowly for in the clay muds of recent salt lakes the composition of the suspended mud shows no noticeable changes (DROSTE, 1961). Yet the examples investigated by DROSTE were soda or Na sulphate-rich continental salt lakes whose chemistry is not comparable with that of oceanic salt seas. In them the disequilibrium between the salt solution and the continental clay detritus is much smaller so that the tendency towards recrystallization is also lower. Conversely, in young deep sea muds the formation of chlorite at the expense of illite appears possible, as was reported in the San Diego borehole (180 m deep from the ocean floor at a depth of 4000 m) (lecture, E. GOLDBERG, 1961). Further, the distribution of talc in dark bands parallel to the bedding in anhydrite rocks cannot be explained as secondary (FÜCHTBAUER and GOLDSCHMIDT, 1959; DREIZLER, 1962).

Starting from muscovite and Mg in the salt solution as the initial materials for chlorite formation, free K and SiO_2 result if it is supposed that all the Al is fixed in the newly formed chlorite. As in the salt clays (T 1–T 4) the authigenic formation of potash felspar declines in contrast to the clay residues from sylvite rocks, etc., the major part of the K must remain in solution. However, the assumption of the complete fixation of Al, as was previously realized, is probably not warranted. The muscovite itself, of course, is not particularly representative of continental detritus. Furthermore a partial recrystallization of the illite of continental origin is not improbable.

There appears to be a dependence of the clay minerals on palaeogeographic position, at least, in the present state of knowledge. Towards the center of the basin the quantity of the clay minerals diminishes and talc appears to become more abundant, while the micas disappear and the chlorite minerals (including corrensite) become less common. The silicic acid required for the formation of talc was probably dissolved from the continental detritus and transported in this form independently of the detritus. In the interior of the basin precipitation from solution is possible. The solubility of silicic acid in salt solutions can be established petrographically, for example, by the evidence of solution on originally idiomorphic quartz. It is not uncommon to find authigenic quartz showing signs of solution on the prism faces while the rhombohedral surfaces are unaffected. Hour-glass forms may develop in this way (Hoffmann, 1961) and may eventually separate into two parts. The effect of the solubility anisotropy of the different crystal forms of quartz should be regarded as indicative of very small fluctuations in the properties of the circulating solutions (temperature, common ions etc.). It also follows that the formation and alteration of the insoluble residues are long-term effects which do not always proceed in the same direction. Present knowledge does not extend to linking these processes with the formational sequence of salt rocks.

In comparison with normal clay sediments, the quantity of newly formed minerals in the salt clays is much greater. In the sense of the earlier definition (p. 92), it is not easy to distinguish whether the minerals were formed in the diagenetic stage (alteration and reaction in unchanged conditions, that is, removal of the disequilibria) or in the metamorphic stage (low-temperature metamorphism of Eckhardt, 1958). Yet the distribution of many of the newly formed minerals shows that a rise in temperature is not an important factor, but rather a suitable medium in which the establishment of stable equilibria is established or facilitated. As yet no account can be given of the kinetics of these reactions, ionic diffusion velocities and so on.

An attempt can be made to estimate the necessary amount of Mg in the gray salt clay. The approximation is made possible by the use of Niemann's (1960) data for Königshall-Hindenburg. As the salt clay there has an above average thickness of some 30 m, the estimate provides an upper value for the whole area which may be three to five times the average figure.

On average the gray salt clay contains about 12% chlorite by weight, which corresponds to about 2% Mg. A Q solution at 25° C contains about 6.7% Mg, therefore 0.3 kg Q solution/1 kg gray salt clay is required, or about 38 % Q solution by volume in the gray salt clay, to provide the necessary amount of Mg. Naturally only a part of the Mg present will be fixed in the clay minerals, but the volume of Q solution should suffice so long as the terrestrial detritus is not Mg-free. Then, from Tables 10 to 12 the approximate water depth at which carnallite precipitation occurs can be estimated; this is most simply done from the theoretical thickness of the bischofite layer. As an adequate, practical rule of thumb:

$$\text{Solution depth} = 1.6 \cdot \text{thickness of the bischofite layer}.$$

In the case of complete deficiency in $MgSO_4$ (Table 12, IV), which to a good approximation is the case in Königshall-Hindenburg, for a thickness of about 15 m of carnallite (after allowing for the rock salt precipitated from influxing brine) we have $1.6 \cdot 35.3\,\text{m} \cdot \frac{15\,\text{m}}{8.6\,\text{m}} \approx 100\,\text{m}$ water depth. This figure is a minimum value, since the calculation was based on static evaporation and complete carnallite precipitation, so that at the end of the process the solution composition reached Z. The effect of influxing brine on this model is to increase the amount of solution without raising the absolute Mg content of the solution very much. With normal $MgSO_4$-bearing salt sequences, the depths obtained for the solution vary considerably according to whether the primary precipitation occurred with or without reaction with an initial kainite layer. In the case of complete reaction of initial kainite, after the precipitation of the carnallite the water depth would be about twice the thickness of the total carnallite layer, but even in this extreme case there would be excess Mg in the residual solution available for diagenetic new formations.

The gray salt clay was deposited over the Stassfurt seam as the result of a massive invasion of new brines, but in all probability the Q solution was not driven from the basin because of the stable layering of dilute over concentrated solutions. There was, in any event, sufficient Q solution for authigenic clay mineral formation. In the case of clay intercalations in chloridic salt rocks, excess Mg is always present in the salt solutions.

d) Filtration Effects in Salt Clay

The great importance of pore spaces in sediments was briefly described in a comprehensive account by ENGELHARDT (1960). He details the special properties of clay vis à vis brine currents, e.g. the marked reduction of anion mobility (p. 42 et seq.), and the alteration in the chemical composition of the brine as a result of the authigenic formation of chlorite and dolomite (pp. 49, 164) and cation exchange with the clay minerals (p. 165). These changes in the composition of pore solutions were clearly distinguished from the commonly observed increase in the total salt content with depth (p. 49, 158).

The clays seem to serve a double function in the chemical alteration of salt solutions. Through the authigenic formation of Mg silicates they produce a relative impoverishment in Mg and relative enrichment in Na and Ca in the pore solution. The concentration increase in the pore solution is still poorly understood. A new angle on this problem has been pointed out, which must be explained by further research as soon as possible. It concerns the so-called "filtration effect" (KORSHINSKY, 1947), that is, a more rapid migration of the solvents than of the ions in solution. It thus produces a more concentrated solution in the filter. To a certain degree the clays act as semipermeable membranes. Experimentally the effect was first discovered in ultrafiltration through colloidal membranes (ERSCHLER, 1934; HACKER, 1941) and later was investigated in a porous quartz flour ($\sim 2\,\mu$ particle size) using various dilute electrolytes (OVCHINNIKOV and MAKSENKOV, 1949). The filtration effect varies very much according to the electrolyte used, and is dependent upon the filtration conditions, but not on the thickness of the filter. It falls with increasing concentration and rising temperature, no doubt as a consequence of the higher diffusion rate which operates in the opposite sense. Presumably the filtration effect also depends upon reduced anion mobility (v. ENGELHARDT, pers. comm.) and is produced by the negative surface charge of the clay minerals. Experimentally an increase in the electrolytic concentration of pore solutions was noted when a montmorillonite clay was compressed at above 800 atmospheres (v. ENGELHARDT and GAIDA, 1963), and salt solutions were forced through up to 0.5 cm thick layers of Wyoming bentonite and a "clay" of quartz, montmorillonite, illite, and kaolinite. The effect was much smaller in the latter case (MCKELVEY and MILNE, 1962). Up to the present, experiments have been carried out with relatively dilute solutions. They do, however, appear to provide a plausible if entirely qualitative explanation for the increasing concentration of salt solutions with increasing porosity as a result of the squeezing out of fluid. Because of the different operation of the filter with different electrolytes, a change in solution composition may be anticipated. In recent marine clays a higher salt content of the pore solution in comparison with the overlying seawater has been found and also attributed to the filtration effect of clay (SIEVER et al., 1961).

It is not possible to conclude from present knowledge whether the concentration due to this effect can lead to the precipitation of salts. Yet it must be remembered that the salt clays contain appreciable amounts of salt which can hardly have been derived directly from the evaporation of the salt solution. The large amount of clay present, loosely called "gray salt clay", indicates an important influx of dilute solution which was not saturated with respect to halite etc. However, the greater part of the "clay" was already flocculated and had sunk through the more concentrated basal solution when precipitation of evaporites recommenced, first carbonate (= upper third of the gray salt clay) and later anhydrite (= Hauptanhydrit A 3), associated with influxes of new solution (p. 257). This sequence is a reliable criterion for the influx of new brine and one which, like NIEMANN's (1960) sedimentary petrographic data, argues against an aeolian origin of the gray salt clay. Below a dilute surface layer water cannot evaporate from the more concentrated basal solution in the pores of the salt clay. It is conceivable, however, that with the later diagenetic reduction of the pore space a filtration effect could bring about a concentration of the already saturated pore solution up to the point of salt precipitation. Whether this hypothesis can also explain certain tachhydrite occurrences, such as the enrichment in the clay-rich layers of the Stassfurt seam of Neustassfurt (KNAK, 1960), can only be established by further investigations which include that of the "clays."

Signs which identify solutions from the gray salt clay have already been mentioned, namely the red coloration of the adjacent salt rocks (p. 230). The fibrous salts (halite, carnallite) and the koenenite in joints are also precipitates from the pore solutions from the salt clay.

Even today the salt clays are not dry, but still contain recognizable amounts of highly concentrated liquid inclusions. In particular, the inclusions of $CaCl_2$ solutions (Upper Rhine potash layer, GÖRGEY, 1912) belong to this category. There could also be liquid inclusions in the gray salt clay of Königshall-Hindenburg (NIEMANN, 1960) since NIEMANN did not directly observe tachhydrite or carnallite (excluding fibrous carnallite in joints).

Both these cases involve $MgSO_4$-free salt parageneses (p. 162) and the possibility arises that these may be conserved, and further concentrated, primary solutions. The degree of change in the relative ionic proportions in the solution as a consequence of the new formation of chlorite cannot be estimated, but clearly a closed system is unlikely. The relative Mg impoverishment and Ca enrichment is characteristic of all pore solutions with more than a certain minimum salt content ($\gtrsim 0.1\%$ Cl, LOEWENGART, 1962).

The filtration effect has already been suggested by RICOUR (1960) as an explanation for the diagenetic precipitation of salts in some French Triassic salt occurrences. But so long as there is no adequate experimental basis, all attempted explanations are tentative. Perhaps this effect may also provide the key to the understanding of the genesis of the problematic clastic salt clays (Tonbrockensalze) of the Middle Muschelkalk of Southern Germany (SCHACHL, 1954).

e) On the Formation of Koenenite

The crystal chemistry of the oxychloride koenenite is still not well known, but it does provide a clear indication of intensive material exchange between salt solutions and clays. Its formation requires $MgCl_2$ and Al in a form (soluble) susceptible to reaction (with NaCl, H_2O, etc.).

According to KÜHN (D'ANS and KÜHN, 1960 and earlier), a distinction must be made between primary and secondary koenenite. The secondary form occurs in joints and nests in anhydrite and particularly good occurrences are also found in salt clay. Here the Al is obviously mobilized from the salt clay while the Mg could be derived from the concentrated pore solution in the salt clay. The primary koenenite, according to KÜHN (1961), occurs in a finely divided form but is moderately abundant only from Zechstein 3, whereas in Zechstein 2 it is found in a few of the mining areas in the central part of the basin. In principle it is possible for this koenenite to originate by exactly the same process as the secondary koenenite, as with progressive evaporation of the solution there may be a sufficient enrichment in the Al dissolved out of the clay. It is clear that the Al content of seawater itself is quite insufficient for the formation of koenenite. On the other hand, it is not necessary to postulate continental lateritic products (D'ANS and KÜHN, 1960) as a source of Al. As the gray salt clay is fairly thick and accumulated slowly, only a small uptake of Al by the basal solution would explain the observed koenenite content (average 0.1% koenenite in Zechstein 3 and 4, KÜHN, 1960). The discussion of the chemical peculiarities of salt clay has already demonstrated their relative impoverishment in aluminum. This Al deficiency should be of the right order of magnitude to explain the observed amount of koenenite.

According to this interpretation, the koenenite represents a different stage in the exchange reactions between detritus and salt solution than the talc, as the Al from the detritus was transported preferentially in solution. Up till now the special factors which induce Al and Si to behave differently in saturated salt solutions are not known. Koenenite and talc do not appear to be mutually exclusive (FÜCHTBAUER and GOLDSCHMIDT, 1958, p. 325 onwards) but, in the few known cases where they occur together, it is not known whether they were formed at the same time. This seems quite possible however, since the regions of formation of koenenite and talc overlap or are adjacent.

f) Carbonates, Fluorides, Phosphates

In anhydrite rocks there are always varying amounts of carbonates, usually in the form of dolomite, although in the Basalanhydrit calcite occurs. Generally the enrichment occurs in layers parallel to the bedding (dolomitic fine striping of JUNG in JUNG and KNITZSCHKE, 1960 and others). There are some very reliable average figures from bores in the S. E. Harz foreland (see Table 35).

Table 35. *Dolomite and calcite content in the Zechstein anhydrites in the S.E. Harz foreland (rounded-off from* JUNG *and* KNITZSCHKE, *1960, 1961)*

	m average thickness	% dolomite		% calcite	
		average	spread	average	spread
Hauptanhydrit (A 3)	44.2	10.4	1.3–39.5	5.9	0 –17.3
Basalanhydrit (A 2)	2.7	1.1	0.9– 1.9	13.7	10.6–22.6
Werraanhydrit (A 1)	57.2	7.0	1.0–13.3	1.8	0 – 4.2

However, nothing at all is known about variations within the Zechstein basin. The cause of the predominance of calcite in the Basalanhydrit has also not been explained. Of interest is the confirmation of a characteristic magnesite content in the Hauptanhydrit, particularly near the base. It contains, in the three bores examined from the A 3 α zone, an average of

$$19.4\%\ MgCO_3 \text{ over a thickness of } 2.6\text{ m}.$$

Calculated for the total thickness of the Hauptanhydrit this amounts to about 1.5% $MgCO_3$. The high magnesite content of the basal part of the Hauptanhydrit is known, qualitatively, in many other outcrops. LANGBEIN (1961) established that the magnesite is not restricted to the base of the Hauptanhydrit, but occurs in two further somewhat calcareous layers although in smaller amounts, which indicates a cyclic structure of the Hauptanhydrit.

The content of magnesite, particularly at the base of the Hauptanhydrit, appears to confirm the supposition made in consideration of the Mg balance of the gray salt clay that there were large volumes of residual solutions from the Stassfurt cycle. That it did not lead to the formation of polyhalite in the newly instituted precipitation of anhydrite (as in the model on p. 100) is the result of the almost complete precipitation of K and $MgSO_4$ in the Stassfurt seam. In solutions with a slightly higher $MgCl_2$ content than at point *Q*, polyhalite is not stable at any temperature.

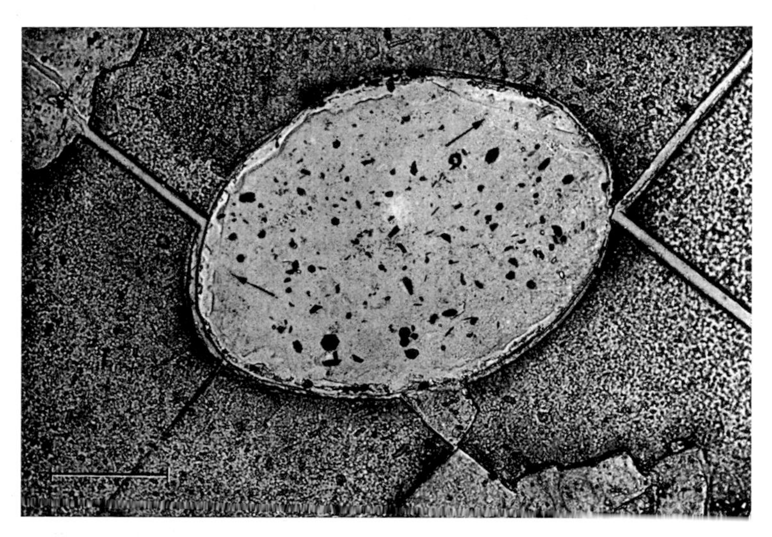

Fig. 45. Magnesite in sylvite: it contains idiomorphic haematite inclusions (Arrows indicate the rhombohedral planes of magnesite). The sylvite contains essentially finer haematite inclusions, most not recognizably single crystals. They show partial enrichment fringing relict carnallite (lower right). Scale 0.2 mm. Königshall-Hindenburg mine

In the chloride salt rocks the amount of dolomite diminishes, calcite is mostly absent, while magnesite is the dominant carbonate. It generally is found as tabular crystals after (0001); the rhombohedral planes are very small but may occasionally determine the form. The elongate habit is only occasionally observed without clear crystal form or with occasional acute rhombohedra, as for example close to basalt dykes in the Werra district. Similar forms have been found in the insoluble residues from the Choctaw rock salt, Louisiana (TAYLOR, 1937).

In individual cases the petrographic data provide clues to the relative ages of the authigenic minerals. For example, the thin tabular haematite inclusions in magnesite (Fig. 45) from a sylvite at the transition to carnallite show that the magnesite was probably formed after the incongruent alteration of carnallite, as such haematite is characteristic of carnallite. Other magnesites may be older than, for example, kieserite. In any event the occurrence of one mineral as inclusions in another is often an uncertain age criterion.

There are practically no quantitative data on the magnesite content of chloride salt rocks. There are approximate figures for Königshall-Hindenburg (averages from semiquantitative X-ray and microscopic examinations of the same samples as in Figs. 39, 40) in Table 36.

Table 36. *Average magnesite content from the Königshall-Hindenburg mine (determinations by the author)*

	% magnesite in solution residues	% magnesite in the total rock
Stassfurt seam (K 2)	5– 3	~0.05
Stassfurt rock salt (upper part Na 2 γ)	10–15	~0.05

The magnesite content in the Werra potash seams is about an order of magnitude less, but there are still insufficient data. Quantitative data are lacking for other outcrops (HOFFMANN, 1961, determined magnesite but only in the fraction $>60\ \mu$). Nevertheless, high carbonate contents are known in individual clay intercalations, e.g. in the Upper Rhine salts (GÖRGEY, 1912).

The geochemistry of fluorine and phosphorus in salt deposits is still poorly known (cf. KORITNIG, 1951). In the Königshall-Hindenburg mine the occurrence of wagnerite, isokite and apatite, and occasionally lueneburgite, has been confirmed, the apatite often occurring as oriented growths on detrital nuclei. Fluorite occurs relatively commonly in the first precipitated carbonate rocks (CORRENS, 1939; p. 238; FÜCHTBAUER, 1958; ABRAMOVICH and NECHAYEV, 1960). In chloridic salt rocks it is certainly very rare, and until now has only been found secondarily with anhydrite and lueneburgite in a small thrust zone (BRAITSCH, 1960). In this case it can be assumed that the B, F, and P were introduced by subsequent infiltrating solutions. According to HERRMANN (1961 b), salt solutions in the South Harz contain 0.0011–0.11% B (7 natural occurrences) and 0.15 to 0.9 g P/ton, while in other regions of the Zechstein 2 amounts of 0.006–3.4 g P/ton of salt solution occur. Sellaite is known at very few outcrops (p. 21).

In total, the fluorides and phosphates are too rare in salt deposits to be used as index minerals. Exceptionally, sellaite may be locally important, as in the Zechstein 3 at Salzdetfurth (SIEMEISTER, 1961), and apatite, for example, in the Cambrian salts of Hormuz island. The phosphate content found at Königshall-Hindenburg is generally less than 0.1% (but occasionally as much as 0.3%) of the residues. For the whole rock this amounts to 1–3 g P_2O_5/ton with in addition 0.3–1 g F/ton. The same order of magnitude for P_2O_5 was found by KORITNIG (1951). However, the observed fluorine contents are many times the amount found in phosphates, and in the Stassfurt seam a sample with up to 300 g F/ton has been found. Some of this might be contained in the clay minerals where it would replace (OH). KORITNIG found the highest fluorine contents in fine clay layers and salt clay. In four additional analyses of Salzdetfurth clay samples there was a considerable spread of values from 0.03 to 0.23% F (SIEMEISTER, 1961). Special investigation of this question is necessary, particularly because no generalizations can be made about the mines in question because of their special genetic and geochemical peculiarities.

On thermodynamic grounds SAHAMA (1945) deduced that sellaite is stable with anhydrite at 500° C but, in contrast, not fluorite with kieserite. With dolomite, both fluorite and selaite are possible, with fluorite perhaps preferred at low temperatures. The few observations in salt deposits are in accord with these statements.

F. Geological Conclusions and Problems

1. Normal and Altered Seawater

The division of fossil salt deposits according to the differences in their $MgSO_4$ content is a natural consequence of the clear distinction in their chemistry and petrography. All these types trace their origin back to the evaporation of seawater, for only in seawater is there the inexhaustible supply of Na, K and Mg etc. sufficient to account for the enrichment in salt deposits. In purely terrestrial salt lakes the precipitation of the chlorides of potassium and magnesium is scarcely ever attained and there is always a predominance of sodium and other carbonates or sulphates. The occurrence of borates and nitrates, on the evidence of the prevailing anions, is a clear indication of non-marine origin. The extent and salt content of terrestrial salt lakes is small (for details, see LOTZE, 1938).

It is thus necessary to explain the alteration of normal seawater. There are two principal hypotheses:

a) The deficit is not so much of $MgSO_4$ but of SO_4^{--}, resulting from the bacterial reduction of SO_4^{--} to H_2S, most of which subsequently escapes into the atmosphere (BORCHERT, 1940, 1959). The cation excess must, however, be balanced and this can only occur through $(OH)^-$ ions so that the system must have a strongly alkaline reaction[55]. This would produce precipitates quite different from those which result from seawater, in particular, brucite ($Mg(OH)_2$). This is not found. In addition, the possible extent of bacterial SO_4^{--} reduction is quite restricted, as will be more clearly explained on p. 263.

b) SO_4^{--} is precipitated as $CaSO_4$ through the influx of water carrying Ca in solution (STURMFELS, 1943, and many predecessors; see LOTZE, 1938; VALYASHKO, 1958; LUKJANOWA et al., 1958 and several other Russian authors). The influx of calcium hydrocarbonate in solution would, for instance, give the following reaction.

$$Ca^{++} + 2\,HCO_3^- + Mg^{++} + SO_4^{--} + 2\,H_2O \rightarrow CaSO_4 \cdot 2\,H_2O + Mg(HCO_3)_2\,.$$

The precipitates are here assumed to be gypsum and a hypothetical magnesium hydrocarbonate. The primary precipitation of magnesium

[55] The bacterial reduction of sulphates produces H_2S and carbonate ions, but not OH ions.

is not at all well known. Observations on recent to subrecent salt lagoons suggest it might be protodolomite, hydromagnesite

$$[Mg(OH)_2 \cdot 4\,MgCO_3 \cdot 4\,H_2O]\,,$$

or even magnesite (with a somewhat enlarged lattice) (ALDERMANN and SKINNER, 1957; GRAF et al., 1961; ALDERMAN and BORCH, 1961), or perhaps a metastable H_2O-bearing carbonate. A possible connection between these observed sediments and metastable aragonite, obtained in crystallization experiments, has not been confirmed. The end product of diagenetic alteration in any event is dolomite or magnesite. In its end result secondary dolomitization also produces Mg impoverishment and Ca-enrichment of the solution, as dolomitization must to a significant extent be associated with Ca/Mg exchange. The balance of charges is not upset. The fixation of equivalent amounts of Mg^{++} and SO_4^{--} thus results in the formation of two different solid phases. As their solubilities are quite different, it follows that the precipitation is not simultaneous but successive and the two phases are more or less independent of one another. They are not, therefore, to be expected together in one exposure, but on a larger scale (several kilometers) a primary transition from the precipitation of carbonate to sulphate is possible.

This hypothesis, whether in the form of secondary dolomitization, or of alteration through influxes from the continent, corresponds best to the conditions found in Nature (p. 162). It is very important in the Russian literature and is referred to as "metamorphization" of the solution (KURNAKOV, cited by VALYASHKO, 1958). As an index of the alteration of a solution the relation ("metamorphization coefficient")

$$K_{MC} = \frac{MgSO_4}{MgCl_2}$$

is applied and has often been used to characterize the water of recent brine pools. As ions are present in the solution, the calculation of the coefficient depends upon the type of compounds assumed. Furthermore, commencing with the precipitation of Mg-bearing sulphates, the coefficient loses its uniqueness even when the calculation of the compounds is referred to a norm, because it decreases even with precipitation from unaltered seawater. The not infrequent application in the Russian literature of this coefficient to the waters of the Kara Bugas Gulf and other terrestrial salt lakes, which in comparison with seawater contain far too much SO_4^{--}, is conducive to error.

As a special case, BENTOR (1961) attributes the composition of the Dead Sea brine, a $CaCl_2$-bearing solution (Fig. 24), in principle to influxes of $CaCl_2$-rich solutions into water (from the Jordan) similar in composition to seawater. In this case, only $CaSO_4$ would precipitate. BENTOR's calculation automatically yields complete agreement for the Mg content, and also for Na. On the other hand, there are relative errors of 50% in the quantitative estimates of potassium and bromine. In addition, the calculation is based upon $CaCl_2$ solutions, from which two-thirds of the Dead Sea salts are derived, which do not occur at the surface but are rather pore solutions from deep borings. It is not known in what manner such solutions could get into the Dead Sea. BENTOR explains the much higher observed bromine content as due to fossil residual solutions from Tertiary phases of salt precipitation, so that a correspondingly large proportion of the $MgCl_2$ would also have been derived from the same source. This, however, is not taken into account in the calculation so it cannot be accepted as evidence. Moreover, the introduction of $CaCl_2$-bearing solutions merely begs the real question, the explanation of the Dead Sea $CaCl_2$ content (Fig. 24). Finally, the calculation takes no account of salt precipitation, ion exchange and similar effects. More importantly, the available pore space in the Dead Sea basin subsurface is insufficient to contain the necessary volume of $CaCl_2$ solution (LOEWENGART, 1962). In contrast, LOEWENGART (1962) attributes the salt content to seawater (predominantly as cyclic "airborne" salt) and explains the change in composition as due to Ca-Mg exchange (in association with the precipitation of NaCl and $CaSO_4$ in the Dead Sea itself). He shows that a relatively small amount of dolomitization of limestone produces a sufficient change in the Ca/Mg ratio and that this reaction, together with a partial absorption of potash, proceeds in the main in the accumulated groundwater around the Dead Sea. The Jordan and salt water springs are thus only secondary transport media, and the salt water springs are in the main more or less dilute residual solutions of an earlier salt precipitation phase, but with a strongly displaced Ca/Mg ratio.

This interpretation is superior to BENTOR's and has the merit that it can be considered as a concrete example of the $MgSO_4$ impoverishment of seawater and even of the extreme case of alteration within the $CaCl_2$ region (Fig. 28). At the same time LOEWENGART's hypothesis serves to explain the relative K deficiency of the Dead Sea in comparison with pure seawater. The spatial separation of dolomite and gypsum formation is also noteworthy.

The alteration of seawater thus need not occur only in the basin where precipitation is taking place. Under special conditions the saline solutions from a lagoon can infiltrate underlying or surrounding carbonates, particularly those from the first evaporation phases, there be altered by dolomitization and later return to the parent water body. In this way a Ca excess can develop in the solution persisting until gypsum saturation is reached, when all the excess Ca is precipitated as gypsum as long as sufficient SO_4^{--} ions are available. After the total sulphate is removed in gypsum, then the solution moves into the $CaCl_2$ region.

Theoretically this situation can be reached at the beginning of NaCl precipitation. The extent of actual $MgSO_4$ impoverishment differs from one solution to the next. According to this hypothesis, it is to be expected wherever thick carbonate precipitates underlie salt deposits. In the case of the German Zechstein, this would apply to the Hauptdolomit Ca 2d. The latter is particularly well-developed in the neighbourhood of Königshall-Hindenburg in the South Harz, just in that region where $MgSO_4$ impoverishment has been recorded in the overlying Stassfurt Series (RICHTER-BERNBURG, 1955). No Ca hydrocarbonate influx from continental sources is, therefore, required to explain the alteration of the brine, which can be much more satisfactorily explained by early dolomitization of the Hauptdolomit. Naturally through diagenetic compaction, the circulation of solutions and communication with the main water body may be restricted or halted before dolomitization is completed. This can continue with the Mg from the trapped pore solutions with a consequent further increase in the pore fluid Ca/Mg ratio, so providing a possible explanation for the highly concentrated $CaCl_2$ solutions found in pores and joints of the Hauptdolomit (see, for example, HERRMANN, 1961 b).

In the central Zechstein basin the Hauptdolomit is represented by a thin bituminous shale (Ca 2 st) so that no significant dolomitization could occur. There, too, no noteworthy impoverishment in $MgSO_4$ is found. What happened was rather the alteration of the brine chiefly due to the syngenetic formation of polyhalite as a consequence of the influx of fresh seawater.

The types of salt in a sequence will depend upon the underlying bed only if the surrounding waters are not mixed by currents. This cannot be assumed without some qualification. However, by the time brine has reached the stage of saturation with respect to NaCl, the viscosity of the solution is already so great that currents and other mixing processes have a greatly reduced efficiency. More important in this context are such palaegeographic factors as dividing ridges and the like, although these cannot be demonstrated in the particular case of the South Harz basin as against the Stassfurt Deep on the basis of present outcrops.

From the preceding discussion it should be clear that there is no strong interdependence between the types of carbonate and chloride precipitation facies. There are so many examples of variation in fossil salt deposits, that it is impossible to generalize as to whether the alteration of the original brine solution can be traced to dolomitization or to the precipitation of primary Mg-bearing carbonates (as a result of the influx from continental streams). The larger the salt deposit, the smaller the probability of a sufficient continental influx.

A complete understanding of the processes involved is still far off. Apart from the uncertainty about the details of the dolomitization or dolomite precipitation mechanism[56], it is uncertain to what extent conditions in the Dead Sea can be compared with the significantly larger salt lagoons of the geological past. One important difference lies in the much greater influx of seawater (p. 251), which in the larger salt deposits certainly brought the greater part of the salt content into the lagoon only after the stage of $CaSO_4$ precipitation. Each influx carried with it a certain amount of SO_4^{--} so that even during the NaCl precipitation phase, the alteration of the solution by the said mechanism must have continued.

In spite of many unanswered questions two general statements can be made:

1. Case b (p. 246) can be accepted as the mechanism of primary $MgSO_4$- impoverishment (in the extreme case with the displacement of the brine composition into the $CaCl_2$ region), as a sufficient number of examples of primary (as defined on p. 92) anhydrite-bearing $MgSO_4$-free halite-carnallite deposits are known.

2. It is certain that for the impoverishment of a solution in $MgSO_4$ the addition of calcium carbonate is necessary, whether in the form of an influx from the land surface or in form of thick, earlier deposits which are later dolomitized. The Ca content of pure seawater influxes would permit only limited alteration of the solution in a roundabout way via (early diagenetic) polyhalite precipitation.

The particular case of Zechstein 2 shows that both sides who have been arguing for years for and against primary facies differentiation were correct. The various forms of carnallite rock are primary and reflect varying degrees of $MgSO_4$ impoverishment (conditioned indirectly by palaeogeographic controls rather than by vertical differences in concentration of the solution). On the other hand, all economically important sylvite deposits are secondary, developed as a result of solution metamorphism, whilst the primary anhydrite-bearing sylvinitic halite occurs only in special cases, and only with non-workable amounts of sylvite. However, the solution metamorphism is probably not early diagenetic or sub-primary (RICHTER-BERNBURG, 1953) as the $MgSO_4$-bearing transition types of the South Harz should then contain kainite and not kieserite and sylvite.

[56] But see the very important monograph by USDOWSKI, H. E.: „Die Genese von Dolomit in Sedimenten“, Mineralogie und Petrographie in Einzeldarstellungen, Vol. 4, Berlin-Heidelberg-New York: Springer 1967.

2. *Rhythmic Bedding and Influxes*

The even rhythmic bedding of many anhydrite and halite rocks often for great thicknesses, is a characteristic feature. LOTZE (1957) has discussed in detail the bedding and the related general problem of its formation so that only one important special case will be touched upon here.

There is a general tendency to regard the regular rhythmic banding in the German Zechstein 2 as annual banding. Meanwhile it can be demonstrated that in the case of the Stassfurt halite series, at least,

1. the bedding is conditional upon the influx of fresh seawater and
2. the concentration of brine had strong, periodic variations.

Assumption 1. is common to all hypotheses which postulate a barrier of some kind. This is indispensable in explaining the quantity of salt precipitated and requires no further discussion. The desert hypothesis (J. WALTHER, 1910, p. 250/251) favoured earlier, for instance, by JÄNECKE (1923, p. 86) is not applicable to oceanic salt deposits for it explains neither the great salt thicknesses, nor the regular salt successions and in particular neglects the qualitative composition. It must be noted that the influx of fresh seawater onto a so-called "saturation shelf" in the absence of a barrier, as occasionally presented (RICHTER-BERNBURG, 1953, Fig. 2), does not lead to a stable increase in salt concentration because of the reflux of concentrated solutions.

Assumption 2. is common to all hypotheses which suppose periodic influxes. This idea first appeared in ZIMMERMANN (1915) and was based upon the fact than in fine banding a more and a less soluble salt alternate regularly, both in the Stinkschiefer (Ca 2 st) and in the Stassfurt halite (Na 2), and in the Basalanhydrit (A 2). In such cases there is usually a thin coating of clay, which sometimes contains carbonates, at the bottom of the less soluble layer.

According to LÜCK (1913), the rhythms in the anhydrite region in the Stassfurt anticline are on average 10 cm thick, the anhydrite "threads" being about $^1/_2$ cm. The halite grain size within the rhythm often increases upwards (cf. p. 155). In the polyhalite region the rhythms average 7 cm in thickness. The rhythms may also be irregular, particularly through the introduction of a thicker sulphate layer, or through the absence of bedding over a thickness of several meters. The halite layers also contain small quantities of loosely disposed sulphates. In hand-samples the separation into sulphate-rich (fine-grained, dull white, gray or dark) and halite-rich layers is clearly distinguishable. Between the anhydrite and polyhalite zones according to LÜCK (1913), there is a transition layer with about 1% glauberite. This zone was found by RÜHLE (1913) on the Bernburg anticline and later observed in several other localities in the central Zechstein basin.

In the Stassfurt section, the quantitative ratios in the anhydrite and halite layers correspond significantly enough to the theoretical quantities calculated for the static evaporation of seawater [theoretical: 3 mm anhydrite (4%) upon 10 cm NaCl out of $8^1/_2$ m seawater, or out of $2^1/_2$ m of a brine already saturated with respect to gypsum; observed: 4–5% anhydrite, Fig. 38]. This shows not only a regular change in the brine concentration, but, more importantly, that the formation of each couplet runs through the complete concentration range, at least from the beginning of gypsum saturation up to NaCl precipitation. Each couplet thus forms a small static evaporation cycle.

The alternative is to assume some form of continuous influx, as in Ochsenius' Bar theory; this, as so far presented, is physico-chemically and sediment-petrographically inadequate to explain the Zechstein 2. Neither the assumption of rhythmic precipitation (a detailed discussion in LOTZE, 1957) nor the assumption of temperature fluctuations can explain the double layers, especially as in an already saturated NaCl brine there is at most 0.7 g $CaSO_4$ per 100 g NaCl + $CaSO_4$ instead of the observed 4–5% (and in some places up to 9%). It is indeed possible to precipitate NaCl by cooling (p. 104) but this will not change the absolute amount of $CaSO_4$ in one and the same solution.

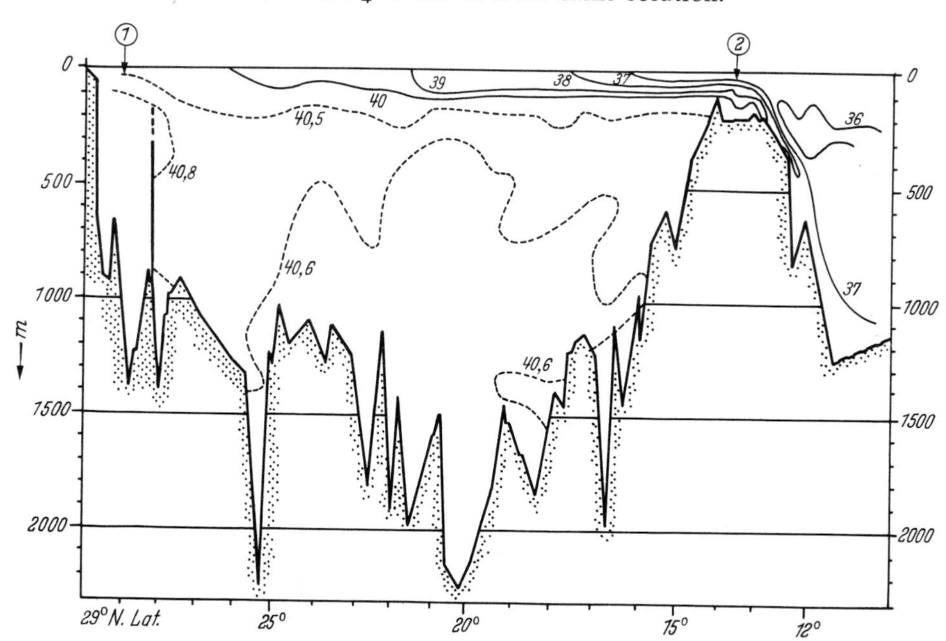

Fig. 46. The salt content of the Red Sea in ‰, June 1958 (simplified after NEUMANN and MACGILL, 1961). *1* Gulf of Aquaba; *2* Straits of Bab el Mandel in the Gulf of Aden

Assumption 2, that of periodic strong fluctuations in concentration in the upper brine layer, can be maintained even with continual brine influx. This possibility, which does not appear to have been discussed until now, will be developed below, using the Red Sea as an example.

The Red Sea, according to recent research (NEUMANN and MCGILL, 1961), shows a fairly continuous increase in salinity from about 36‰ at the bar in the Straits of Aden to over 40‰ in the Gulf of Suez (Fig. 46). The deep water throughout the Red Sea possesses the same salinity, about 40.6‰, up to the level of the bar (about 125 m below sea level). Immediately behind the bar there is thus a steep vertical concentration gradient (about 1‰/25 m) whilst the horizontal concentration gradient is very small (about 1‰/400 km). On account of the clear seasonal variation in evaporation (PRIVETT, 1959), upon which is superimposed a further variation due to latitude, a seasonal variation in the concentration gradient may be expected. In the Mediterranean such conditions are known to exist. In winter the 37‰ salinity line extends to about 2° E and in summer to 18° E (MCGILL, 1961, Fig. 3). In transposing these conditions to the Zechstein salt lake, a significantly greater horizontal gradient must first be assumed, which would be possible with a considerably shallower bar and longer or narrower straits than in the Red Sea, and in the case of the Kara Bugas Gulf is known to exist (SEDELNIKOV, 1958). As a consequence of seasonal fluctuations in the concentration gradient, the limit of gypsum or halite saturation in the surface layers at different times of the year must have lain at varying distances from the bar. The saturation limits must then have moved according to the season. There would thus be a transition zone in which there would be an alternation of gypsum and halite precipitation. All that is required is a pronounced variation in the horizontal concentration gradient in the surficial brine layers. From the data in Table 8 the magnitude of the necessary concentration variation from the beginning of gypsum precipitation to the precipitation of halite is evident, and this argues against the model. Should an annual rhythm, however, be demonstrable, it would appear to be most suitable as an explanation of anhydrite-halite banding and for that reason will not be totally rejected.

It is also possible to consider a continual seawater influx with a periodic influx from terrestrial sources containing Ca^{++} and HCO_3^- ions. This would lead to the precipitation of SO_4^{--} as calcium sulphate over a clay layer and at the same time bring about an impoverishment of the brine in $MgSO_4$ (p. 246). Although this can be excluded in the case of Zechstein 2, as then $MgSO_4$ depletion would occur in the central Zechstein basin, too, the model cannot be excluded in other cases.

The immediate cause of rhythmic bedding is certainly related to similar rhythmic fluctuations in the brine composition. The original bar theory model, which postulates that brine composition is more or less constant, or at least at any one spot in the salt lagoon excludes short-period variations in composition, cannot explain the results obtained from the Stassfurt Series, irrespective of whether or not a barrier is assumed to have separated subsidiary basins. If an annual rhythm is considered as the cause of the periodicity, then the available theories are either ZIMMERMANN'S (1915) hypothesis in which the seasonal influx occurred only during the Monsoon, or the rather more plausible hypothesis of the seasonal horizontal displacement of gypsum or NaCl saturation concentrations in the upper brine levels. In both cases, climate is the controlling influence. However, the second hypothesis is more consistent with oceanographic data than the first, and does not require a tongue of land as a barrier which has to be broken and reformed every year in the change from Monsoon to counter-Monsoon.

Yet the linking of the periodicity to an annual rhythm is an uncertain hypothesis. Apart from LOTZE'S (1955) views on the geologically short duration of salt formation, the following points must also be considered:

1. The normal thickness of the anhydrite layers (<1 mm) is quite compatible with the maximum annual evaporation, but this is not true of the not uncommon anomalies of 5 mm and more (up to >20 mm) in a single layer. To precipitate 5 mm anhydrite requires the evaporation of 14 m of normal seawater or 4 m of brine saturated in gypsum. At the present time the maximum evaporation from large water surfaces in very arid regions is 2 m (e.g. Red Sea; J. NEUMANN, 1952; PRIVETT, 1959). In more concentrated solutions this figure is never reached, but falls by about 1% for a 1% increase in salt concentration. In halite-saturated solutions the fall in evaporation is about 30% (HARDING, 1949, p. 65). Furthermore, in the absence of wind, halite precipitates at the water surface, reducing evaporation still further, so that the present-day evaporation from halite-saturated solutions may be taken as about $1^1/_2$ m per year. It is quite incompatible with well established physical principles to assume that a slight increase above an average temperature boundary will bring about much greater precipitation of $CaSO_4$ (RICHTER-BERNBURG, 1953). Only a marked negative temperature coefficient of gypsum solubility, which seems unlikely from the results of experiments so far carried out, under conditions of strong heating in shallow water could bring about the precipitation of gypsum from inflowing brines. This would require evaporation in the main basin to bring the brine to gypsum saturation, yet with the precipitation of gypsum itself concentrated over a much smaller area than the evaporation.

2. The precipitation of the observed anhydrite-NaCl couplets in Zechstein 2 from normal seawater likewise requires a higher annual

evaporation rate (about $8^1/_2$ m of normal seawater or 2.6 m of seawater concentrated to the point of gypsum evaporation). The Stassfurt halite cannot be directly compared with the average 10 cm halite precipitation per annum found in modern salt lakes such as Kara Bugas (FIVEG, 1954), as in these lakes the rhythmic banding with $CaSO_4$ is absent. The precipitation of a pure halite layer does not require strong periodic concentration variations nor does the precipitation of 10 cm NaCl demand greatly increased evaporation.

3. Regarding the anhydrite layer which sometimes replaces the Werra halite, JUNG (1959) established that it has 10 times as many "annual rings" as the halite itself. Of course, it could be argued that the halite was deposited only in the troughs without any equivalent on the ridges.

4. An attempt has been made to demonstrate statistically an 11-year cycle upon the basis of the anomalies in the fine bedding and to refer it to the sunspot cycles (RICHTER-BERNBURG, 1950, and later). Such a relationship is not yet well established. The anomalies of the sunspot cycles are a periodic fluctuation between a maximum and minimum, whilst the bedding thickness anomalies usually occur abruptly. One way of characterizing this is the determination of the periodicity from: $\frac{\text{number of varves}}{\text{number of anomalies}}$. Thus, even qualitatively, another rhythm may be suspected. Quantitatively the relationships are again not well established, quite apart from subjective factors in thickness measurements and counting and variations in the thickness of the actual beds. Even assuming that the measurements are reliable, the periodicity itself must be ascertained by harmonic analysis, as applied to all periodic geophysical processes (magnetic variations, sunspot numbers, etc.). Such an objective test is indispensable as, according to RICHTER-BERNBURG (1960), in counts of several hundred bands "no eleven-unit period is recognizable" or else the periods are of variable length. It remains to be seen whether under these circumstances there is a convincing dependence of the anomalies upon the rather constant solar periods. Apart from these methodological considerations, no effect of sunspot activity on the gross weather pattern has yet been detected (BAUR, 1961)[57], although the climate of the whole Earth has been recorded at a sufficient number of stations only over the last two or three sunspot cycles. Nevertheless,

[57] Page 65: "whilst sunspots and solar flares exhibit a clear, average 11-year variation, there is no 11-year periodicity in the general weather pattern. Rather in the rhythmic analysis of certain weather elements an 11-year cycle can be seen, but the amplitudes are always so small that they in no way explain weather anomalies. The 11-year period is thus well within the range of chance".

a connection between sunspot activity and the climate of the Earth is not impossible.

5. The enrichment of the clay layers in organic matter is regarded as a consequence of the plankton dying off in autumn (RICHTER-BERNBURG, 1960). This is open to more than one interpretation. Assuming a periodic influx of fresh seawater, a rich planktonic flora could be introduced from the adjacent open sea, but the increasing salinity due to evaporation would soon become too great for the majority of stenohaline plankton and their death would consequently be independent of season. The organic enrichment of the clay layers can thus be seen as a direct indication of the influx of fresh brine. This would involve evaporation on a scale so large it could not be explained by annual rhythm. Convenient and reasonable as the annual banding hypothesis may appear, it leads to difficulties at many points.

To be sure, the assumption of periodic influxes, resembling minor marine transgressions, runs into similar difficulties. In the first place it fails to explain the regularity of the influxes, and the same time applies to possible endogenic causes of rhythmic banding (LOTZE, 1957). Eustatic changes in sea level have also been suggested (HAM, 1961). Less probable is WILFAHRT'S great flood hypothesis, for as LOTZE (1957) noted it is highly improbable that the necessary amount of water could enter a basin as large as the Zechstein during a single high tide. The hypothesis requires that the evaporation surface of the salt sea should lie below sea level. This demands in turn an old depression, completely blocked off, with a depth at least equal to the salt thickness. This is quite feasible.

In this context the so-called "descendence" theory must be considered. It was introduced by EVERDING (1907, p. 31) for "salt beds which, immediately after precipitation of the sequence of mother salts and still during Zechstein time, were formed by redeposition or transformation of the mother material"[58]. This was clearly refuted by SCHÜNEMAN (1913) (see p. 178) for the Stassfurt area. For the younger Zechstein series (Z 3 and Z 4) it has been variously defended (D'ANS and KÜHN, 1940–1960), in particular as an explanation for the decreasing bromine curves. These arguments were disposed of on p. 142 upon the basis of a model with periodic seawater influx, although it was conceded that the D'ANS model (even with the solution of bromine-rich Stassfurt rock salt in rainwater) leads to a more rapid decrease in the bromine curve. KÜHN (in KÜHN and SCHWERDTNER, 1959; SCHAUBERGER and KÜHN, 1959) sought to use the large single crystals of NaCl (1–2 cm) occurring in the "Jüngeres Steinsalz" especially in zones Na 3 γ and Na 3 δ, as well as in Alpine

[58] Quoted from EVERDING as follows: "Salzablagerungen, die alsbald nach Abscheidung der Muttersalzfolge noch während der Zechsteinzeit aus der Umlagerung oder Umbildung des Muttermaterials hervorgegangen sind".

salt deposits, as an independent proof of the mechanical redeposition of coarse-grained older salts. In this context the so-called "Augensalze" were regarded as fragmentary salts or salt conglomerates. Excluding the sedimentary-petrographical weaknesses [59] of this explanation, geological objections have been raised, particularly as regards the amount of "descendence" (LOTZE, 1957; BORCHERT, 1959). It is certainly true that the character of the sedimentation of the younger halite horizons is disturbed with numerous progressive and recessive epochs. Only, however, in a fully isolated basin which is periodically flooded is basin shrinkage possible with, at least in places and at times, the redissolving of older salt precipitates in the dried-out marginal areas, either by a fresh influx of seawater or (less likely) by continental run-off or rain-water. On the other hand, the mechanical redeposition of sylvite and carnallite is quite certainly impossible, and that of NaCl is probably also excluded, so that this hypothesis is inapplicable to potash deposits. Yet, insofar as the level of the brine always lay at sea level, the redissolving of older salts is altogether unlikely, and the significance of such a process in the Zechstein salt sequences would generally have been slight.

The model of a completely isolated basin can only have developed late, if at all, either during or towards the end of chloride precipitation, possibly even following carbonate or gypsum precipitation in the bar region. If a return flow of concentrated brine was already excluded at the time of gypsum precipitation in the main basin, the amounts of halite and potash salts precipitated would have been much greater. There must thus have been an appreciable potential return flow of concentrated brines at the beginning of the saline phase, as is the case in the water layer above the barrier zone in the smaller enclosed seas at the present time (Fig. 46). To be sure, no quantitative calculation is ever likely to be possible for Zechstein 2, the total salt content of the basin being far from well known, because of the great variations in thickness and lateral changes in the facies.

In contrast, during the formation of the Hauptanhydrit, the greater part of the NaCl-solutions must have flowed back. In the present state of knowledge it appears that the Hauptanhydrit has a fairly uniform thickness over the whole basin while the thickness of the overlying "younger halite" (Jüngeres Steinsalz) is far too small in relation to the anhydrite thickness. Yet it is not known whether the highly concentrated

[59] From his examination of erosion in the salt stocks in the Persian Gulf, the author concludes that the transport of material is by pure solution weathering with the formation of sharp-angled solution channels. The only solids transported in salt water run-off streams are the insoluble residues. As well as the depositional mechanism required, the necessary flow rate is improbable within a salt lake; furthermore, such coarse-grained salt rocks occur too rarely in the Stassfurt halite and almost always as the result of secondary recrystallization.

residual solutions of the Stassfurt time were also washed out of the basin. This seems unlikely on account of the abnormally high bromine content of the "younger halite" (Jüngeres Steinsalz) immediately above the Hauptanhydrit (p. 204). It seems more likely that there was a very stable layering of the dilute influxes over a highly concentrated basal solution, so that the return flow did not carry away this old, residual brine. The characteristic radial structures of the Hauptanhydrit also suggest special precipitation conditions which can perhaps be explained by recrystallization of this concentrated basal solution. Admittedly, the Hauptanhydrit might, at least in part, be interpreted as a sedimentary precipitate (due to large influxes of $Ca(HCO_3)_2$ solutions). This would be consistent with the facies transition from Hauptanhydrit to Plattendolomit indicated by ZIMMERMANN (1915).

Weighing all the observed facts, it appears certain that during the formation of the fine, rhythmically banded carbonate and anhydrite rocks the surface level of the salt basin must have been the same as that of the open ocean, for otherwise the reflux of concentrated brine solutions would have been possible. This in turn would suggest periodic variations in the composition of the brines and be a point in favour of the annual banding hypothesis for the fine-banded layers. Yet whether the anomalously thick anhydrite layers could form in a single year remains very doubtful. Using the proposed mechanism of periodic horizontal displacements of the saturation limits for a given solid phase, the absence of an "annual ring" can be explained by the fact that, during the formation of these anomalously thick beds, the horizontal displacement was smaller and hence the surface brine layer remained saturated in the same solid phase over a longer period[60]. The same explanation could also be applied to the halite series, particularly where no strict regularity is found in the rhythmic bedding. It is certain that at the end of the halite precipitation, at least in the Stassfurt series, the salt basin became completely isolated. There were, nevertheless, several ingressions of fresh seawater, as witnessed by the intercalated rock salt banks (e.g. the Unstrutbänke, p. 212). This was early recognized by D'ANS (1915, p. 269).

It therefore seems right to adopt a compromise view of the bedding question, as LOTZE (1957) also does, since part of the sediments appear to be correlated with an annual rhythm. It is certainly an error to try to ascertain the duration of deposition simply by counting the bands, as an individual band may not uncommonly represent more than one year. Conversely, it is still uncertain whether individual bands always represent a complete year.

[60] BRAITSCH, O.: „Die Entstehung der Schichtung in rhythmisch geschichteten Evaporiten", Geol. Rundschau **52**, 405–417 (1962).

3. Temperature of Primary Precipitation[61]

VAN'T HOFF used the first appearance and the disappearance of solid phases, and occasionally also changes in paragenesis, as indicators of temperature (see Figs. 19, 20). As most primary parageneses have disappeared from fossil salt deposits, the determination of the temperature of formation is uncertain. Further, primary metastable parageneses also occur. D'ANS (1947) estimated the increase in temperature necessary to maintain constant evaporation from the surface of a concentrating solution. He arrived at a figure of 30–35° C for the beginning of NaCl saturation, 34–39° C for the beginning of polyhalite precipitation and 41–47° C for that of carnallite. Even if the assumption of equal evaporation and equal annual precipitation is questionable, the temperatures obtained below 50° C are feasible.

In $MgSO_4$-bearing salt deposits, the kieserite-carnallite paragenesis offers a good starting point for temperature determination. If the quantitative ratio of these two minerals is very sensitive to temperature, at constant temperature it is independent of influxes. Very high NaCl- and anhydrite-contents are therefore without effect. However, the region over which it can be applied is restricted to temperatures over 25° C because at lower temperatures hexahydrite and epsomite can occur. Unfortunately there are no new data for the critical temperature range between 35° and 55° C, so that the temperature data are still quite crude. The calculation gives (in weight %; remainder = NaCl).

Temperature	25°	35°	55° C
kieserite	40	29.5	6.5
carnallite	48	59	86.5
ratio $\frac{\%\ \text{kieserite}}{\%\ \text{carnallite}}$	0.83	0.5	0.075

The graphic representation of the kieserite/carnallite ratio against temperature shows a shallow curve. On account of the steep gradient of the curve against the temperature axis, a determination to $\pm 3°$ should be possible. Temperature variations about an average figure are not an important source of error so long as the kieserite/carnallite ratio is determined from samples averaged over a large section of the beds. For, although carnallite is preferentially precipitated during cooling, in the succeeding evaporation phase the corresponding equi-

[61] See also: BRAITSCH, O.: "The Temperature of Evaporite Formation. Problems in Palaeoclimatology". Ed. by NAIRN, A. E. M., Interscience Publishers London-New York-Sydney, 479–531 (1963).

librium amount of kieserite must be precipitated. An important source of error, however, lies in the early diagenetic alteration of metastable hexahydrite-carnallite primary precipitates to the stable paragenesis kieserite-carnallite. As this alteration occurs in the presence of saturated solutions, one obtains from the temperature determined for the kieserite/carnallite ratio the temperature of transformation to the stable equilibrium forms, which lies an unknown number of degrees above the temperature of formation of the primary precipitates.

The possibility of the formation of primary kainite (or metastable epsomite-sylvite) with subsequent reaction in solution R is a further possible source of error. In this case, the temperature dependence has not yet been investigated. Above all, the kieserite/carnallite ratio does not remain constant even within the carnallite layer (see Fig. 27) so that a temperature determination is only possible with an exact knowledge of the individual stages in formation. In practice, this is generally not possible. At any event the last stage in the kieserite-carnallite precipitation (between R and Z) can be found with the help of careful bromine analyses. Certainly the average value of the kieserite/carnallite ratio over all sections of the seam resulting from the reaction gives a crude approximation to the kieserite/carnallite ratio in the last section ($R-Z$), so that an approximate figure is obtained from the average of the whole seam. The error limits must be at least $\pm 10°$ C.

The application of this method to certain adequately known Stassfurt occurrences results in the following values:

Stassfurt anticline Stassfurt seam (after Table 24)					Hildesia (Hildesheim)		
					after ENGEL from LOTZE (1938)		after KOKORSCH (1960) p. 180
	I–O	I–O	I–O	K–M			
$\frac{\text{kieserite}}{\text{carnallite}}$	0.31	0.38	0.33	0.22	0.47	0.29	0.32
temperature °C	42	39	41	47	36	43	41

The temperature values found in this way are thus about 40° C. There is no significant difference between the values from the two occurrences.

In completely $MgSO_4$-free potash salts, some indication of temperature can be obtained from the bromine content of the sylvite and halite immediately at the base of the potash layer (Fig. 35), if the bromine content of the original brine is known or can be estimated. As this possibility has not previously been considered, there is still no systematic research on this aspect. The initial bromine content of the brine can be estimated from the bromine present in the halite at the base of the halite precipitation, but only where secondary recrystallization can be excluded by petrographic examination. The bromine curve should also rise steadily within the salt sequence, recessive phases due to seawater dilution being without effect. The bromine content of the halite and sylvite at the base of the potash layer must be ascertained from the pure phases by careful separation and extrapolation, and the paragenetic ratio, within the limits of analytical accuracy, should be 1 : 10.

As in $MgSO_4$-free parageneses no metastable phases (excluding a small gypsum admixture, whose early diagenetic alteration is without significant effect) occur, this method therefore yields the temperatures of primary precipitation. The error limit is determined by analytical errors, by the uncertainty of the distribution factors and the models as well as by the uncertainty in the determination of the initial bromine content of the brine. It should be less than $\pm 10°$ C.

In the rock salt below the "Unteres Lager" in the Upper Rhine Valley a progressive increase is seen in the bromine content, the initial value corresponding to that of normal seawater (p. 203). At the base of the potash layer no determinations are available, moreover, the halite is not from the sylvinitic region but derived from the sylvite-halite double layers. For orientation we first take the minimum content of 0.32% Br/KCl in the potash layer and the maximum content in the halite immediately below the potash layer (0.03% Br/NaCl or 0.02_6% Br/NaCl); we then find from Fig. 35 a primary evaporation temperature between 40° (from the sylvite value) and 30° or 8° C (from the two NaCl values). In the absence of new research data, we cannot say which temperature is the more probable, possibly the higher. The question can be answered by the examination of carefully collected samples from the lowest sylvinite horizon[62]. The usefulness of the method cannot be judged from data in the literature, as in earlier researches samples were selected with quite different objects in view. BAAR and KÜHN (1962, p. 330) selected on palaeontological grounds, basded their calculations on a temperature of 15° C.

In contrast to VAN'T HOFF'S methods of determining temperatures, both methods (interpretation of the paragenetic kieserite/carnallite ratio; bromine content at the change of paragenesis) permit determination of temperature within the stability field of the parageneses, and do not merely define its limits. Both methods are complementary as they can be applied to the two basic potash types: $MgSO_4$- free and $MgSO_4$-bearing.

Geological estimates of temperature lie between 10° and 72° C. On the basis of solution equilibrium data, the formation of potash

[62] See footnote p. 142.

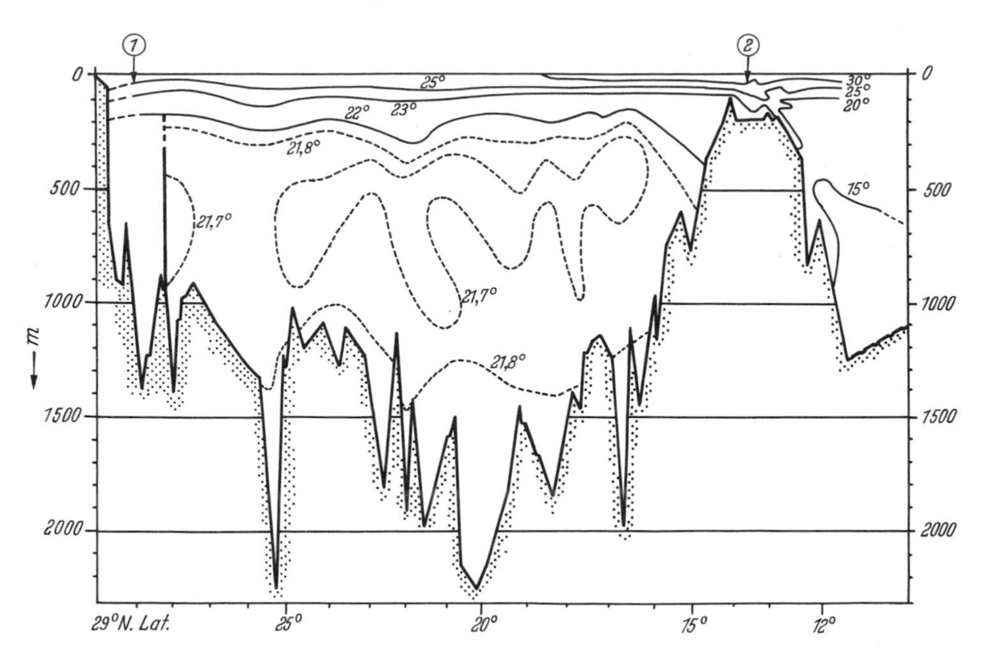

Fig. 47. Temperature distribution in °C in the Red Sea, June 1958 (simplified after NEUMANN and MACGILL, 1961). *1*, *2* as in Fig. 46

salts in this wide temperature range is indisputable. It appears that observations on Recent salt lakes (cf. LOTZE, 1957, p. 92) are not directly applicable to the Zechstein and the other large salt-basins of the geological past. A better indication is probably obtained from the temperature distribution today in minor seas such as the Red Sea (Fig. 47). In more isolated seas higher temperatures of up to 50° C are not excluded. The actual temperature fluctuations and the strong temperature gradient occur only in the upper water layers, from which it follows that in large salt seas the greater part of the salts, particularly the chlorides, are precipitated as a consequence of the cooling which occurs as the saturated solution sinks into the basinal solution.

4. Redox Potential and Hydrogen Ion Concentrations

The insoluble residues give an indication not only of the redox potential (E_h) but of the hydrogen ion concentration (pH) during the formation and alteration of salt deposits, although the salt parageneses are relatively less sensitive to E_h and pH (KRUMBEIN and GARRELS, 1952). Unfortunately, there have been no experimental investigations of saturated salt solutions where in any case the ionic activities differ from those in dilute solutions, which are those usually studied. Conclusions drawn by analogy with normal marine sediments are not permissible without severe restrictions, and so only qualitative discussion is possible. A complication is introduced by the widespread secondary alteration which occurs under anomalous E_h–pH conditions.

The Fe-free carbonates, phosphates, and probably also borates are independent of E_h but strongly dependent upon pH; the Mg silicates are less markedly pH-dependent. The occurrence of carbonate and phosphate is not possible in an acid medium. It can therefore, be concluded from the presence of magnesite and certain Ca–Mg phosphates that the Zechstein salts, for example, were formed in a neutral to weakly alkaline medium. Chlorites are also an indication of a neutral to weakly alkaline pH region. On the other hand, the absence of brucite ($Mg(OH)_2$) speaks against strongly alkaline conditions, that is to say a $pH < 9$.

Organic matter is strongly dependent upon E_h but to a large extent independent of pH. Under positive E_h values (aerobic conditions) organic substances are completely oxidized to CO_2, but with negative E_h values they remain in various stages of anaerobic decomposition (bitumen). The occurrence of organic matter in the clay residues from salt and anhydrite rocks in many formations indicates the prevalence of reducing conditions. There are, of course, differing opinions on the extent and geological significance of the reducing conditions (p. 246). As a measure of reduction, the capacity of the reducing system is more important than the potential (E_h), and the capacity depends upon the concentration of the reducing medium. In the rock itself there is scarcely any indication of this, although the total carbon content provides a starting point. In the insoluble residues of salt rocks it amounts to about 1% by volume, that is to say about 0.01% of the total rock. In contrast the total organic carbon in true sapropelites amounts to about 10%. Although the sapropel facies usually occurs at the beginning of salt precipitation, the decrease of organic matter during salt precipitation indicates a corresponding decrease in the reducing capacity. Free hydrogen sulphide is no index of exclusive reduction of SO_4^{--}, as anaerobic decomposition of organic matter (albumin etc.) gives H_2S as a by-product. Even in the Black Sea, SO_4^{--} reduction is quantitatively unimportant, as more than 97% of the original SO_4^{--} remains (SKOBINZEV et al., 1958).

The iron minerals are strongly dependent upon both E_h and pH. The results obtained from normal marine sediments again cannot be applied to salt deposits without qualification. Between the stability fields of haematite (at pH 8 somewhat above $E_h = -0.05$ V) and pyrite (at pH 8 somewhat below $E_h = -0.2$ V, see KRUMBEIN and GARRELS, 1952) siderite ($FeCO_3$) should occur. It does not, however, seem to occur in salt deposits. Only in the Upper Rhine potash salts does ankerite (p. 22) play a certain role. On the other hand, in the salt residues of Königshall-Hindenburg and of the South Harz, which have been investigated in detail, in Fe-poor to Fe-free magnesites, pyrite + haematite with subordinate magnetite occur together in the same sample. It is true that in these cases the haematite is found only in carnallite and sylvite, whilst in contrast pyrite and magnetite are found in the insoluble clay residues both in solvents and in small clay inclusions. As a result of secondary impoverishment haematite is replaced by pyrite, although the E_h–pH mechanism for this is still obscure. It can perhaps be assumed that the reducing conditions existed only within the clay residues. In these too are the organic remains. As the clay minerals as well as the salt minerals are re-formed and recrystallized, there is no detectable time difference between the formation of pyrite and haematite. The secondary red coloration of the salt beds in contact with the clay (p. 230) is suggestive of a different redox potential in the salts.

Apart from these unexplained problems, the occurrence of haematite is, according to CORRENS (1960), the most reliable criterion against the assumption of large-scale SO_4^{--} reduction. Even when the haematite is supposed to have resulted from the subsequent alteration of $FeCl_2$ (p. 223), it still indicates that the S^{--} content of the evaporating solution must have been several orders of magnitude less than the SO_4^{--} content. In the least favourable case, the activity ratio, which is approximately proportional to the concentration, is $a_{SO_4^{--}}/a_{S^{--}} \approx 10^{12}$. Whenever the S^{--} content rises above this, all the Fe^{++} must precipitate as the sulphide. In the Stassfurt series this is certainly not the case. Here pyrite and haematite (possibly derived from primary $FeCl_2$) occur in approximately equal amounts.

Only in the Werra series does the S^{--} concentration appear to have been adequate to allow nearly all the iron to precipitate as pyrite (together with much smaller amounts of magnetite). The clay residues there appear to contain a higher organic content. Yet in the Fulda district near the basin margin some haematite occurs. There can thus have been no great difference in order of magnitude between the SO_4^{--} reduction within the Werra series in comparison with the Stassfurt series.

All the indications point to the same conclusion, namely that the German Zechstein salts developed under weakly reducing conditions near the neutral point. Reduction of SO_4^{--} to S^{--} plays no part in the precipitation of the salts as the requisite E_h conditions never occurred. Their main effect should have occurred in the diagenetic phase, without, however, involving any significant amount of the total sulphates.

5. The Origin of the Metamorphic Solutions

Many reactions in salt deposits produce solutions, and the most important metamorphic changes are brought about through the alteration of existing salts by penetrating solutions which are unsaturated with respect to the salts with which they are in contact. Unfortunately there are relatively few published analyses of saline solutions from salt rocks (BAUMERT, 1928; STORK, 1954; HARTWIG, 1955; HERRMANN, 1961 b) as the influxing of solutions in potash workings is generally kept secret. In view of this, no specialized data are available in this context. Only pure chemical systems can be considered, such as those used by HERRMANN (1961 b). Here, three groups can be distinguished, namely:

1. $CaCl_2$-bearing solutions:
a) with low Mg content,
b) with high Mg content (> 50 mg $MgCl_2$/l) usually saturated in NaCl;
2. predominantly $MgCl_2$ bearing solutions, in part highly concentrated with a composition near point Z (Table 6);
3. predominantly NaCl-bearing solutions, mostly saturated in NaCl.

Type 1 a) solutions are not peculiar to salt deposits, for they appear to be the commonest type of pore solutions in sediments, although often with a low total salt content (v. ENGELHARDT, 1960, and literature references; KREJCI-GRAF, 1962). They are in part the result of enclosed, diagenetically altered seawater (p. 247), but in part also derive from the decomposition of organic matter (KREJCI-GRAF, 1962), in particular, plankton which themselves are more than 90% water.

From the technical and mining standpoint the solutions are classified according to the extent to which they endanger mining operations. The only solutions which are not dangerous are those least saturated with carnallite ($MgCl_2$ concentration $\geqq$ the $MgCl_2$ concentration of Q solutions, i.e. group 2); moreover, their composition must remain constant with time. A classification of solutions according to risks is to be found in HARTWIG (1955).

There are still many problems in trying to systematize the genesis of salt solutions, and it would be a mistake to try and trace them back to a single process of formation. Three main categories may be considered:

1. Residual solutions from the evaporation of seawater, the so-called "Urlaugen" of the older literature. Definite examples are unknown. They should be characterized by a high lithium and iodine content, with bromine present to more than half a weight percent and relatively low rubidium. As only small cavities or pore spaces remain after diagenetic recrystallization, large quantities of residual fluid are unlikely to occur. At most, the saliferous clays (Salztone) may contain altered residuals of such solutions.

2. Authigenic solutions resulting from dehydration reactions and thermal metamorphism. Thermal metamorphism is, however, not very important. Possibly certain highly concentrated $MgCl_2$ solutions belong here in so far as they are derived from the thermal metamorphism of carnallite (p. 112). They could, however, be derived from the decomposition of carnallite in foreign solutions at high temperature (HERRMANN, 1961b). The most important authigenic solutions arise from the dehydration of gypsum; their significance, however, is problematic as will be shown below.

3. Foreign solutions, those which infiltrate the salt deposits from outside. Indisputably in this category are the solutions associated with the top of salt diapirs and the gypsum and kainite cap solutions, plus the post-volcanic solutions consequent upon basalt intrusions and so on. In addition, there are appreciable quantities of foreign solutions in joints and in porous salt horizons which are undoubtedly very important in the explanation of metamorphism. These are not completely or not continuously isolated from the solution and gas-bearing horizons of the enclosing beds. This is particularly true of thin salt beds (for example, in the South Harz region of the German Zechstein) or for strongly tectonically, deformed salt rocks or diapirs. In the latter case, fracturing permits ready penetration of foreign solutions.

All solutions on their way through a salt body reach equilibrium with the salts with which they are in contact, and their chemical peculiarities, for example, the content of characteristic trace elements, are a result of this complicated history. It is thus not always possible to establish the genesis of a solution, and furthermore, solutions of mixed origin are not rare.

BORCHERT (1959) and his school are inclined to attribute solution metamorphism predominantly to the gypsum-anhydrite dehydration solutions. Such a post-diagenetic gypsum-anhydrite dehydration is improbable. In the presence of NaCl-rich salt solutions the transition temperature (Fig. 6) is so low that the persistence of metastable gypsum over geological time is unlikely. From the practical standpoint the most interesting solution metamorphism is the formation of kieserite + sylvite from kieserite + carnallite, which occurs only above 72° C. This is too large a temperature interval to permit the preservation of metastable gypsum. The accumulation of the newly formed solutions in the pores of the anhydrite rock is restricted by its low porosity. This would further involve the assumption that all anhydrite rocks were dehydrated at a correspondingly late stage, thus excluding any possibility of finding even mere traces of potash salts in the surrounding beds, e.g. in the Werra anhydrites and the Hauptanhydrit. Most important of all, the minor components in these brines indicate that they are not dehydration solutions, for their relatively high content of heavy metals, rare elements and particularly gases could not have been derived from the dehydration and dissolution of salts, with the possible exception of small amounts of hydrogen (see SAVCHENKO, 1958) and helium (see BORN, 1934/35). One characteristic is the enrichment in biophilic elements, such as boron and iodine, also strontium and barium, particularly in $CaCl_2$-bearing solutions; these, in fact, are the elements which characterize the most important types of pore solution in sediments (often described as petroleum water or oilfield water). The gases present along with the solutions are probably an important source of energy for fluid migration, as can be directly proved in cases where sufficient data are available the time dependence of the amounts of solution afterflow. This behaviour conforms to the Boyle-Mariotte law (BAAR, pers. comm.), time dependence being seen as a measure of the pressure drop. This can, however, only be related to the expansion of gases[63].

From the preceding discussion it thus seems more correct to look upon most of the solutions which play a part in solution metamorphism as foreign solutions.

[63] For data on the content of gases in salt rocks see, for instance: ACKERMANN, G., SCHRADER, R., HOFFMANN, K.: Untersuchungen an gashaltigen Mineralsalzen, II. Teil: Methodik und Ergebnisse der gasanalytischen Untersuchungen. Bergakademie **16**, 676–679 (1964).

G. Review

In closing the following points may be recapitulated:

1. Based upon the substances they contain, different primary types of marine salt deposits may be distinguished according to the degree of their impoverishment in $MgSO_4$. Evaporites from completely unaltered seawater are unknown. In the present state of knowledge two principal groups can be discerned, namely, deposits in which there is little $MgSO_4$ impoverishment (e.g. Zechstein 1 and Zechstein 2 but only in the center of the basin; the Sarmatian of Sicily) and the highly, often completely $MgSO_4$-deficient deposits (e.g. the Devonian salts of Saskatchewan, Canada; the upper Devonian salts in the Pripyat basin, White Russia; the Permian salts of Solikamsk; the lower Oligocene salts of Alsace). Transition types, up to the present, are known only in minor amounts.

2. In considering the conditions of formation, in all compositional types a distinction must be made between primary precipitates, and those diagenetically and metamorphically alterated forms. Usually metastable solid phases precipitate first, particularly in the Mg and Ca sulphates and borates. In these cases the change from metastable to stable paragenesis occurs in the (early) diagenetic phase. Unfortunately no details are known, apart from the fact that pseudomorphs after gypsum occur. Also, the significance of the reaction of older precipitates with the concentrated brines of later evaporation stages is still not sufficiently clear. The most important case of this kind is the reaction of original kainite to kieserite-carnallite after carnallite saturation has been reached. This probably occurred in the Stassfurt seam in the centre of the basin. Metamorphic processes can frequently be demonstrated. Thermal metamorphism is unimportant and is only to be expected in $MgSO_4$-bearing salt successions. The so-called "Flockensalz" in the Hesse seam (Zechstein 1) is a probable example. Solution metamorphism is very important and of great practical significance as many sylvite deposits, particularly in the German Zechstein salts, originated from the incongruent alteration of carnallite. On account of the great sensitivity of salt deposits to temperature change and infiltrating solutions, most metamorphic reactions occur at crustal depths. In their susceptibility to and in their rate of reaction, evaporites exceed all other rocks (probably including coal).

3. The kieserite-sylvite paragenesis cannot result from primary precipitation. It can occur either through the thermal metamorphism of kainite + carnallite or, more probably, through the solution metamorphism of kieserite-bearing carnallite. The anhydrite-sylvite paragenesis can result from the solution metamorphism of anhydrite-bearing carnallite (as in some examples from the marginal zone of the Stassfurt seam), but in many cases it can be a primary precipitate, namely, in $MgSO_4$-free salt series.

4. Within a given salt basin, primary facies differences are widespread. In this category belong the great differences in the thickness of the anhydrite and halite sediments as between the margin and the centre of the basin. These are probably explained to a great extent by the contrasting temperature dependence of the solubility, $CaSO_4$ (as gypsum) precipitating as the water warms up at the shallower basin margin, and halite as the water cools in the deeper central region of the basin. Even in potash salts primary facies differences can be demonstrated. There exist, particularly in the German Zechstein 2, large areas where there is a compositional trend from the $MgSO_4$-containing basin region to the $MgSO_4$-free marginal zone. However, these primary facies differences in the potash beds are themselves overprinted by massive solution metamorphism which has led to the present-day assemblage of carnallite, sylvite, KCl-poor or KCl-free rocks in such an extraordinarily small area. Only in the carnallite rocks are the primary facies differences still clearly preserved.

5. The reactions of enclosed detritus or of older carbonate sediments (through secondary dolomitization) with saline solutions are of great importance. They bring about (in addition to reactions with $Ca(HCO_3)_2$ solutions) the $MgSO_4$ depletion of the brine. This is also seen in regional differences in the minor components. One indication of oceanic salt deposits is the formation of authigenic Mg silicates. In the German Zechstein basin this is characterized in the marginal regions by the formation of chlorites, not only in the clayey residues but also in large amounts in the gray salt clay (Salzton), and in the centre of the basin (although this has not yet been systematically investigated) the formation of talc and/or koenite. Strontium, at least during chloride precipitation, shows an enrichment in the marginal regions of the basin, while boron concentrates in the centre of the basin. Both elements show a pronounced and regular dependence upon the stage of evaporation. A preferential enrichment of talc is observed in anhydrite rocks. Boron is found both in distinct mineral phases and in anhydrite and clay residues.

6. Among the minor components, bromine permits the most reliable conclusions to be drawn about progressive and recessive salt precipitation, concentration stage, and in favourable circumstances the initial

bromine content of the brine, as well as the evaporation temperature in $MgSO_4$-free salt series. Bromine is also sensitive to secondary alterations, although in this case different kinds of disturbance may lead to the same end result. The bromine content in different examples of primary precipitation or metamorphism can be described by means of simple quantitative models. The original depth of the salt sea cannot, however, be deduced from the bromine content.

7. Thick salt deposits cannot be explained without the influx of seawater. Recessive salt precipitation may also be explained by means of seawater influx.

8. In special cases the composition of salt deposits may be modified by the introduction of material by metamorphic solutions. Under this heading $CaCl_2$-rich solutions probably play an important part by the secondary formation of tachhydrite, and on a smaller scale and in individual cases, by the possible formation of secondary anhydrite through reaction with kieserite. Part of the strontium, boron and other minor components may be introduced later by such solutions or at any rate redistributed. The formation of rinneite always requires the introduction of $FeCl_2$-rich metamorphic solutions.

9. The finely banded rhythmic carbonate and anhydrite beds may represent an annual rhythm, in which case an approximate duration of the sedimentation can be obtained. With coarse rhythms, and especially in the rhythmic anhydrite/halite sequence, a yearly rhythm is questionable and a determination of the duration of sedimentation highly unreliable. In potash salts, an estimation of the length of time of sedimentation is impossible in most cases.

10. The influence of tectonic processes on the composition of salt deposits can only be indicated. It depends not only upon the creation of circulation channels for solutions, but also upon the considerable differences in plasticity which, when there is strong deformation, can cause partial demixing. In contrast, the effect of inhomogeneous pressure distribution on diagenetic recrystallization and salt precipitation in cavities and joints was treated, although only qualitatively.

To these special points a few general remarks must be added:

VAN'T HOFF summarized his research into salt systems in his publications of 1905–1909. At that time the Stassfurt beds were the best known, and were considered the normal type of oceanic salt deposits. Their complex tectonic structure (p. 178) and the extent of metamorphic overprinting had not yet been appreciated. VAN'T HOFF attempted to explain the deposit as purely primary, and all his experimental work was directed to that end. As at 25° C even qualitative differences were found between theory and observation, he also investigated the 83° isotherms. Although the agreement was better, in that kainite was absent and the kieserite-sylvite paragenesis was found, the temperature was unacceptable on general geological grounds, although defenders of the results were not lacking. Particularly in geological works, right up to the present time there has been a tendency to regard primary precipitation as the only salt-forming process and to attribute everything else to mechanical deformations. Yet it was shown a long time ago that the composition of many salt deposits cannot be explained by a single formation process. The first significant step towards the necessary theoretical understanding was taken by JÄNECKE (1923 and earlier) when he introduced thermal metamorphism. This metamorphism is certainly of lesser importance than solution metamorphism which, although long ago recognized qualitatively, was formulated quantitatively by D'ANS and KÜHN only in 1940.

Quite recently BORCHERT (1959) in his book on oceanic salt deposits emphasized and described in detail the many diverse transformations and metamorphic changes which can occur in salt deposits; nevertheless, there seemed some justification for taking up these questions again in the present work. It seemed possible, on the basis of present knowledge of solution equilibria, to make real progress not only towards the quantitative understanding of primary precipitation but also of metamorphism. This was the primary objective of the present book, involving the construction of models which could be directly compared with observations made in nature.

Of equal importance was the testing of the models against natural occurrences. To this end it was necessary to collect and examine quantitative data on the composition of salt deposits. There is certainly a great deal of data about the potash deposits themselves, but what is required for the present purpose is a continuous series of samples through the whole salt succession. Unfortunately, in this respect the study of salt deposits is very backward when compared with the petrography of igneous rocks, and there exists a shocking disparity between theoretical knowledge and the quantitative compositional survey of the salt deposits themselves. Most of the data on the composition of salt deposits come from literature published between 1910 and 1930! The only important new data published in recent years concern the minor components, in particular bromine and strontium. This is only significant in the context of complete compositional data. Today we can and must demand of geological investigations, insofar as they have pretensions to pose and answer questions concerning the genesis of salt deposits, adequate documentation of their qualitative and quantitative composition and, over and above this, an understanding of the underlying physico-chemical principles. When all is said and done, salt deposits are chemical sediments and they can only be reliably interpreted upon the basis of chemical principles. This is true despite petrographic conclusions which in many cases, e.g. with respect to secondary carnallitization, are in open contradiction to chemical principles. It is, however, clear that the contradictions are due to the sheer inadequacy of the criteria for determining age in a crystallization succession.

The comparison of models and observations leads in most cases to qualitative, and in many cases also to passable quantitative agreement. It is abundantly clear that a great deal of work, both theoretical and experimental, remains to be done on the petrographic as well as the chemical aspects if the objective, a quantitative derivation of the history of the formation of salt deposits, is to be satisfactorily achieved.

The study of salt deposits is not an isolated chapter in the study of rocks. The same general principles which apply to the formation of magmatic and metamorphic rocks are valid here, in particular, a knowledge of chemical equilibria, the phase rule, the principles of thermodynamics, Nernst's partition law, etc. Salt deposits form under rather simple and easily understood conditions, and they thus form useful models to approach the much more difficult and complex problems of the formation of magmatic and metamorphic rocks, as RINNE clearly realized more than half a century ago. Indicative of this is the fact that BOEKE, stimulated by the pioneer work of VAN'T HOFF and building upon his research into salt deposits, went on to become one of the founders of physico-chemical methods of investigation and hence of the exact scientific treatment of rock formation.

References

ABRAMOVICH, YU. M., NECHAYEV, YU. A.: Authigenic fluorite in Kungurian deposits of the permian Preurals. Doklady Akad. Nauk S.S.S.R. **135**, 414—415 (1960); (Engl. Transl. p. 1288—1289).

ADAMS, L. H.: Equilibrium in binary systems under pressure. I. An experimental and thermodynamic investigation of the system NaCl–H_2O at 25°. J. Am. Chem. Soc. **53**, 3769—3813 (1931).

— GIBSON, R. E.: The melting curve of sodium chloride dihydrate. An experimental study of an incongruent melting at pressures up to twelve thousand atmospheres. J. Am. Chem. Soc. **52**, 4252—4264 (1930).

AHLBORN, O.: Die Flöze „Thüringen" und „Hessen" der Werraserie und ihre wechselseitigen Beziehungen. Z. deut. geol. Ges. **105**, 664—673 (1953).

ALDERMANN, A. R., BORCH, C. C. v. D.: Occurence of Magnesite-Dolomite Sediments in South Australia. Nature **192**, Nr. 4805, 861 (1961).

— SKINNER, H. C. W.: Dolomite sedimentation in the south-east of South Australia. Am. J. Sci. **255**, 561—567 (1957).

D'ANS, J.: Untersuchungen über die Salzsysteme ozeanischer Salzablagerungen. Experimentell bearbeitet mit A. BERTSCH u. A. GESSNER. Kali **9**, 148, 161, 177, 193, 217, 229, 245, 261 (1915).

— Die Lösungsgleichgewichte der Systeme der Salze ozeanischer Salzablagerungen. 254 S. Kali-Forschungsanstalt. Berlin: Verl. Ges. f. Ackerbau 1933.

— Ein neues Syngenitvorkommen (vorl. Mitt.). Kali **38**, 206 (1944).

— Über die Bildung und Umbildung der Kalilagerstätten. Naturwissenschaften **34**, 295—301 (1947).

— Über die Bildungsmöglichkeiten des Tachhydrits in Kalisalzlagerstätten. Kali u. Steinsalz **3**, 119—125 (1961).

— Über die Entwässerung der Magnesiumchloridhydrate und des Carnallites. Kali u. Steinsalz **3**, 126—129 (1961).

— AUTENRIETH, H., BRAUNE, G., FREUND, H. E.: Nonvariante Temperaturen mit drei Hydraten eines Salzes als Bodenkörper. Bestimmung d. unt. Bildungstemp. von Magnesiumsulfathydraten. Z. Elektrochem. **64**, 521—525 (1960).

— BEHRENDT, K. H.: Über die Existenzbedingungen einiger Magnesiumborate. Kali u. Steinsalz **2**, 121—137 (1957).

— BRETSCHNEIDER, D., EICK, H., FREUND, H. E.: Untersuchungen über die Calciumsulfate. Kali u. Steinsalz **1**, Heft 9, 17—38 (1955).

— FREUND, H. E.: Versuche zur geochemischen Rinneitbildung. Kali u. Steinsalz **1**, Heft 6, 3—9 (1954).

— FREUND, H. E., KAUFMANN, E. K.: Über binäre Systeme mit einem Quadrupelpunkt mit drei festen binären Verbindungen und einer Gasphase. Z. Elektrochem. **61**, 546—549 (1957).

— KÜHN, R.: Getrübter Sylvin. Kali **32**, 152—155 (1938).

— — Über den Bromgehalt von Salzgesteinen der Kalisalzlagerstätten. Kali **34**, 42, 59, 77 (1940).

— — Bemerkungen zur Bildung und zu Umbildungen ozeaner Salzlagerstätten. Kali u. Steinsalz **3**, 69—84 (1960).

ARMSTRONG, G., DUNHAM, K. C., HARVEY, C. O., SABINE, P. A., WATERS, F.: The paragenesis of sylvine, carnallite, polyhalite and kieserite in Eskdale borings nos. 3, 4 and 6, north-east Yorksh. Mineral. Mag. **29**, 667–689 (1951).

ARRHENIUS, S.: Über die physikalischen Bedingungen bei den Salzablagerungen zur Zeit ihrer Bildung und Entwicklung. Kali **6**, 361—365 (1912).

ASSARSSON, G.: Equilibria in aqueous systems containing K^+, Na^+, Ca^{+2}, Mg^{+2} and Cl^-. J. Am. Chem. Soc. **72**, 1433, 1437, 1442 (1950).

— Kristallisationserscheinungen und Paragenese in den Systemen der Alkalichloride-Erdalkalichloride-Wasser. Sveriges Geol. Undersökn. **556**, Ser. C, 1—17 (1957).

— BALDER, A.: The poly-component aqueous systems containing the chlorides of Ca^{2+}, Mg^{2+}, Sr^{2+}, K^+ and Na^+ between $18+93°$. J. Phys. Chem. **59**, 631 to 633 (1955).

AUTENRIETH, H.: Neuere Entwicklungen auf dem Gebiet d. graph. und rechnerischen Behandlung der Vorgänge bei der Gewinnung von Kalidüngemitteln aus Rohsalzen. Kali u. Steinsalz **1**, H. 2, 3—17 (1953).

— Die stabilen und metastabilen Gleichgewichte des reziproken Salzpaars $K_2Cl_2 + MgSO_4 \rightleftharpoons K_2SO_4 + MgCl_2$ ohne und mit NaCl als Bodenkörper, sowie ihre Anwendung in der Praxis. Kali u. Steinsalz **1**, H. 7, 3—22 (1954).

— Neue, für die Kalirohsalzverarbeitung wichtige Untersuchungen am quinären, NaCl-gesättigten System der Salze ozeanischer Salzablagerungen. Kali u. Steinsalz **1**, H. 11, 18—32 (1955).

— Untersuchungen am Sechs-Komponenten-System $K^{\cdot}$, $Na^{\cdot}$, $Mg^{\cdot\cdot}$, $Ca^{\cdot\cdot}$, SO_4'', (Cl'), H_2O mit Schlußfolgerungen für die Verarbeitung der Kalisalze. Kali u. Steinsalz **2**, 181—200 (1958).

— BRAUNE, G.: Ein neues Salzmineral, seine Eigenschaften, sein Auftreten und seine Existenzbedingungen im System der Salze ozeanischer Salzablagerungen. Naturwissenschaften **45**, 362—363 (1958).

— — Das Sechskomponentensystem $K^{\cdot}$, $Na^{\cdot}$, $Mg^{\cdot\cdot}$, $Ca^{\cdot\cdot}$, SO_4'', (Cl') H_2O bei 90° C und seine Anwendung auf Schlammprobleme der Kalisalzverarbeitung. Kali u. Steinsalz **2**, 395—405 (1959).

— — Die Lösungsgleichgewichte des reziproken Salzpaares $Na_2Cl_2 + MgSO_4 + H_2O$ bei Sättigung an NaCl unter besonderer Berücksichtigung des metastabilen Bereichs. Kali u. Steinsalz **3**, 15—30 (1960a).

— — Weitere Untersuchungen im stabilen und metastabilen Gebiet des reziproken Salzpaares $Na_2Cl_2 + MgSO_4 + H_2O$ bei Sättigung an NaCl. Kali u. Steinsalz **3**, 85—97 (1960b).

BAAR, A.: Entstehung und Gesetzmäßigkeiten der Facieswechsel im älteren Kalilager am westlichen Südharz unter besonderer Berücksichtigung des Kaliwerkes Bismarckshall. Kali **38**, 175, 189, 207 (1944); **39**, 3 (1945).

— Entstehung und Gesetzmäßigkeiten der Facieswechsel im Kalilager am Südharz (II). Bergakademie **4**, 138—150 (1952).

— Untersuchungen des Bromgehalts im Zechsteinsalz. Bergbautechnik **4**, 284 bis 288 (1954).

— Über die facielle Entwicklung der Kalilagerstätte des Staßfurtflözes. Neues Jahrb. Geol. Paläontol., Abh. **111**, 111—135 (1960).

— Über gleichartige Gebirgsverformungen durch bergmännischen Abbau von Kaliflözen bzw. durch chemische Umbildung von Kaliflözen in geologischer Vergangenheit. Freiberger Forschungsh. A **123**, 137—157 (1958).

— KÜHN, R.: Der Werdegang der Kalisalzlagerstätten am Oberrhein. Neues Jahrb. Mineral. Abhandl. **97**, 289—336 (1962).

BAILEY, R. K.: Talc in the salines of the potash-field near Carlsbad, Eddy-County. Am. Mineralogist **34**, 757—759 (1949).
BALK, R.: Structure of grand saline salt dome, Van Zandt County, Texas. Bull. Am. Assoc. Petrol. Geologists **33**, S. 1791 (1949).
BARTH, T. F. W., CORRENS, C. W., ESKOLA, P.: Die Entstehung der Gesteine. 422 S. Berlin-Göttingen-Heidelberg: Springer-Verlag 1939.
BAUMERT, B.: Über Laugen- und Wasserzuflüsse im deutschen Kalibergbau. Diss. Aachen, 90 S. (1928).
BAUR, F.: Die wissenschaftlichen Grundlagen und Probleme der langfristigen Witterungsvorhersage. Naturwissenschaften **48**, 61—66 (1961).
BEEVERS, C. A., STEWART, F. H.: p-Veatchite from Yorkshire. Mineral. Mag. **32**, 500—501 (1960).
BEHNE, W.: Untersuchungen zur Geochemie des Chlor und Brom. Geochim. et Cosmochim. Acta **3**, 186—214 (1953).
BENTOR, Y. K.: Some geochemical aspects of the Dead Sea and the question of its age. Geochim. et Cosmochim. Acta **25**, 239—260 (1961).
BESSERT, F.: Geologisch-petrographische Untersuchungen der Kalilager des Werragebietes. Arch. Lagerstättenforsch. **57**, 1—45 (1933).
BILTZ, W., MARCUS, E.: Über die chemische Zusammensetzung der Staßfurter Salztone. Z. anorg. Chem. **68**, 91—101 (1910).
— — Über die Verbreitung von borsauren Salzen in den Kalisalzlagerstätten. Z. anorg. Chemie **72**, 302—312 (1911).
— — Über die chemische Zusammensetzung des roten Salztones. Z. anorg. Chemie **77**, 119—123 (1912).
BLOCH, M. R., SCHNERB, J.: On the Cl^-/Br^--ratio and the distribution of Br-Ions in liquids and solids during evaporation of bromide-containing chloride solutions. Bull. Research Council Israel **3**, 151—158 (1953/54).
BOEKE, H. E.: Über das Kristallisationsschema der Chloride, Bromide, Jodide von Natrium, Kalium und Magnesium, sowie über das Vorkommen des Broms und das Fehlen von Jod in den Kalisalzlagerstätten. Z. Krist. **45**, 346—391 (1908).
— Über die Borate der Kalisalzlagerstätten. Centr. Mineral., 531—539 (1910a).
— Übersicht der Mineralogie, Petrographie und Geologie der Kalisalzlagerstätten. Illustr. Jb. Wirtsch. u. Technik im deutsch. Kalisalz-Bergbau. Herausg. A. STANGE Berlin 1910b.
— Ein Schlüssel zur Beurteilung des Kristallisationsverlaufes der bei der Kalisalzverarbeitung vorkommenden Lösungen. Kali **4**, 271—284, 300—307 (1910c).
— Über die Eisensalze in den Kalisalzlagerstätten. Neues Jahrb. Mineral. Geol. **1911 I**, 48—76.
BORCHERT, H.: Die Vertaubungen der Salzlagerstätten und ihre Ursachen. I. Kali **27**, 97, 105, 124, 139, 148 (1933).
— Die Vertaubungen der Salzlagerstätten und ihre Ursachen II: Das dynamisch-polytherme System der Salze der ozeanen Salzablagerungen. Kali **28**, 290, 301 (1934); **29**, 1 (1935).
— Die Metamorphose ozeaner Salzablagerungen. Kali **32**, 132, 143, 169 (1938).
— Die Salzlagerstätten des deutschen Zechsteins. Arch. Lagerstättenforsch. H. **67**, 196 S. (1940).
— Ozeane Salzlagerstätten. 237 S. Berlin: Gebr. Bornträger 1959.
— Deszendenz und synsedimentäre Senkung oder primäre Lagerbildung und Metamorphose bei den Kalilagerstätten des Oberrheintal-Grabens. Neues Jahrb. Mineral. Abh. **94**, 636—661 (1960).
— BAIER, E.: Zur Metamorphose ozeaner Gipsablagerungen. Neues Jahrb. Mineral. Abh. **86**, 103—154 (1954).

BORCHERT, W.: Zusammenhang zwischen Verfärbung und Luminiszenz bei Bestrahlung von Steinsalz. Kali u. Steinsalz **2**, 223—224 (1958).

BORN, H. J.: Der Bleigehalt der Norddeutschen Salzlager und seine Beziehungen zu radioaktiven Fragen. Chemie der Erde **9**, 66—87 (1934/35).

— Zur Frage der geochemischen Folgerungen aus den Hahnschen Arbeiten über Mitfällungen. Beiträge zur Physik u. Chemie des 20. Jahrhunderts S. 130—134. Braunschweig: Fr. Vieweg 1959.

BRADLEY, W. F., WEAVER, C. E.: A regularly interstratified chlorite-vermiculite clay mineral. Am. Mineralogist **41**, 497—504 (1956).

BRAITSCH, O.: Über den Mineralbestand der wasserunlöslichen Rückstände von Salzen der Staßfurtserie im südlichen Leinetal. Freiberger Forschungsh. A **123**, 160—163 (1958).

— 1 Tc-Strontiohilgardit $(Ca, Sr)_2$ $[B_5O_8(OH)_2Cl]$ und seine Stellung in der Hilgarditgruppe X_2^{II} $[B_5O_8(OH)_2Cl]$. Beitr. Mineral. u. Petrogr. **6**, 233—247 (1959a).

— Über p-Veatchit, eine neue Veatchit-Varietät aus dem Zechsteinsalz. Beitr. Mineral. u. Petrogr. **6**, 352—356 (1959b).

— Über Strontioginorit, eine neue Ginorit-Varietät aus dem Zechsteinsalz. Beitr. Mineral. u. Petrogr. **6**, 366—370 (1959c).

— Mineralparagenesis und Petrologie der Staßfurtsalze in Reyershausen. Kali u. Steinsalz **3**, 1—14 (1960).

— Neue Daten für Lüneburgit und Sulfoborit. Beitr. Mineral. u. Petrogr. **8**, 60—66 (1961a).

— Ein einfacher Vorlesungsversuch des inkongruenten Schmelzens. Beitr. Mineral. u. Petrogr. **8**, 67—68 (1961b).

— Zur Entstehung der Staßfurter Salzfolge. Naturwissenschaften **48**, 402 (1961c).

— Zur Entstehung der Schichtung in rhythmisch geschichteten Evaporiten. Geol. Rundschau **52**, 405—417 (1962).

— HERRMANN, A. G.: Zur Bromverteilung in salinaren Salzsystemen bei 25° C. Naturwissenschaften **49**, 346 (1962).

— — Zur Geochemie des Broms in salinaren Sedimenten. Teil I.: Experimentelle Bestimmung der Br-Verteilung in verschiedenen natürlichen Salzsystemen. Geochim. et Cosmochim. Acta **27**, 361—391 (1963).

BREDIG, M. A.: Isomorphism and allotropism in compounds of the type A_2XO_4. J. Phys. Chem. **46**, 746—764 (1942).

BÜCKING, H.: Sulfoborit, ein neues kristallisiertes Borat von Westeregeln. S.-B. Akad. Wiss. Berlin **1893**, 967.

BUDZINSKI, H., LANGBEIN, R., STOLLE, E.: Danburit im Zechstein des Südharzes. Chemie d. Erde **20**, 53—70 (1959).

BURKSER, E. S., PONIZOVSKIJ, A. M., MELESHKO, E. P.: Brom in Salzwässern der Krim, am Schwarzen und Asowschen Meer. Akad. Nauk Ukr. S.S.R., Kiev, 152—160 (1958).

CLABAUGH, P. S.: Petrofabric Study of Deformed Salt. Science **136**, 389—391 (1962).

CONLEY, R. F., BUNDY, W. M.: Mechanism of gypsification. Geochim. et Cosmochim. Acta **15**, 57—72 (1958).

COOPER, L. H. N.: Oxidation-reduction potential of seawater. J. Marine Biol. Assoc. United Kingdom **22**, 167—176 (1937).

CORRENS, C. W.: Introduction to Mineralogy (Crystallography and Petrology) 484 pages. Berlin-Heidelberg-New York: Springer 1969.

— Zur Geochemie der Diagnose. Geochim. et Cosmochim. Acta **1**, 49—54 (1950).

— The geochemistry of the halogens. Phys. and Chem. Earth **1**, 181—233 (1956).

CORRENS, C. W.: Buchbesprechung: H. BORCHERT, Ozeane Salzlagerstätten. Neues Jahrb. Mineral. Mh., 190—192 (1960).

— STEINBORN, W.: Experimente zur Messung und Erklärung der sogenannten Kristallisationskraft. Z. Krist. **101**, 117—133 (1939).

DALY, R. A.: First calcareous fossils and the evolution of the limestones. Bull. Geol. Soc. Am. **20**, 153 (1909).

DANA's System of Mineralogy. 7. Aufl. Bd. II, 1124 S. New York: Wiley & Sons 1951.

DOMMERICH, S.: Festigkeitseigenschaften bewässerter Salzkristalle VI. Z. Physik **90**, 189—196 (1934).

DREIZLER, I.: Mineralogische Untersuchungen zweier Gipsvorkommen aus der Umgebung von Göttingen. Beitr. Mineral. Petrogr. **8**, 323—338 (1962).

DROSTE, J. B.: Clay minerals in sediments of Owens, China, Searles, Panamint, Bristol, Cadiz and Danby Lake basins, Calif. Geol. Soc. Am. Bull. **72**, 1713—1722 (1961).

DUBININA, W. N.: Zur Mineralogie und Petrographie des Kalisalz-Lagerstättenbezirkes an der oberen Kama. Trudy Vsesojuz. Nauch.-Issledovatel. Inst. Galurgii S.S.S.R. **29**, 3—128 (1954); – Ref.: A. G. HERRMANN, „Geologie" **7**, 1093—1097 (1958).

DZENS-LITOWSKIJ: Die hydrogeologischen Bedingungen zur Bildung von Salzlagerungen auf Eis. Trudy Vsesojuz. Nauch.-Issledovatel. Inst. Galurgii **30**, 224—252 (1955); — Ref.: Montanwissenschaftl. Literaturberichte Geo 58/751.

ECHLE, W.: Mineralogische Untersuchungen an Sedimenten des Steinmergelkeupers und der Roten Wand aus der Umgebung von Göttingen. Beitr. Mineral. u. Petrogr. **8**, 28—59 (1961).

ECKHARDT, F.-J.: Über Chlorite in Sedimenten. Geol. Jahrb. **75**, 437—474 (1958).

ENGELHARDT, W. v.: Kreislauf und Entwicklung in der Geschichte der Erdrinde. Nova acta leopoldina. Abhandl. Akad. Naturforscher Leopoldina **143**, 83—99 (1959).

— Der Porenraum der Sedimente. 207 S. Berlin-Göttingen-Heidelberg: Springer 1960.

— FÜCHTBAUER, H., ZEMANN, J.: Heidornit $Na_2Ca_3[Cl(SO_4)_2B_5O_8(OH)_2]$ ein neues Bormineral aus dem Zechsteinanhydrit. Beitr. Mineral. u. Petrogr. **5**, 177—186 (1956).

— GAIDA, K. H.: Concentration changes of pore solutions during the compaction of clay sediments. J. Sedimentary Petrology **33**, 919—930 (1963).

ERSCHLER: Über die Grenzen der Anwendung der Ultrafiltration zum Nachweis des kolloiden Zustandes. Kolloid-Z. **68**, 289—298 (1934).

EVERDING, H.: Zur Geologie der deutschen Zechsteinsalze. Deutschl. Kalibergbau, Kgl. Geol. L.-A. Berlin, 25—133 (1907), (Festschr. X. Allg. Bergmannstag Eisenach).

FABIAN, H.-J., GAERTNER, H., MÜLLER, G.: Zwei neue Funde von Danburit im Zechstein Nordwestdeutschlands. Erdöl u. Kohle **14**, 597—599 (1961).

FINDLAY, A.: Die Phasenregel und ihre Anwendungen. 368 S. Weinheim: Verl. Chemie 1958.

FIVEG, M. P.: Über die Bildungsdauer der Salzablagerungen. Trudy Vsesojuz. Nauch.-Issledovatel. Inst. Galurgii **29**, 341—350 (1954).

— Bedingungen für die Entstehung der Kalisalz-Lagerstätten der UdSSR. Vortrag Kali-Symposium Merseburg, Okt. 1961.

FLINT, R. F., GALE, W. A.: Stratigraphy and radiocarbon dates at Searles Lake, California. Am. J. Sci. **256**, 689—714 (1958).

FOURNIER, R. O.: Regular interlayered chlorite-vermiculite in evaporite of the salado formation, New Mexico. U.S. Geol. Survey, Prof. Paper 424-D, 323—327 (1961).

FREDERICKSON, A. F., REYNOLDS, R. C., JR.: Geochemical method for determining paleosalinity. Clay and Clay Minerals, Proc. 8. nat. Conf. 203—213 (1960). Pergamon Press 1960.

FRIEDRICH, K.: Gefüge und Tektonik im Hartsalz des Werragebietes. Z. deut. geol. Ges. **111**, 502—524 (1959).

FÜCHTBAUER, H.: Die petrographische Unterscheidung der Zechsteindolomite im Emsland durch ihren Säurerückstand. Erdöl u. Kohle **11**, 689—693 (1958).

— GOLDSCHMIDT, H.: Ein Zechsteinanhydrit-Profil mit Einlagerungen von Montmorillonit und einer abweichenden Serpentinvarietät. Heidelberger Beitr. Mineral. u. Petrogr. **5**, 187—203 (1956).

— — Die Tonminerale der Zechsteinformation. Beitr. Mineral. u. Petrogr. **6**, 320—345 (1959).

FULDA, E.: In: Erläuterungen z. Geol. Karte von Preußen, Blatt Vienenburg Nr. 2231, 2. Aufl. Berlin 1931.

— Handbuch der vergleichenden Stratigraphie Deutschlands. Zechstein. 409 S. Berlin: Bornträger 1935.

GAERTNER, H. R. v.: Petrographie und paläogeographische Stellung der Gipse vom Südrande des Harzes. Jahrb. preuß. geol. Landesanstalt (Berlin) **53**, 655—694 (1932).

GAERTNER, H., ROESE, K.-L., KÜHN, R.: Fabianit = $CaB_3O_5(OH)$, ein neues Mineral. Naturwissenschaften **49**, 230 (1962).

GAHM, J., NACKEN, R.: Skelettkristallbildung bei den Alkalihalogeniden, bes. beim Steinsalz. Neues Jahrb. Mineral. Abh. **86**, 309—366 (1954).

GARRELS, R. M., THOMPSON, M. E.: A chemical model for sea water at 25° C and one atmosphere total pressure. Am. J. Sci. **260**, 57—66 (1962).

GATTOW, G., ZEMANN, J.: Über Doppelsulfate vom Langbeinit-Typ $A_2^+B^{2+}(SO_4)_3$. Z. anorg. allgem. Chemie **293**, 233—240 (1958).

GLEMSER, O.: Ergebnisse und Probleme von Verbindungen der Systeme Oxyd-Wasser. Angew. Chem. **73**, 785—805 (1961).

GLÖCKNER, F.: Ein Vorkommen von Kupferkies in Kalisalzen. Kali **8**, 307—308 (1914).

GODLEVSKY, M. N.: Mineralogical investigation of the Inder borate deposits. Zapiski Vsesoyuz. Mineral. Obshchestva **66**, 345—368 (1937).

GÖRGEY, R.: Salzvorkommen aus Hall in Tirol. Tschermak's mineral. petrog. Mitt. **28**, 334—346 (1909).

— Zur Kenntnis der Kalisalzlager von Wittelsheim im Ober-Elsaß. Tschermak's mineral. petrog. Mitt. **31**, 339—468 (1912).

— Über die Kristallform des Polyhalit. Tschermak's mineral. petrog. Mitt. **33**, 48—102 (1915).

GOTTESMANN, W.: Zur Trümmercarnallitbildung auf Menzengraben (Rhön). Geologie **11**, 51—82 (1962) [Berlin].

GRAF, D. L., EARDLEY, A. J., SHIMP, N. F.: A preliminary report on magnesium carbonate formation in glacial lake Boneville. J. Geol. **69**, 219—223 (1961).

GREENWOOD, N. N., THOMPSON, A.: The reactions of boron halides with anhydrous sulfuric acid: Boron tri(hydrogen sulphate) and tetra(hydrogen sulphato) boric acid. J. Chem. Soc. (London) p. 3643—3645 (1959).

GUNZERT, G.: Die „obere bituminöse Zone“ im Bereich der Kalisalzlagerstätte von Buggingen (Baden). Kali u. Steinsalz **3**, 111—118 (1961).

GUNZERT, G.: Zur Frage des Diapirismus in der tertiären Salzlagerstätte des südlichen Oberrheintals. Neues Jahrb. Geol. Paläont. Abh. **116**, 69—88 (1962).
HACKER, W.: Die Konzentrationsänderung in Elektrolytlösungen bei der Filtration durch Kollodiummembranen. Kolloid-Z. **94**, 11—29 (1941).
HAHN, O.: Über Blei und Helium in ozeanischen Alkalihalogeniden. Naturwissenschaften **20**, 86 (1932).
— Applied Radiochemistry. Ithaca: Cornell Univ. Press 1936.
HAM, W. E.: Middle Permian evaporites in southwestern Oklahoma. Rept. Intern. Geol. Congr. XXI, part XII, 138—151, Copenhagen (1960).
— MANKIN, C. J., SCHLEICHER, J. A.: Borate minerals in permian gypsum of West-Central Oklahoma. Oklahoma Geol. Survey, Bull. **92**, 77. S. (1961).
HARBORT, E.: Über zonar in Steinsalz und Kainit eingewachsene Magnetkieskristalle aus dem Kalisalzbergwerk Aller-Nordstern. Kali **9**, 250—253 (1915).
— Über Zirklerit. Kali **22**, 157—161 (1928).
HARDER, H.: Beitrag zur Geochemie des Bors, Teil II: Bor in Sedimenten. Nachr. Akad. Wiss. Göttingen II math.-naturw. Kl. Nr. **6**, 123—183 (1959).
— Einbau von Bor in detritische Tonminerale. Geochim. Cosmochim. Acta **21**, 284—294 (1961).
HARDING, S. T.: Evaporation from free water surfaces. In: Physics of the earth **9** (Hydrology) New York. 56—82 (1949). Herausg. O. E. MEINZER.
HARTWIG, G.: Zur Petrographie und Transversalschieferung der tieferen Stufen der Zechstein-Großfolge 2 im Untergrund von Solling-Elfas und Dün-Hainleite-Eck mit Ausblicken auf die Verhältnisse unter der östlichen Randhochfläche des Göttinger Leinetals. Kali u. Steinsalz **1**, H. 8, 8—29 (1955).
HEALD, M. T.: Cementation of Simpson and St. Peter sandstones in parts of Oklahoma, Arkansas and Missouri. J. Geol. **64**, 16—30 (1956).
HEIDE, F., WALTER, G., URLAU, R.: Zur Kristallchemie des Borazits. Naturwissenschaften **48**, 97—98 (1961).
HEIDORN, F.: Über ein Vorkommen von Sellait (MgF_2) in Paragenese mit Bitumen aus dem Hauptdolomit des mittleren Zechsteins bei Bleicherode. Centr. Mineral. **1932**, 356—364.
HENTSCHEL, J.: Die Faciesunterschiede im Kaliflöz Staßfurt im Kalisalzbergwerk „Königshall-Hindenburg“ bei Nörten-Hardenberg. Diss. Mainz 1958 [gekürzt in Kali u. Steinsalz **3**, 137—157 (1961)].
HERRMANN, A. G.: Geochemische Untersuchungen an Kalisalzlagerstätten im Südharz. Freiberger Forschungsh. **C 43** (Berlin) 1—112 (1958).
— Zur Geochemie des Strontiums in den salinaren Zechsteinablagerungen der Staßfurt-Serie des Südharzbezirkes. Chemie der Erde **21**, 137—194 (1961a).
— Über das Vorkommen einiger Spurenelemente in Salzlösungen aus dem deutschen Zechstein. Kali u. Steinsalz **3**, 209—220 (1961b).
— Über die Einwirkung Cu-, Sn-, Pb- und Mn-haltiger Erdölwässer auf die Staßfurt-Serie des Südharzbezirkes. Neues Jahrb. Mineral. Mh. **1961c**, 60—67.
— Eine quantitative Bestimmung des in Tachhydrit diadoch eingebauten Strontiums. Neues Jahrb. Mineral. Mh. **1961d**, 141—143.
— HOFFMANN, R. O.: Zur Genese einiger Borate in den Salzablagerungen der Staßfurt-Serie des Südharzbezirkes einschließlich der Grube Königshall-Hindenburg. Neues Jahrb. Mineral. Mh. **1961**, 52—60.
HERRMANN, R.: Die Entstehung der Verdoppelungen des Kalilagers im Gebiet von Staßfurt und Bernburg. Z. prakt. Geol. **47**, 150ff. (1939).
HILMY, M. E.: Structural crystallographic relation between sodium sulfate and potassium sulfate and some other synthetic sulfate minerals. Am. Mineralogist **38**, 118—135 (1953).

VAN'T HOFF, J. H.: Zur Bildung der ozeanischen Salzablagerungen. 85 u. 90 S. Braunschweig: Fr. Vieweg 1905, 1909.
— Untersuchungen über die Bildungsverhältnisse der ozeanischen Salzablagerungen. 374 S. Leipzig: Akad. Verl. Ges. 1912; Herausg.: PRECHT u. COHEN. (Nachdruck der i. d. S.-B. Kgl. Preuß. Akad. veröff. Originalarbeiten von 1897—1908.)
HOFFMANN, R. O.: Die Mineralzusammensetzung der in Wasser schwer löslichen Rückstände von Filterschlämmen und Rohsalzen einiger mitteldeutscher Kaliwerke. Bergakademie **13**, 237—248 (1961).
HOPPE, W.: Die Bedeutung der geologischen Vorgänge bei der Metamorphose der Werra-Kalisalzlagerstätte. Freiberger Forschungsh. **A 123**, 41—60 (1958).
HOWARD, C. L. H., KERR, P. F.: Blue halite. Science **132**, 1886—1887 (1960).
HURLBUT, C. S., JR.: Parahilgardite, a new tricline-pedial mineral. Am. Mineralogist **23**, 765—771 (1938).
— TAYLOR, R. E.: Hilgardite, a new mineral species from Choctaw salt dome, Louisiana. Am. Mineralogist **22**, 1052—1057 (1937).
ITO, T., MORIMOTO, N., SANDANAGA, R.: The crystal structure of boracite. Acta Cryst. **4**, 310—316 (1951).
JÄNECKE, E.: Gesättigte Salzlösungen vom Standpunkt der Phasenlehre. Halle a. d. Saale: W. Knapp 1908.
— Eine graphische Darstellung der Gewichtsverhältnisse bei den ozeanischen Salzablagerungen. Kali **6**, 255—258 (1912).
— Die Entstehung der deutschen Kalisalzlager. 2. Aufl. 111 S. Braunschweig: Fr. Vieweg 1923.
DE JONG, W. F., BOUMAN, J.: Das reziproke und das Bravaissche Gitter von Gips. Z. Krist. **100**, 275—276 (1938).
JOHNSEN, A.: Beiträge zur Kenntnis der Salzlager, I. Centr. Mineral. Geol. Paläont. **1909**, 168—173.
JONES, C. L.: The occurence and distribution of potassium minerals in Southeastern New Mexico. Guidebook of Southeastern New Mexico, 5th Field Conf. New Mexico Geol. Soc., Oct. 21—24 S. 107—112 (1954).
JUNG, W.: Zur Feinstratigraphie der Werraanhydrite (Zechstein 1) im Bereich der Sangerhäuser und Mansfelder Mulde. Geologie (Berlin) **7**, Beih. 24, 1—88 (1958).
— Das Steinsalzäquivalent des Zechstein 1 in der Sangerhäuser und Mansfelder Mulde und daraus resultierende Bemerkungen zum Problem der „Jahresringe". Ber. geol. Ges. (Berlin) **4**, 313—325 (1959).
— Zur Feingliederung des Basalanhydrits (Z 2) und des Hauptanhydrites (Z 3) im SE-Harzvorland. Geologie (Berlin) **9**, 526—555 (1960).
— KNITZSCHKE, G.: Kombiniert-feinstratigraphisch-geochemische Untersuchungen der Anhydrite des Zechstein 1 im SE-Harzvorland. Geologie (Berlin) **9**, 58—72 (1960).
— — Kombiniert feinstratigraphisch-geochemische Untersuchungen des Basalanhydrits (Z 2) und des Hauptanhydrits im SE-Harzvorland. Geologie (Berlin) **10**, 288—301 (1961).
KARLIK, B.: Der He-Gehalt von Steinsalz und Sylvin. Mikrochem. Mikrochim. Acta **27**, 216ff. (1939).
KARSTEN, O.: Diagramme der Lösungsgleichgewichte des quinären Systems Na_2Cl_2, K_2Cl_2, $MgCl_2$, $MgSO_4$ u. H_2O unter bes. Berücksichtigung metastabiler Zustände. Z. anorg. Chemie **263**, 292—304 (1950).
KAUFMANN, D. W., SLAWSON, C. B.: Ripple mark in rock salt of the salina formation. J. Geol. **58**, 24—29 (1950).

KELLEY, K. K., SOUTHARD, J. C., ANDERSON, C. T.: Thermodynamic properties of gypsum and its dehydration products. U.S. Bur. Mines Tech. Papers **625** (1941).

KIPPER: Die Zechsteinformation zwischen dem Diemel- und Ittertale am Ostrande des rheinisch-westfälischen Schiefergebirges unter besonderer Berücksichtigung der Kupfer-, Gips-, Eisen-, Mangan-, Zink-, Blei-, Cölestin- und Schwerspat-Vorkommen. Glückauf **44**, 1029—1036, 1065—1075, 1101—1110, 1137—1148 (1908).

KLAUS, W.: Über die Sporendiagnose des deutschen Zechsteins und des alpinen Salzgebirges. Z. deut. geol. Ges. **105**, 776—788 (1953).

KLING, P.: Tachhydritvorkommen in den Kalisalzlagerstätten der Mansfelder Mulde. Centr. Mineral. Geol. Paläont. **1915**, 11 und 44.

KNAK, I.: Feinstratigraphische Aufnahme des Kalilagers Staßfurt auf den Schachtanlagen Neustaßfurt VI/VII, Berlepsch-Maybach und Ludwig II. Freiberger Forschungsh. **C 90**, 7—51 (1960).

KOKORSCH, R.: Zur Kenntnis von Genesis, Metamorphose und Faciesverhältnissen des Staßfurtlagers im Grubenfeld Hildesia-Mathildenhall, Diekholzen bei Hildesheim. Beih. Geol. Jahrb. **41**, 140 S. (1960).

KORITNIG, S.: Ein Beitrag zur Geochemie des Fluor. Geochim. Cosmochim. Acta **1**, 89—116 (1951).

KORSHINSKY, D. S.: Der Filtrationseffekt in Lösungen und seine geologische Bedeutung. Izvest. Akad. Nauk S.S.S.R. Ser. geol. Nr. **2**, 35—48 (1947).

KREJCI-GRAF, K.: Über Ölfeldwässer. Erdöl u. Kohle **15**, 102—109 (1962).

KRUMBEIN, W. C., GARRELS, R. M.: Origin and classification of sediments in terms of pH and oxidation-reduction-potential. J. Geol. **60**, 1—33 (1952).

KÜHN, R.: Die Mikroskopie der Kalisalze; 1. Teil: Untersuchungsmethoden und Mineralien. Herausgegeben von der Kaliforschungsstelle, Empelde (Hannover) 1950.

— Nachexkursion im Kaliwerk Hattorf, Philippsthal — als Beitrag zur Kenntnis der Petrographie des Werra-Kaligebietes. Fortschr. Mineral. **29/30**, 101—114 (1950/51).

— Sellait als salinares Mineral. Fortschr. Mineral. **30**, 390 (1951).

— Reaktionen zwischen festen, insbesondere ozeanischen Salzen. Heidelberger Beitr. Mineral. Petrogr. **3**, 147—168 (1952).

— Tiefenberechnung des Zechsteinmeeres nach dem Bromgehalt der Salze. Z. deut. geol. Ges. **105**, 646—663 (1953a).

— Petrographische Studien an Jahresringen in Steinsalz. Fortschr. Mineral. **32**, 90—92 (1953b).

— Mineralogische Fragen der in den Kalisalzlagerstätten vorkommenden Salze. Kalium-Symposium 1955 (Internat. Kali-Inst., Bern) 51—105 (1955a).

— Über den Bromgehalt von Salzgesteinen, insbesondere die quantitative Ableitung des Bromgehaltes nichtprimärer Hartsalze oder Sylvinite aus Carnallit. Kali u. Steinsalz **1**, H. 9, 3—16 (1955b).

— Führung durch das Kalibergwerk Neuhof-Ellers, obere Sohle, nebst einigen Beiträgen zur Petrographie des Werra-Fulda-Kalireviers. Fortschr. Mineral. **35**, 60—81 (1957).

— Die Mineralnamen der Kalisalze. Kali u. Steinsalz **2**, 331—345 (1959).

— Zur Frage des Auftretens und Ausmaßes von Deszendenzen in der Leine- und Aller-Serie des deutschen Zechsteins. (Vortrag salzgeol. Koll. TH Hannover 16. 12. 1960, Manuskript.)

— Die chemische Zusammensetzung des Koenenits nebst Bemerkungen über sein Vorkommen und über Faserkoenenit. Neues Jahrb. Mineral., Abh. **97**, 112—141 (1961).

KÜHN, R., BAAR, A.: Ein ungewöhnliches Vorkommen von Danburit. Kali u. Steinsalz **1**, Heft 10, 17—21 (1955).

— RITTER, K. H.: Der Kristallwassergehalt von Kainit und Löweit. Kali u. Steinsalz **2**, 238—240 (1958).

KULP, J. L.: The geological time scale. Proceedings XXI Intern. Geol. Congr., Part III, 18—27, Copenhagen (1960).

LACMANN, R.: Über die Kristallisation auf Unterlagen. Die Keimbildungsarbeiten. Z. Krist. **116**, 13—26 (1961).

LAMCKE, K.: Gefügeanalytische Untersuchungen am Anhydrit nebst einem Beitrag zu den optischen und röntgenoptischen Methoden der Gefüge-Analyse. Schriften a. d. mineralog.-petrogr. Inst. Univ. Kiel, H. 4 (1937).

LANGBEIN, R.: Zur Petrographie des Hauptanhydrits (Z 3) im Südharz. Chemie der Erde **21**, 248—264 (1961).

LEONHARDT, J., BERDESINSKI, W.: Zur laugenfreien Synthese von Salzmineralien. Fortschr. Mineral. **28**, 35—38 (1949/50).

— — Semisalinare (assimilierte) Mineralkomponenten der Salzlagerstätten. Chemie der Erde **16**, 22—26 (1952).

— TIEMEYER, R.: Sylvin mit gesetzmäßig eingelagertem Eisenglanz. Naturwissenschaften **26**, 410—411 (1938).

LEPESCHKOW, I. N.: Untersuchungen der Schule N.S. KURNAKOWs zur physikalischen Chemie der natürlichen Salze und Salzsysteme. Freiberger Forschungsh. **A 123**, 105—116 (1958).

LOBANOVA, V. V.: A new borate-Strontioborite. Doklady Akad. Nauk S.S.S.R. **135**, 173—175 (1960), (Engl. Transl. p. 1285—1287).

— KHURSHUDYAN, E. KH.: Untersuchungen von Sulfoborit-Kristallen von der Inder-Lagerstätte. Zapiski Vsesoyuz. Mineral. Obshchestva **88**, 701—705 (1959).

— YARSHEMSKY, YA. YA.: Zur Mineralogie der Inder-Bodenerhebung. Voprosy Miner. Osadochnykh Obraz. Kniga **5**-j, Lvov, 177—190 (1958).

LOEWENGART, S.: The geochemical evolution of the Dead sea basin. Bull. Research Council Israel, **11 G**, 85—96 (1962).

LÖFFLER, J.: Die Carnallitgesteine des Raumes Aschersleben-Schierstedt. Freiberger Forschungsh. **C 87**, 1—63 (1960).

LOHSE, H. H.: Erfahrungen bei der röntgenogr. Identifizierung semisalinarer und nichtsalinarer Minerale der Salzlagerstätten. Diss. Kiel (1958).

LOTZE, F.: Steinsalz und Kalisalze, Geologie. 936 S. Berlin: Gebr. Borntraeger 1938.

— Steinsalz und Kalisalze 1. Teil, 2. Aufl. 465 S. Berlin: Gebr. Borntraeger 1957.

— Der englische Zechstein in seiner Beziehung zum deutschen. Geol. Jahrb. **73**, 135—139 (1958) [Hann.].

LÜCK, H.: Beitrag zur Kenntnis des älteren Salzgebirges im Berlepsch Bergwerk bei Staßfurt, nebst Bemerkungen über die Pollenführung des Salztones. Diss. Leipzig 1913.

LUKJANOWA, E. I., LUSHNAJA, N. P.: Über die Prozesse der Metamorphisation natürlicher Salzlösungen vom Meerestyp. Freiberger Forschungsh. **A 123**, 61—69 (1958).

MACDONALD, G. J. F.: Anhydrite-gypsum equilibrium relations. Am. J. Sci. **251**, 884—898 (1953).

— Thermodynamics of solids under non-hydrostatic stress with geological applications. Am. J. Sci. **255**, 266—281 (1957).

MADGIN, W. M., SWALES, D. A.: Solubilities in the system $CaSO_4$-NaCl-H_2O at 25° and 35°. J. Appl. Chem. (London) **6**, 482—487 (1956).

MARR, U.: Zur Verteilung der Fe-Gehalte in Salzgesteinen des Staßfurt-Zyklus. Geologie (Berlin) **6**, 41—70 (1957).

Mayer, F.: Geologisch-mineralogische Studien aus dem Berchtesgadener Land. Geognost. Jahresh. **25**, 121—161 (1912).

Mayrhofer, H.: Über ein Langbeinit- und Kainit-Vorkommen im Ischler Salzgebirge. Karinthin **30**, 94—98 (1955).

— Schauberger, O.: Pseudomorphosen von Talk nach Steinsalz als stratigraphisches Leitmineral im Hallstädter Salzberg. Berg- u. Hüttenmänn. Mh. **98**, H. 6, 111 (1953).

McGill, D. A.: A preliminary study of the oxygen and phosphate distribution in the Mediterranian Sea. Deep-Sea Research **8**, 259—269 (1961).

McKelvey, J. G., Milne, I. H.: The flow of salt solutions through compacted clay. Clay and Clay Minerals **11**, 248—259 (1962). Pergamon Press.

Morris, R. C., Dicky, P. A.: Modern evaporite deposition in Peru. Am. Assoc. Petrol. Geologists Bull. **41**, 2467—2474 (1957).

Mügge, O.: Über die Kristallform und Deformationen des Bischofit und der verwandten Chlorüre von Kobalt und Nickel. Neues Jahrb. Mineral. **1906 II**, 91—112.

— Über die Minerale im Rückstand des roten Carnallits von Staßfurt und des schwarzen Carnallits von der Hildesia. Kali **7**, 1—3 (1913).

— Über die Entstehung fasriger Minerale und ihrer Aggregationsformen. Neues Jahrb. Mineral. Geol. Paläont. Beil. Bd. **A 58**, 303—348 (1928).

Müller, A., Schwartz, W.: Über das Vorkommen von Mikroorganismen in Salzlagerstätten (geomikrobiologische Untersuchungen III). Z. deut. geol. Ges. **105**, 789—802 (1953).

Müller, G.: Eine sedimentäre Cölestin-Lagerstätte im Oberen Malm NW-Deutschlands. Fortschr. Mineral. **38**, 189 (1960).

— Die Löslichkeit von Cölestin ($SrSO_4$) in wäßrigen NaCl- und KCl-Lösungen. Neues Jahrb. Mineral. Mh. **1960**, 237—239.

— Zur Geochemie des Strontiums in ozeanen Evaporiten unter besonderer Berücksichtigung der sedimentären Cölestin-Lagerstätte von Hemmelte-West (Süd-Oldenburg). Geologie **11**, Beiheft 35, 1 (1962).

— Puchelt, H.: Die Bildung von Cölestin ($SrSO_4$) aus Meerwasser. Naturwissenschaften **48**, 301—302 (1961).

Murray, G. E.: Geology of Gulf Coast Salt Domes Northern Ohio Geol. Soc., Sympos. on Salt. 3.–5. Mai 1962 (Abstract).

Mursajev, P. M.: Über die genetischen Beziehungen von Gips und Anhydrit. Zapiski Vsesoyuz. Mineral. Obshchestva **75**, 339—341 (1947).

Naumann, M.: Die Entstehung des „konglomeratischen" Carnallitgesteins und des Hartsalzes sowie die einheitliche Bildung der deutschen Zechsteinlager ohne Deszendenzperioden. Kali **7**, 87—92 (1913).

Neumann, A. C., McGill, D. A.: Circulation of the Red Sea in early summer. Deep-Sea Research **8**, 223—235 (1961).

Neumann, J.: Evaporation from the Red Sea. Israel Exploration J. **2**, 153—162 (1952). (Jerus.)

Niemann, H.: Untersuchungen am Grauen Salzton der Grube „Königshall-Hindenburg", Reyershausen b. Göttingen. Beitr. Mineral. u. Petrogr. **7**, 137—165 (1960).

Nikolaev, A. V.: Über das Schema d. Genesis der Borlagerstätten vom Inder. Doklady Akad. Nauk S.S.S.R. **51**, 285—286 (1946) (Russ.).

— Chelishcheva, G.: On the primary deposition of borates from sea water. Compt. rend (Doklady) acad. Sci. U.R.S.S. **28**, 502—504 (1940).

Noellner, C.: Über Lüneburgit. S.-B. bayr. Akad. Wiss. München **1870**, 291.

Noll, W.: Zur Genesis porphyrischer Struktur in Gipsgesteinen. Chemie der Erde **9**, 1—21 (1934/35).
— Geochemie des Strontiums. Chemie der Erde **8**, 507—595 (1934).
Ochsenius, C.: Magnesit im Carnallit von Douglashall. Chem. Ztg. Nr. **19**, S. 304 (1890).
Ogniben, G.: Studio analitico della regola di orientazione del gesso primario di sedimentazione chimica. Periodico di Mineralogia **24**. 331—347 (1955).
Ogniben, L.: La "Regola di Mottura" di orientazione di gesso Periodico di Mineralogia (Roma **23**, 53—72 (1954).
— Inverse graded bedding in primary gypsum of chemical deposition. J. Sediment. Petrol. **25**, 273—281 (1955).
— Petrografia della seria solfifera siciliana e considerazione geologiche relative. Mem. descrit. carta geol. Italia **33**, 275 p. (1957a).
— Secondary gypsum of the Sulphur Series, Sicily, and the so-called integration. J. Sediment. Petrol. **27**, 64—79 (1957b).
Ovchinnikov, L. N., Maksenkov, V. G.: Experimentelle Untersuchung des Filtrationseffektes in Lösungen. Izvest. Akad. Nauk S.S.S.R. Ser. Geol. Nr. **3**, 82—94 (1949).
Oxburgh, U. M., Segnit, R. E., Holland, H. D.: Coprecipitation of Strontium with calcium carbonate from aqueous solutions. Bull. Geol. Soc. America **70**, 1653—1654 (1959).
Parchow, G.: Über den Gehalt des Carnallits an Eisenoxyd und Magnesia. Kali **4**, 95—96 (1910).
Pei-Keng-Leng: Gefügeuntersuchung des Trümmercarnallits von Krügershall zu Teutschenthal, Halle, mit besonderer Berücksichtigung des Carnallitgefüges. Diss. TH Berlin 1945 (Ref.: Zbl. Mineral. Geol. **1949 II**, 241—244).
Pitrovskaya, Z. N.: Goyazite in the breccia of the Romny and Issachki salt domes. Doklady Akad. Nauk S.S.S.R. **25**, 502 (1939).
Posnjak, E.: Deposition of calcium sulfate from sea water. Am. J. Sci. **238**, 559—568 (1940).
Privett, D. W.: Monthly charts of evaporation from the N. Indian Ocean (including the Red Sea and the Persian Gulf). Quart. J. Roy. Meteorol. Soc. **85**, 424—428 (1959) London.
Rawitsch, M. I.: Die heterogenen Gleichgewichte in Wasser-Salz-Systemen bei hohen Temperaturen. Freiberger Forschungsh. **A 123**, 269—286 (1958) [Berlin].
Richter, A.: Die Rotfärbung in den Salzen der deutschen Zechsteinlagerstätten. Chemie der Erde **22**, 508—546 (1962); **23**, 179—203 (1964).
Richter-Bernburg, G.: Zur Frage der absoluten Geschwindigkeit geologischer Vorgänge. Naturwissenschaften **37**, 1—8 (1950).
— Über salinare Sedimentation. Z. deut. geol. Ges. **105**, 593—645 (1953a).
— Stratigraphische Gliederung des Deutschen Zechsteins. Z. deut. geol. Ges. **105**, 843—854 (1953b).
— Zeitmessung geologischer Vorgänge nach Warven-Korrelationen im Zechstein. Geol. Rundschau **49**, 132—148 (1960).
Ricke, W.: Ein Beitrag zur Geochemie des Schwefels. Geochim. Cosmochim. Acta **21**, 35—80 (1960).
Ricour, J.: Hypothèse sur le milieu de formation des niveaux salifères du Trias français. Proceedings XXI. Internat. Geol. Congr., Part **XXI**, 215—222, Copenhagen (1960).
Riedel, O.: Chemisch-mineralogisches Profil des älteren Salzgebirges im Berlepschbergwerk bei Staßfurt. Z. Krist. **50**, 139—173 (1912).

RINNE, F.: Metamorphosen von Salzen und Silikatgesteinen. J.-Ber. Niedersächs. Geol. Ver. **7**, 252—269 (1913).

— Die geothermischen Metamorphosen und die Dislokationen der deutschen Kalisalzlagerstätten. Fortschr. Mineral. **6**, 101—136 (1920).

— Gesteinskunde 10. u. 11. Aufl. Leipzig: Verl. Dr. M. Jänecke 1928 (erstmalig 1908).

RONOV, A. B.: On the post-precambrian geochemical history of the atmosphere and hydrosphere. Geochemistry (Geokhimiya) **1959**, 493—506.

— KHLEBNIKOVA, Z. V.: Chemical composition of the main genetic clay types. Geochemistry (Geokhimiya) **1957**, 527—552.

ROTH, H.: Ausbildung und Lagerungsformen des Kaliflözes „Hessen" im Fuldagebiet. Z. deut. geol. Ges. **105**, 674—684 (1953).

RÓZSA, M.: Die Entstehung des Hartsalzes und die sekundären Umwandlungen der Zechsteinsalze im Zusammenhang mit den Gleichgewichtsschemata VAN'T HOFFs. Z. anorg. Chemie **91**, 299—319 (1915).

RUBEY, W. W.: Geologic history of sea water. Bull. Geol. Soc. Am. **62**, 1111—1148 (1951).

RÜHLE, C.: Der Aufbau der Kalisalzlagerstätte des Bernburger Sattels, insbesondere d. „alteren Lagers" von „Solvay in Preußen". Jahresber. Niedersächs. geol. Ver. 116—147 (1913).

RUSSEL, G. A.: Crystal growth and solution under local stress. Am. Mineral. **20**, 733—737 (1935).

SAHAMA, TH. G.: Abundance relation of fluorite and sellaite in rocks. Ann. Acad. Sci. Fennicae, ser. A III No. **9**, 20 S. (1945).

SAVCHENKO, V. P.: The formation of free hydrogen in the earth's crust, as determined by the reducing action of the products of radioactive transformations of isotopes. Geochemistry (Geokhimiya) p. 16—25 (1958).

SCHABUS, J.: Über die Krystallformen des Bleichlorides PbCl, des Eisenchlorürs FeCl · 4 HO und des Eisenchlorür-Kaliumchlorides KCl, FeCl, 2 HO. Ber. Kais. Akad. Wien; math.-naturw. Kl. **4**, 456—484 (1850).

SCHACHL, E.: Das Muschelkalksalz in Südwestdeutschland. Neues Jahrb. Geol. Paläontol. Abh. **98**, 309 (1954).

SCHALLER, W. T., HENDERSON, E. P.: Mineralogy of drill cores from the potash field of New Mexico and Texas. Bull. U.S. Geol. Surv. Bull. **833** (1932).

— MROSE, M. E.: The naming of the hydrous magnesium borate minerals from Boron, California — a preliminary note. Am. Mineralogist **45**, 732—733 (1960).

SCHAUBERGER, O., RUESS, H.: Über die Zusammensetzung der alpinen Salztone. Berg- u. Hüttenmänn. Monatsh. (Leoben) **96**, 187—195 (1951).

SCHELLMANN, W.: Experimentelle Untersuchungen über die sedimentäre Bildung von Goethit und Haematit. Chemie der Erde **20**, 104—135 (1959).

SCHMIDT, E.: Über die Auflösung von Boracit (Staßfurtit) in konzentrierten Magnesiumchloridlösungen. Monatsber. deut. Akad. Wiss. Berlin **1**, 546—553 (1959).

SCHMIDT, R.: Über die Beschaffenheit und Entstehung parallelfaseriger Aggregate von Steinsalz und Gips. Kali **8**, 161, 197, 218, 239 (1914).

SCHNEIDER, W.: Neubestimmung der Kristallstruktur des Mangan-Leonits $K_2Mg(SO_4)_2 \cdot 4\,H_2O$. Acta cryst. **14**, 784—791 (1961).

— ZEMANN, J.: Die kristallographischen Konstanten von Löweit. Beitr. Mineral. u. Petrogr. **6**, 201—202 (1959).

SCHOTT, G., KIBBEL, H. U.: Über Sulfatoborate. Z. anorg. allg. Chemie **314**, 104—112 (1962).

SCHULZE, G.: Stratigraphische und genetische Deutung der Bromverteilung in den mitteldeutschen Steinsalzlagern des Zechsteins. Freiberger Forschungsh. C **83**, 114 S. Berlin 1960.

SCHÜNEMANN, F.: Vorläufige Mitteilungen über einzelne Ergebnisse meiner Untersuchungen auf den Kaliwerken des Staßfurter Sattels. Z. prakt. Geologie **21**, 205—216 (1913).

SCHWERDTNER, W.: Korngefügeuntersuchungen an Anhydritgesteinen im Benther Salzstock (Werk Ronnenberg) bei Hannover. Kali u. Steinsalz **3**, 173—182 (1961).

SDANOVSKY, A. B., LYAKHOVSKAYA, E. I., SHLEYMOVITCH, R. E.: Handbuch über die Löslichkeit der Salzsysteme. Verlag Goskhimizdat (Leningrad). Pt. I, 1953; Pt. II, 1954; 954 S.

SEDELNIKOW, G. S.: Über die hydrochemischen Verhältnisse bei der Salzbildung in der Karabugas-Bucht. Freiberger Forschungsh. A **123**, 166—174 (1958).

SEIDL, E.: Beiträge z. Morphologie und Genesis d. permischen Salzlagerstätten Mitteldeutschlands. Z. deut. geol. Ges. **65**, 124—144 (1913).

SENFT: Der Gipsstock bei Kittelsthal mit seinen Mineraleinschlüssen. Z. deut. geol. Ges. **16**, 166—177 (1861).

SHCHERBINA, V. N.: Rhythms and cycles of sedimentation in the section of the salt series of the Starobinsk Potash deposit. Doklady Akad. Nauk S.S.S.R. **131**, 398—401 (1960).

SIEGEL, F. R.: Variations of Sr/Ca Ratios and Mg-contents in recent carbonate sediments of the northern Florida Keys area. J. Sediment. Petrol. **31**, 336—342 (1961).

SIEMEISTER, G.: Primärparagenese und Metamorphose des Ronnenberglagers nach Untersuchungen im Grubenfeld Salzdetfurth. Diss. Clausthal 1961.

SIEVER, R., GARRELS, R. M., KANWISHER, J., BERNER, R. A.: Interstitial waters of recent marine muds off Cape Cod. Science **134**, Nr. 3485, S. 1071—1072 (1961).

SKINNER, H. C. W.: Formation of modern dolomitic sediments in South Australian lagoons. Prog. Annual meetings, Denver, S. 208 (1960).

SKOPINZEV, B. A., GUBIN, P. A., VOROBYEVA, R. V., VERCHININA, O. A.: Salzzusammensetzung des Wassers des Schwarzen Meeres. Doklady Akad. Nauk S.S.S.R. **119**, 121—124 (1958).

SOURIRAJAN, S., KENNEDY, G. C.: The system H_2O–NaCl at elevated temperatures and pressures. Am. J. Sci. **260**, 115—141 (1962).

STEWART, F. H.: The petrology of the evaporites of the Eskdale No. 2 – boring, east Yorkshire. Part I: the lower evaporite bed. Mineral. Mag. **28**, 621—675 (1949).

— The petrology of the evaporites of the Eskdale. Nr. 2 – boring, east Yorkshire. Part II: the middle evaporite bed. Mineral. Mag. **29**, 445—475 (1951a).

— The petrology of the evaporites of the Eskdale. No. 2. – boring, east Yorkshire. Part III: the upper evaporite bed. Mineral. Mag. **29**, 557—572 (1951b).

— Early gypsum in the permian evaporites of north-eastern England. Proc. Geologists' Assoc. (Engl.) **64**, pt. 1, 33—39 (1953).

STÖCKE, K., BORCHERT, H.: Fließgrenzen von Salzgesteinen und Salztektonik. Kali **30**, 191, 204, 214 (1936).

STORCK, U.: Die Entstehung der Vertaubungen und des Hartsalzes im Flöz Staßfurt im Zusammenhang mit regelmäßigen Begleiterscheinungen auf dem Kaliwerk Königshall-Hindenburg. Kali u. Steinsalz **1**, Nr. 6, 21—31 (1954).

STRUNZ, H.: Mineralogische Tabellen. 3. Aufl. 448 S. Leipzig: Akad. Verl. Ges. Geest & Portig 1957. (4. Aufl. 1966).

— Kristallographie von D'Ansit, ein auf marinsedimentären Lagerstätten zu erwartendes Salz. Neues Jahrb. Mineral. Mh. **1958**, 152—155.

Sturmfels, E.: Das Kalisalzlager von Buggingen (Südbaden). Neues Jahrb. Mineral. Abh. Beil. Bd. **78 A**, 131—216 (1943).

Sugawara, K., Kawasaki, N.: Sr and Ca distribution in W Pacific, Indian and antarctic oceans. Records of oceanographic works in Japan. (Special No. 2). March 1958.

Sverdrup-Johnson-Fleming: The oceans, their physics, chemistry and general biology. New York: Prentice Hall 1942.

Sydow, W.: Die Ausbildung des Ronnenberg-Lagers unter besonderer Berücksichtigung des petrographischen Aufbaus und seiner sekundären Veränderung. Kali u. Steinsalz **2**, 406—418 (1959).

Tammann, G., Seidel, K.: Zur Kenntnis der Kohlensäureausbrüche in Bergwerken. Z. anorg. Chem. **205**, S. 209 (1932).

Taylor, R. E.: Water-insoluble residues from the rock-salt of Louisiana salt plugs. Bull. Am. Assoc. Petrol. Geologists **21**, 1268—1310 (1937).

Tinnes, A.: Die ältere Salzfolge Mitteldeutschlands unter besonderer Berücksichtigung des Unstrutgebietes. Arch. Lagerstättenforsch. H. **38** (1928).

Tittel, M.: Thermodynamische Berechnungen zu Untersuchungen über eine einstufige Carnallitentwässerung. Freiberger Forschungsh. **A 123**, 457—463 (1958).

Todes, O. M.: Die Kinetik der Salzkristallisation. Freiberger Forschungsh. **A 123**, 341—363 (1958).

Trusheim, F.: Über Halokinese und ihre Bedeutung für die strukturelle Entwicklung Norddeutschlands. Z. deut. geol. Ges. **109**, 111—151 (1957).

— Mechanism of salt migration in northern Germany. Bull. Am. Assoc. Petr. Geol. **44**, 1519—1540 (1960).

Tschermak, M. G.: Beitrag zur Kenntnis der Salzlager. Sitzber. Akad. Wiss. (math. nat.) Wien, **63**, 305—324 (1871).

Tschoepke, R., Karl, F.: Gefügeanalytische Untersuchung der Faltungsrichtungen im Tonlöser des Flözes Hessen in der Grube Neuhof-Ellers. Kali u. Steinsalz **2**, 96—100 (1957).

Turner, F. J., Verhoogen, J.: Igneous and metamorphic petrology. McGraw-Hill Book Comp. New-York, Toronto, London. 2. ed. 694 pp. 1960.

Valeton, J. J. P.: Kristallform und Löslichkeit. Ber. Verhandl. sächs. Akad. Wiss. Leipzig, **67 I**, 1—59 (1915).

Valyashko, M. G.: Geochemistry of bromine in the processes of salt deposition and the use of the bromine content as a genetic and prospecting criterion. Geochemistry (Geokhimiya) 570—589 (1956).

— (Waljaschko) Die wichtigsten geochemischen Parameter für die Bildung der Kalisalzlagerstätten, Freiberger Forschungsh. A **123**, 197—233 (1958).

— Nechaeva, A. A.: Experimentelle Untersuchungen der Bildungsbedingungen des Polyhalits. Handb. Mineralogie der geol. Ges. Lvov (Lemberg) Nr. **6** (1952) [Russ.]; cited by Valyashko 1958.

Vinogradov, A. P., Ronov, A. B.: Evolution of the chemical composition of clays of the Russian Platform. Geochemistry (Geokhimiya) 123—139 (1956).

Vinokurov, V. M.: Über das blaue Steinsalz der Solikamsk-Lagerstätte. Zapiski Vsesoyuz. Mineral. Obshchestva **87**, 504—507 (1958).

Volmer, M.: Kinetik der Phasenbildung. 220 S. Dresden u. Leipzig: Th. Steinkopff 1939.

Wagner, W.: Die tertiären Salzlagerstätten im Oberrheintal-Graben. Z. deut. geol. Ges. **105**, 706—728 (1953).

Waljaschko: see Valyashko.

Walther, H. W.: Orogen-Struktur und Metallverteilung im östlichen Zagros (Südost-Iran). Geol. Rundschau **50**, 353—374 (1960).

WALTHER, JOH.: Das Gesetz der Wüstenbildung in Gegenwart und Vorzeit. 2. Aufl. 342 S. Leipzig: Verlag Quelle & Meyer (spez. S. 250/251) 1912.

WEBER, KLAUS: Untersuchungen über die Faciesdifferenzierungen, Bildungs- und Umbildungserscheinungen in den beiden Kalilagern des Werra-Fulda-Gebietes unter besonderer Berücksichtigung der Vertaubungen. Diss. Clausthal 1961.

WEBER, KONR.: Geologisch-petrographische Untersuchungen am Staßfurt-Egelner Sattel unter besonderer Berücksichtigung der Genese der Polyhalit- und Kieserit-Region. Kali **25**, 17, 33, 49, 65, 82, 97 (1931).

WHITMAN, W. G., RUSSEL, R. P., DAVIS, G. H. B.: The solubility of ferrous hydroxide and its effect upon corrosion. J. Am. Chem. Soc. **47**, 1, 70—79 (1925).

WILFARTH, M.: Sedimentationsprobleme in der germanischen Senke zur Perm- und Triaszeit. Geol. Rundschau **24**, 349—377 (1933).

WIMMENAUER, W.: Petrographische Untersuchungen über das Ankaratrit-Vorkommen im Kalisalzlager von Buggingen in Baden. Mitt.-Bl. bad. geol. Landesamt **1951**, 117—128.

YARZHEMSKY, YA. YA.: On the questions related to the origin of Inder Borates. Doklady Akad. Nauk S.S.S.R. **47**, 642—645 (1945).

— Kurgantait – ein neues Bormineral (russ.). Mineral. Sbornik, Lvovsk. Geol. Obshchestva Nr. **6**, 169—175 (1952).

— Zur Genese des Hydroborazites in Salzgesteinen. Doklady Akad. Nauk S.S.S.R. (2) **88**, 1051—1054 (1953).

— Zur Frage der Genese des Polyhalits in Kalisalzlagerstätten. Trudy Vsesoyuz. Nauch.-Issledovatel. Inst. Galurgii **29**, 223—259 (1954) [russ.]. Ref.: Geologie **6**, 861—864 (1957).

— Preobrashenskit, ein neues Borat der salzführenden Schichten der Inder-Erhebung. Doklady Akad. Nauk S.S.S.R. **111**, 1087—1090 (1956).

ZÄHRINGER, J.: Altersbestimmung nach der K–Ar-Methode. Geol. Rundschau **49**, 224—237 (1960).

ZEMANN, J.: Beitrag zum Verständnis von Zustandsdiagrammen in Systemen mit überwiegend heteropolarer Bindung. Beitr. z. Mineral. u. Petrogr. **6**, 89—95 (1958).

ZERNIKE, J.: Chemical Phase Theory. E. Kluwer, Deventer 493 S. (1957).

ZIMMERMANN, E.: Über den Pegmatitanhydrit. Kali **3**, 309—312 (1909).

— Der thüringische Plattendolomit und sein Vertreter im Staßfurter Zechsteinprofil, sowie eine Bemerkung zur Frage der „Jahresringe". Z. deut. geol. Ges. **65**, Mber. 357—372 (1915).

Author and Subject Index

Minerals, Rocks and Inorganic Materials

Already published:

Vol. 1: W. G. ERNST: Amphiboles. With 59 figures, X, 134 pages. DM 27,20; US $ 6.80 (Subseries "Experimental Mineralogy")

Vol. 3: B. R. DOE: Lead Isotopes. With 24 figures, IX, 137 pages. DM 36,—; US $ 9.90 (Subseries "Isotopes in Geology")

Vol. 4: O. BRAITSCH: Salt Deposits, Their Origin and Composition. With 47 figures. XIV, 297 pages. DM 72,—; US $ 19.80.

In press:

Vol. 2: E. HANSEN: Strain Facies. With 78 figures and 21 plates. Approx. 200 pages. DM 58,—; US $ 16.00.

Volumes being planned:

K. KEIL: Quantitative Microprobe Analysis
F. LIPPMANN: Carbonate Minerals of Sedimentary Rocks
N. PETERSEN and A. SCHULT: Magnetic Properties of Rockforming Minerals and Rocks
A. RITTMANN et al.: The Mineral Association of Volcanic Rocks

Subseries "Experimental Mineralogy":

P. M. BELL, S. W. RICHARDSON, and M. CH. GILBERT: Synthesis and Stability of the Aluminum Silicate Minerals
G. M. BROWN and D. H. LINDSLEY: Synthesis and Stability of Pyroxenes
W. G. ERNST: Blueschist Facies Metamorphism, Petrology and Tectonics
H. P. EUGSTER and D. R. WONES: Synthesis and Stability of Micas
J. J. FAWCETT: Synthesis and Stability of Chlorite and Serpentine
I. KUSHIRO: Experimental Petrology of Gabbro
D. H. LINDSLEY: Iron-Titanium Oxides: Crystallography, Crystal Chemistry, and Synthesis
M. S. PATERSON: Experimental Rock Deformation

Subseries "Isotopes in Geology":

M. L. JENSEN: Biochemistry of Sulfur Isotopes
J. L. POWELL and G. FAURE: Strontium Isotope Geology